Climate Change Adaptation in Practice
From strategy development to implementation

Climate Change Adaptation in Practice

From strategy development to implementation

EDITED BY

Philipp Schmidt-Thomé
Geological Survey of Finland

Johannes Klein
Aalto University, Finland

Part-financed by the European Union (European Regional Development Fund)

A John Wiley & Sons, Ltd., Publication

This edition first published 2013 © 2013 by John Wiley & Sons, Ltd

Wiley-Blackwell is an imprint of John Wiley & Sons, formed by the merger of Wiley's global Scientific, Technical and Medical business with Blackwell Publishing.

Registered office: John Wiley & Sons, Ltd, The Atrium, Southern Gate, Chichester, West Sussex, PO19 8SQ, UK

Editorial offices: 9600 Garsington Road, Oxford, OX4 2DQ, UK
  The Atrium, Southern Gate, Chichester, West Sussex, PO19 8SQ, UK
  111 River Street, Hoboken, NJ 07030-5774, USA

For details of our global editorial offices, for customer services and for information about how to apply for permission to reuse the copyright material in this book please see our website at www.wiley.com/wiley-blackwell.

Library of Congress Cataloging-in-Publication Data

Climate change adaptation in practice : from strategy development to implementation / editors, Philipp Schmidt-Thomé, Johannes Klein.
  pages cm
 Includes index.
 ISBN 978-0-470-97700-2 (cloth)
 1. Climatic changes–Government policy–Europe, Northern.   2. Environmental policy–Europe, Northern.   I. Schmidt-Thomé, Philipp.   II. Klein, Johannes.
  QC903.2.E853C55 2013
  363.738′745610948–dc23

2012048829

A catalogue record for this book is available from the British Library.

Wiley also publishes its books in a variety of electronic formats. Some content that appears in print may not be available in electronic books.

Cover image: Images supplied by iStock
Cover design by Dan Jubb

Set in 8.75/12pt Meridien by Aptara® Inc., New Delhi, India
Printed and bound in Singapore by Markono Print Media Pte Ltd

1   2013

Contents

Contents

List of Contributors

Jussi Ahonen
Geological Survey of Finland (GTK), Espoo, Finland

Tarmo All
Ministry of the Environment, Tallinn, Estonia

Jurga Arustienė
Lithuanian Geological Survey, Vilnius, Lithuania

Birgitta Backman
Geological Survey of Finland (GTK), Espoo, Finland

B. Bedsted
The Danish Board of Technology, Copenhagen, Denmark

Markus Boettle
Potsdam Institute for Climate Impact Research (PIK), Potsdam, Germany

Agrita Briede
University of Latvia, Faculty of Geography and Earth Sciences, Riga, Latvia

Jorge Olcina Cantos
Alicante University, Alicante, Spain

Sven Dahlke
University of Greifswald, Kloster/Hiddensee, Germany

Aldona Damušytė
Lithuanian Geological Survey, Vilnius, Lithuania

Larissa Donges
Leibniz Institute for Baltic Sea Research, Warnemünde, Rostock, Germany

Guntis Eberhards
University of Latvia, Faculty of Geography and Earth Sciences, Riga, Latvia

Mareike Fellmer
Hafen City University Hamburg, Urban Planning and Regional Development, Germany

Christian Filies
EUCC – The Coastal Union Germany, Rostock, Germany

René Friedland
Leibniz Institute for Baltic Sea Research, Warnemünde, Rostock, Germany

Koji Fukuda
Institute for Global Environmental Strategies, Japan

S. Gram
The Danish Board of Technology, Copenhagen, Denmark

Marius Gregorauskas
Vilniaus hidrogeologija Ltd, Vilnius, Lithuania

Inga Haller
EUCC – The Coastal Union Germany, Rostock, Germany

Shinano Hayashi
Institute for Global Environmental Strategies, Japan

Doddy Juli Irawan
Center for Climate Risk and Opportunity Management in Southeast Asia and Pacific, Bogor Agriculture University, Indonesia

Darius Jarmalavičius
Nature Research Centre, Institute of Geology and Geography, Vilnius, Lithuania

Sirkku Juhola
Department of Real Estate, Planning and Geoinformatics, Aalto University; Department of Environmental Sciences, University of Helsinki

Susanna Kankaanpää
Helsinki Region Environmental Services Authority (HSY)

Kiki Kartikasari
Center for Climate Risk and Opportunity Management in Southeast Asia and Pacific, Bogor Agriculture University, Indonesia

List of Contributors

Justas Kažys
Department of Hydrology and Climatology, Vilnius University, Lithuania

Anna-Marie Klamt
Leibniz Institute for Baltic Sea Research Warnemünde, Rostock, Germany

J.E. Klausen
Norwegian Institute for Urban and Regional Research (NIBR), Norway

Māris Kļaviņš
University of Latvia, Faculty of Geography and Earth Sciences, Riga, Latvia

Johannes Klein
Geological Survey of Finland (GTK), Espoo, Finland; Aalto University, Espoo, Finland

Joerg Knieling
Hafen City University Hamburg, Urban Planning and Regional Development, Germany

Jurgita Kriukaitė
Lithuanian Geological Survey, Vilnius, Lithuania

Jürgen P. Kropp
Potsdam Institute for Climate Impact Research (PIK), Potsdam, Germany; University of Potsdam, Institute of Earth and Environmental Science, Potsdam, Germany

Laila Kūle
University of Latvia, Faculty of Geography and Earth Sciences, Riga, Latvia

O. Langeland
Norwegian Institute for Urban and Regional Research (NIBR), Norway

Andris Ločmanis
University of Latvia, Faculty of Geography and Earth Sciences, Riga, Latvia

Samrit Luoma
Geological Survey of Finland (GTK), Espoo, Finland

Matthias Mossbauer
Leibniz Institute for Baltic Sea Research, Warnemünde, Rostock, Germany; EUCC – The Coastal Union Germany, Rostock, Germany

Anika Nockert
Geological Survey of Finland (GTK), Espoo, Finland

Valter Petersell
Geological Survey of Estonia, Tallinn, Estonia

S.V.R.K. Prabhakar
Institute for Global Environmental Strategies, Japan

Gattineni Srinivasa Rao
eeMausam, Weather Risk Management Services, India

Egidijus Rimkus
Department of Hydrology and Climatology, Vilnius University, Lithuania

Jayant K. Routray
School of Environment, Resources and Development (SERD), Asian Institute of Technology (AIT), Bangkok, Thailand

Diego Rybski
Potsdam Institute for Climate Impact Research (PIK), Potsdam, Germany

Daisuke Sano
Institute for Global Environmental Strategies (IGES), Japan

M. Mustafa Saroar
School of Environment, Resources and Development (SERD), Asian Institute of Technology (AIT), Bangkok, Thailand; Urban and Rural Planning, School of Science, Engineering and Technology (SET), Khulna University, Bangladesh

Jonas Satkūnas
Lithuanian Geological Survey, Vilnius, Lithuania

Gerald Schernewski
Leibniz Institute for Baltic Sea Research Warnemünde, Rostock, Germany; Coastal Research & Planning Institute, Klaipeda University, Klaipeda, Lithuania

Philipp Schmidt-Thomé
Geological Survey of Finland (GTK), Espoo, Finland; Potsdam Institute for Climate Impact Research (PIK), Potsdam, Germany

Susanne Schumacher
EUCC – The Coastal Union Germany, Rostock, Germany

Mihkel Shtokalenko
Geological Survey of Estonia, Tallinn, Estonia

Edvinas Stonevičius
Department of Hydrology and Climatology, Vilnius University, Lithuania

Sten Suuroja
Geological Survey of Estonia, Tallinn, Estonia

Ulla Tiilikainen
City of Tampere, Urban Development, Tampere, Finland

Ruusu Tuusa
Department of Real Estate, Planning and Geoinformatics,
Aalto University

Gintaras Valiuškevičius
Department of Hydrology and Climatology, Vilnius University,
Lithuania

Tuire Valjus
Geological Survey of Finland (GTK), Espoo, Finland

Jari Viinanen
Environment Centre, City of Helsinki, Finland

M. Winsvold
Norwegian Institute for Urban and Regional Research (NIBR),
Norway

Tiia Yrjölä
Environment Centre, City of Helsinki, Finland

Gintautas Žilinskas
Nature Research Centre, Institute of Geology and Geography,
Vilnius, Lithuania

About the Editors

Philipp Schmidt-Thomé is a senior scientist and project manager at the Geological Survey of Finland (GTK) and an Adjunct Professor at the University of Helsinki. He is trained as a Geographer (MSc) and holds a PhD in Geology. He leads the Working Group on Climate Change Adaptation under the International Union of Geosciences Commission on Geo-Environment. His scientific focus is on geoscience communication and interdisciplinary cooperation. His recent project work has focused on integrating natural hazards, climate change and risks into land-use planning practices. He is a regular lecturer in several universities and a visiting fellow to the South East Asia Disaster Prevention Institute (SEADPRI).

Johannes Klein works at the Aalto University, Department of Real Estate, Planning and Geoinformatics, Land Use Planning and Urban Studies Group. He graduated from the University of Stuttgart in environmental engineering and is currently a PhD student within the Nordic Centre of Excellence for Strategic Adaptation Research (NORD-STAR). His research focus is on climate change adaptation and urban development. He worked as researcher at the Geological Survey of Finland from 2005 to 2012 and was the coordinator of the BaltCICA project.

Acknowledgements

The Editors acknowledge the significant contribution of Anika Nockert who was largely responsible for the technical and administrative revision process of this book. **Anika Nockert** has a Bachelor of Science in Geography and is an MSc student in 'Physical geography of human-environment-systems' at the Humboldt-University in Berlin. She worked as a research assistant in the 'Climate Impacts & Vulnerability' Research Domain at the Potsdam Institute for Climate Impact Research (PIK) and at the Geological Survey of Finland (GTK) where she primarily supported the BaltCICA project.

1

Communicating Climate Change Adaptation: From Strategy Development to Implementation

Philipp Schmidt-Thomé[1], Johannes Klein[1,2], Anika Nockert[1], Larissa Donges[3] & Inga Haller[4]

[1]Geological Survey of Finland (GTK), Espoo, Finland
[2]Aalto University, Espoo, Finland
[3]Leibniz Institute for Baltic Sea Research Warnemünde, Rostock, Germany
[4]EUCC – The Coastal Union Germany, Rostock, Germany

1.1 Introduction

This book displays climate change adaptation measures that were developed and implemented in the Baltic Sea Region. International and European institutions, such as the Intergovernmental Panel on Climate Change (IPCC) as well as the EU Commission (2009) have identified the necessity of actions to go beyond strategies and called for the implementation of adaptation measures (IPCC, 2007; COM, 2009). Examples that demonstrate the need for the implementation of climate change adaptation measures to be politically pushed towards the local level are the resolution on resilient cities adopted by the Congress of Local and Regional Authorities of the Council of Europe (2012), the position paper on climate change by the Association of Finnish Local and Regional Authorities (Suomen Kuntaliitto, 2010) or the recently published policy document on climate change adaptation by the German Association of Cities (2012). The latter paper lists a number of adaptation measures cities shall take into consideration for future land-use planning.

Consistent with these calls for action, the **Climate Change: Impacts, Costs and Adaptation in the Baltic Sea Region (BaltCICA)** project particularly focused on the implementation of adaptation measures, which are summarised in this book. Representatives of regional and local authorities, municipalities, research institutes of various disciplines and universities from eight countries[1] participated in the project. The BaltCICA project was the third consecutive project on climate change adaptation in the Baltic Sea Region conducted under the Geological Survey of Finland. The first of these projects, SEAREG,[2] focused on awareness raising and structuring of the science-stakeholder dialogue. The second project, ASTRA,[3] identified climate change impacts on regional development and formulated adaptation strategies. The BaltCICA project drew on the experiences of these projects and contributed to the implementation of adaptation measures. It produced new knowledge relating to climate change impacts, costs and benefits and governance of adaptation. It reduced uncertainty in decision-making in relation to adaptation by

[1]Denmark, Estonia, Finland, Germany, Latvia, Lithuania, Norway and Sweden
[2]Sea level change affecting the spatial development in the Baltic Sea Region (2002–2005), www.gtk.fi/slr
[3]Developing Policies & Adaptation Strategies to Climate Change in the Baltic Sea Region (2005–2007), www.astra-project.org

Climate Change Adaptation in Practice: From Strategy Development to Implementation, First Edition.
Edited by Philipp Schmidt-Thomé and Johannes Klein.
© 2013 John Wiley & Sons, Ltd. Published 2013 by John Wiley & Sons, Ltd.

strengthening science-decision maker links and it increased participation of stakeholders and citizens in decision-making on adaptation measures.

Thirteen case studies dealt with a broad range of thematic areas, especially focusing on land-use planning and urban development for adaptation. Interdisciplinary work enabled a multi-faceted approach to these topics. This included modelling of climate change impacts on groundwater and flood-prone areas; the participatory development of adaptation measures with the cooperation of citizens, authorities, scientists and representatives of economic sectors; as well as the assessment of adaptation options with respect to costs, benefits and less tangible criteria such as environmental impacts or aesthetics. These methods were closely interlinked in order to foster climate change adaptation at the local level.

The methodologies to identify and implement adaptation measures were developed on a local level and communicated among project partners via study visits and workshops. These workshops enabled other project partners to both learn about new methodologies and to further develop them according to specific local needs in their respective case studies.

Scenario workshops were designed and employed for direct science-stakeholder cooperation. This methodology was adapted to local circumstances of each case study and applied to identify needs and viabilities of decision-making processes towards implementing adaptation measures. Adapting or changing current land-use plans and underlying regulations, is often a lengthy process. Therefore concrete adaptation actions have been employed in only some of the case study areas, meanwhile in several other municipalities decisions are currently being taken or are high on the political agenda. In any case, the BaltCICA project has had a notable impact in the case studies on developing methodologies on how to take the step ahead from formulating climate change adaptation strategies towards specific adaptation measures.

The project partners have communicated their activities and results beyond the Baltic Sea Region and Europe. In the course of these dissemination activities several new project ideas were born. Some international activities therefore round the book up with examples on how climate change adaptation is perceived and dealt with in areas outside of the Baltic Sea Region.

1.2 Structuring the communication processes

The identification of adaptation necessities and potentials requires interdisciplinary cooperation, not only between scientific disciplines but especially between scientists and stakeholders (including decision makers) (e.g. Adger et al., 2009; Dessai & Hulme, 2004). Therefore the communication process plays a key role. Only if decision makers, scientists and involved citizens agree on local necessities of adaptation options is it possible to develop reasonable and cost-effective options that can be implemented. For decision makers it is usually not practicable to develop measures against impacts that might potentially occur in 100 years. In the daily business of decision makers, the focus is often on current and near future land use patterns. Therefore it is necessary to understand motivations and interests of decision makers in order to find entry points in planning that may respect developments that lie in the farther future. It was shown during the project work that adaptation concepts that can be embedded into current political demands and interests raise the interest and thus also the acceptability among decision makers.

The communication with stakeholders during the BaltCICA project and its predecessors showed that overall 'tool boxes' are difficult to deploy or can even be counterproductive, as every municipality has its own history and special characters. An overall adaptation concept is often received sceptically, so that general concepts, for example, on how to start and endorse communication processes are helpful. But finally each approach for every respective case study has to be completely adapted to the special requirements of each respective case study.

It also turned out that preferred adaptation options are in fact those of no-regret character, that is, those that also offer protection to current hazard patterns. It proved useful to start off with current extreme events (including historical records) rather than using those of potential flood events that might occur in the future. The potential impacts of current extreme events revealed recent developments of local vulnerability patterns. Often it turned out that assets had been constructed in unsuitable, that is, currently hazard prone areas. In the communication process land use developments and future options were then combined with potential changes in sea level and hydro-meteorological phenomena.

The combination of current and potential future land use patterns, climate variables and extreme events then lead to an integrated understanding of present as well as emerging risk patterns. In some case studies adaptation measures were designed to avoid or withstand current impacts, with an outlook on enhancing these measures along with ongoing climate change. In these cases adaptation measures are currently being put into practice. In other cases even more radical approaches of retreat were discussed, which would be implemented and aligned to the life cycles of buildings and infrastructure, and the development of climate impacts.

The examples displayed in this book show that whatever option on climate change adaptation might seem to be important from a scientific perspective, the structure of the communication process with stakeholders is the decisive factor to implement cost effective as well as politically and socially acceptable implementation measures.

1.3 Climate change induced physical impacts on the Baltic Sea Region

Impacts of climate change occur and are perceived differently throughout the Baltic Sea Region. Depending on local circumstances, climate change adaptation processes are in various stages and address different challenges. This section gives an overview on climate change impacts in the Baltic Sea Region, as based on current scientific knowledge. Local impacts are, where necessary, further described and analysed in the respective case studies.

1.3.1 Air Surface Temperature (AST)

Long-term observations of the Baltic Sea Basin mean AST indicates both decadal and seasonal trends. Annual temperature anomaly estimates show stronger fluctuations for the northern areas (north of 60°N) for the investigation period 1961–2001 (Jones & Moberg, 2003; HELCOM, 2007). Negative AST anomalies until

the 1920s were followed by a first warming phase ending in the 1930s (0.274 K/decade). After a period of cooling (−0.156 K/decade) the annual AST anomalies increased steadily since the 1970s, exceeding any previously observed rates in the early 1990s (1977–2001: 0.364 K/decade) (Jones & Moberg, 2003).

For the Baltic Sea Region south of 60°N the AST development is not dramatic. Up until the 1970s, no significant AST trends can be observed. Nevertheless, an even more distinctive AST increase since 1985 (1977–2001: 0.425 K/decade) (Jones & Moberg, 2003), was recorded and was strongest south and east of Tallinn and St Petersburg due to changing patterns of the atmospheric circulation (HELCOM, 2007). The annual linear AST trends for the investigation period 1871–2004 show an overall increase of 0.07 K/decade for latitudes <60°N and of 0.10 K/decade for latitudes >60°N (Heino et al., 2008). With an annual warming trend of 0.08 K/decade, the Baltic Sea ASTs increase faster than global temperatures (0.05 K/decade) (HELCOM, 2007).

For the southern area seasonal trends are significant in spring, autumn and winter, with the highest increase (0.11 K/decade) for spring temperatures (HELCOM, 2007; Heino et al., 2008). In the northern Baltic Sea Basin the most distinct warming trend is also recorded in spring (0.15 K/decade), whereas the development of winter temperatures is insignificant (Heino et al., 2008). Among other consequences, this resulted in a significantly prolonged growing season in the Baltic Sea Region.

Despite certain caveats and uncertainties, all existing projections indicate that atmospheric temperatures in the Baltic Sea Basin may continue to warm during the next decades. Simulations based on the IPCC A2 and B2 emissions scenarios of future AST in 2071–2100 show changes relative to the reference period 1961–90 between 2.8–4.8 K for the Baltic Sea Region (Meier, 2006). There are seasonal differences, indicating a stronger increase in wintertime AST as compared to summertime AST, which are especially high in the northern and eastern sub-regions of the Baltic Sea (Räisänen et al., 2004; HELCOM, 2007). Meier (2006) found the largest monthly mean AST change of 6 K in February (2071–2100). Moreover, the southern parts of the Baltic Sea Region may experience a more pronounced warming in summer than the northern parts (HELCOM, 2007).

1.3.2 Sea Surface Temperature (SST)

As the Baltic Sea is a relatively small and shallow semi-enclosed sea characterized by a low and strongly varying salinity of its surface waters (approximately 20 practical salinity units (PSU) in the Kattegat and 1–2 PSU in the Bothnian Bay and Gulf of Finland) (HELCOM, 2012), changes in SST occur comparatively fast. This holds true for both seasonal and long-term responses of sea temperatures to solar radiation and air temperatures.

Depending on the investigation period, analyses of SST data lead to different results. A reason for that is the long-term variability in the thermal development of the Baltic Sea. For example, the past 100 years were characterized by warming phases in 1920–40 and since the 1970s. These warming phases were interrupted by colder periods, whereby the SST increase rates of 0.65 K/decade since 1985 are unprecedented (Siegel, Gerth & Tschersich, 2008). The warmest years are observed since 1999 when there was a temperature rise of 0.8 K/year (Siegel, Gerth & Tschersich, 2006; HELCOM, 2007), showing strong seasonal and regional variations. The rise of temperatures in summer and autumn mainly determined the positive trend in SST for the Baltic Sea (Siegel, Gerth & Tschersich, 2008). On the other hand, analyses of modelled mean water temperatures for 1970 and 2002, averaged over all depths of the Baltic Sea, showed no trend at all (Heino et al., 2008).

Current simulations of the SST in the Baltic Sea Basin project a positive warming trend for the next decades. Regional coupled atmosphere-ocean models forced by the B2 and A2 emissions scenarios project an increase in annual mean SST between 2 to 4 Kelvin in the period 2071–2100 compared to 1961–1990, which would be most pronounced in the southern and central Baltic Sea (HELCOM, 2007). In comparison, Neumann and Friedland (2011), based their projections on the IPCC B1 and A1B emissions scenarios, assumed an increase in the order of 2–3.5 K until the end of the 21st century.

1.3.3 Precipitation

Compared to other parameters, precipitation varies greatly in time and space. Due to this and the poor data coverage as well as differing measurement techniques, it is difficult to establish long-term trends for the Baltic

Sea Basin. Long-term observations indicate seasonally varying precipitation patterns. For each season, both increasing and decreasing trends can be found for the period 1976–2000 compared to the period 1951–75 (HELCOM, 2007). Nevertheless, an annual increase in precipitation is observed for the period mentioned which however, varies strongly across regions (Heino et al., 2008).

Modelling the development of precipitation under climate change appears to be rather difficult, as the RCM results are still biased, often overestimating winter precipitation (Graham et al., 2008). The general winter precipitation trends may be intensified due to an increasing number and intensity of low-pressure systems from the Atlantic. Changes in summer precipitation may vary regionally with an increase in the northern parts of the Baltic Sea and a decrease in southern parts (HELCOM, 2007; Graham et al., 2008; Neumann & Friedland, 2011). Consequently, precipitation patterns may both shift seasonally and change geographically.

1.3.4 Sea level

Over long-term timescales, the Baltic Sea Region has been subject to dynamic processes, affecting sea level. One of the most important factors is the isostatic effect, due to post-glacial rebound of Fennoscandia, which results in an uplift of the Scandinavian plate with simultaneous lowering of the southern Baltic coast (Heino et al., 2008). Secondarily, since the end of the 19th century there is an eustatic sea-level rise (SLR) of about 1 mm/year (Ekman, 1999). A different factor that is affecting the mean sea level is the salinity gradient (see section on 'salinity') together with a mean west wind component in the Baltic Sea Basin which all cause an increase in the mean sea surface height 'from the Kattegat to the Gulf of Finland and the Bay of Bothnia by about 25 and 32 cm, respectively' as observed in the 20th century (Meier, Broman & Kjellström, 2004). All these factors have to be considered when discussing changes in sea levels along the Baltic coast. Except for the southern Baltic Sea (SLR 1.7 mm/year), post-glacial rebound combined with eustatic SLR results in decreasing sea levels (Ekman, 1996; Heino et al., 2008; Scotto, Barbosa & Alonso, 2009). In the Gulf of Finland, there is a slight net sea level decrease (1–2 mm/year) (Ekman, 1996). The strongest decrease rates can be found in the Gulf of Bothnia (8–9 mm/year) during the 20th century (Ekman, 1996; Heino et al., 2008). Based on observations from the late 19th century Johansson, Boman, Kahma and Launiainen (2004) found a considerable slow-down of sea level decrease or even slight sea level rise in recent decades. For example in Kokkola on the west coast of Finland, the recent SLR led to a less effective post-glacial rebound of the order of 4–5 mm/year compared to a long-term trend in land uplift of 7–8 mm/year (Schmidt-Thomé, Klein & Satkunas, 2010). Additionally, Johansson, Boman, Kahma and Launiainen (2001) found a significant increasing trend in the maximum values in the Baltic Sea nodal area which is at a range of 10 centimetres over half a century (1950–2000) which is more likely to be triggered by larger-scale changes in hydrological and weather conditions than by local storms.

Due to the previously mentioned reasons, an overall assessment of the development of future SLR for the entire Baltic Sea Region is hardly possible. But, it can be expected that some regions that are currently experiencing decreasing sea levels may be confronted with a SLR at the end of the 21st century as well (Fenger, Buch & Jacobsen, 2001).

1.3.5 Salinity

The Baltic Sea is characterized by a decreasing salinity gradient from southwestern to north/northeastern areas determined by the following factors: river runoff, net precipitation and water exchange with the North Sea (Meier, Broman & Kjellström, 2004; Meier, 2006; HELCOM, 2012). The development of salinity during the 20th century is statistically insignificant and no long-term trend can be found (Winsor, Rodhe & Omstedt, 2001; Heino et al., 2008). But because of variations in freshwater inflow and the zonal wind velocity, decadal effects in salinity can be detected (HELCOM, 2007).

Projections of the future development of salinity vary greatly. Nevertheless, it is likely that the salinity of the Baltic Sea might decrease until the end of the 21st century (HELCOM, 2007). On the basis of simulations using the A1B and B1 emissions scenarios, Neumann and Friedland (2011) report a decrease in sea surface salinity in the range of 1.5–2 g/kg or 8–50 % (Meier, 2006) until the end of the 21st century. Due to increasing rainfalls in the northern parts of the Baltic Sea, the river runoff may increase by

up to 15% (Meier, 2006) and may, therefore, influence especially the salinity of the northern and northeastern parts of the Baltic Sea Basin. The decrease in salinity would be more pronounced in the wintertime due to an enhanced river runoff (Neumann & Friedland, 2001).

1.3.6 Sea ice

The development of sea ice in the Baltic Sea is predominantly determined by atmospheric conditions due to the smallness of the basin (Stigebrandt & Gustafsson, 2003). The sea ice concentration 'increases approximately linearly with decreasing temperatures' starting at 1°C (Stigebrandt & Gustafsson, 2003). Currently, half of the Baltic Sea is ice-covered in the wintertime (Meier, 2006). The ice extent is mainly forced by the severeness of a winter season. Since the mid 1990s, the ice winters have been average or milder than average (HELCOM, 2007). Within the Baltic Sea Region, the length of the ice season decreased by 14–44 days in the twentieth century (HELCOM, 2007).

As there is a strong relationship between AST and SST, the future ice coverage is highly dependent on climate change induced warming temperatures. Therefore, the development towards more winters with a comparatively small sea ice extent as well as the decreasing length of the ice season are likely to continue in the Baltic Sea Region. Projections suggest a dramatic retreat of ice cover until the end of the 21st century, which would have a great impact on the Baltic winter climate in general (HELCOM, 2007). Simulations (based on A1B and B1) show that the sea may be covered by only one third of the recent coverage at the end of the 21st century (Neumann & Friedland, 2011). On average, large areas would become ice free and the ice season may decrease by one to two months in the northern parts and two to three months in the central parts of the Baltic Sea (HELCOM, 2007).

1.4 Chapter summaries

The chapters in this book are built on adaptation processes with a clear geographic reference, but touching different thematic aspects. They are grouped as far as possible according to common geographic entities as well as thematic preferences. The book starts off with some of the most important aspects of climate change adaptation, the structuring of communication processes and cost evaluations. In the following, applications of these topics are elaborated for planning aspects in urban and metropolitan areas. The management of resources under climate changes is elaborated in the next section, after which climate change impacts and adaptation measures on larger geographical entities comprising entire coastlines and the tourism sector are assessed. The book closes with four examples, reaching from international insurance approaches towards national activities on natural hazard mitigation and agricultural adaptation projects – towards grass-root level adaptation in cooperation with local people, all exemplarily displaying the wide range of ongoing activities.

Chapter 2 introduces two methods for participatory decision-making for climate change adaptation and their application in Kalundborg, Denmark (Bedsted and Gram). Scenario workshops are presented as a way of addressing uncertainties related to climate change and offering stakeholders the possibility of creating their own development visions for an area or a specific thematic issue. The citizen summit in turn allows citizens to discuss and decide on a set of options for the development of their town. The results provide guidance for political decisions. Chapter 3 and Chapter 4 show two assessment methods for specific adaptation options for flood protection. In Chapter 3 Boettle, Rybski and Kropp test, using the example of Kalundborg, the applicability of cost-benefit analysis (CBA) as a supporting tool for decisions about the level and timing of flood protection measures. In this context they show that the results of the CBA (and hence potential decisions based on the CBA) depend strongly on underlying assumptions of discounting rates and climate sensitivity to greenhouse gases. In Chapter 4 the analysis of flood protection options for two case studies in Northern Germany include a wider set of monetary and intangible criteria. Boettle, Schmidt-Thomé and Rybski show that multi-criteria decision analysis (MCDA) can support decision making and increase the acceptance among stakeholders.

The methods of scenario workshops, CBA and MCDA were adjusted to and applied in a range of other case studies of the BaltCICA project. This is reflected in the following chapters.

The role of climate change in an urban context is discussed on the basis of case studies in six chapters. Tuusa et al. (Chapter 5) look at the development process of the climate change adaptation strategy for the Helsinki Metropolitan Area (HSY, 2012) with a special emphasis on interviews of key stakeholders.

Chapter 6 about Riga, Latvia (Kūle et al.) highlights several aspects of flood protection and climate change. On the background of historical flood events and protection measurements it assesses in detail the communication processes and generation of knowledge related to flooding and introduces Multi Criteria Decision Analysis as a potential support tool for strategic flood risk management in Riga.

Chapter 7 by Knieling and Schaerffer takes a look at developments and new plans along the River Elbe and the harbour area in Hamburg, Germany. These developments are critically reflected upon with respect to the aim of absolute flood safety compared to more flexible concepts of flood risk management.

Bergen is among the most active cities with respect to climate change adaptation in Norway (Dannevig, Rauken & Hovelsrud, 2012). Chapter 8 by Langeland, Klausen and Winsvold focuses on the learning processes, knowledge transfer and coordination that take place among the numerous institutions and project addressing climate change.

Though clearly located in an urban context Chapter 9 by Rimkus, Kažys, Stonevičius and Valiuškevičius illustrates the adaptation process based on specific adaptation measures for flood protection along the Smeltalė River in Klaipėda, Lithuania. The chapter includes the modelling of potential climate change impacts on precipitation and flooding, multi-criteria decision analysis for decision support and the participatory development and assessment of a set of adaptation options. In Chapter 10 Ahonen, Valjus and Tiilikainen explain how geological investigations can help to open up the discussion on climate change adaptation on a broader scale. It starts with a thorough geological investigation of the ridge (esker) and popular housing area that separates the two lakes Näsijärvi and Pyhäjärvi in Tampere, Finland. It then describes the most important impacts of climate change to be expected in the investigated area and points out further issues that should be addressed in municipal planning.

Chapter 11 on water supply in the municipality of Hanko, Finland (Luoma, Klein and Backman) describes how potential climate change impacts on water supply can be assessed with the help of groundwater flow models and how they can provide supporting information for groundwater management. Also, Chapter 12 on groundwater availability in Klaipeda, Lithuania (Arustienė, Gregorauskas, Kriukaitė and Satkūnas) describes the assessment of changing climate conditions on groundwater availability, but also addresses groundwater as an important resource for drinking water in Europe. In Chapter 13 Klamt and Schernewski explore the potential for commercial mussel farming in the southern Baltic with the Blue mussel and the Zebra mussel as two suitable species. Mussel farming is not only seen as an economic potential thanks to higher temperatures and less sea ice in winter, but also as an effective measure to reduce eutrophication in the Baltic Sea.

Sea level rise and changes in flood patterns concern not only urban development but are expected to affect the entire coastal zone. These potential changes are assessed by Petersell, Suuroja, All and Shtokalenko (Chapter 14) for the west Estonian coast. They also investigate potential consequences for groundwater quality, loss of forest and arable land and costs for built-up areas. In Chapter 15 Satkūnas, Jarmalavičius, Damušytė and Žilinskas investigate the geomorphological conditions for Karkle beach in Lithuania, potential consequences of climate change and the indirect effects on long-term plans and investments for tourism in this area. Interestingly, not only can the state of the coastal zone, but also the water conditions in the Baltic Sea have effects on the tourism development in the Baltic Sea Region. Mossbauer, Dahlke, Friedland and Schernewski (Chapter 16) model the potential effects of climate change and the Baltic Sea Action Plan on the growth and development on macroalgae accumulations along the German shore and outline the implications for the maintenance of beaches. Two chapters in this book deal specifically with tourism and climate change. Filies and Schumacher (Chapter 17) identify a wide range of direct, indirect and induced impacts of climate change on tourism at the German Baltic Sea coast. Based on their results they challenge the view that the tourism sector could benefit from climate change thanks to higher temperatures and changes in precipitation.

Additionally, they compare tourism experts' views on adaptation with adaptation requirements suggested by science. Since tourism also depends highly on the satisfaction of the customers (tourists) it seems to be consequent to take tourists' perceptions and opinions into account in strategic planning. In Chapter 18 Donges, Haller and Schernewski show that visitor questionnaires can provide valuable input for short-term decisions and planning, but have restricted benefits for long-term strategies for climate change adaptation.

The outreach section of this book sets off with Chapter 19 in which Olcina analyses recent modifications to Spain's national, regional and local planning regulation to mitigate the impacts of natural hazards. Meanwhile these policies mainly focus on current hazard patterns; initiatives already go further in order to respect potential climate change impacts too.

In Chapter 20 Sano, Prabhakar, Kartikasari and Irawan describe in detail how Indonesia's agricultural sector is planning to adapt to climate changes in order to safeguard food security. Besides the largely positive activities the chapter also explains the difficulties of applying and implementing climate change adaptation on the local level.

In Chapter 21 Saroar and Routray analyse the often-predicted climate change induced mass emigration scenarios at the grass root level. Vulnerability reductions to hinder prospected emigration patterns are explored.

Prabhakar, Rao, Fukuda and Hayashi (Chapter 22) study risk insurances in the Asia-Pacific region. Focusing on India and Japan they identify a set of potential short-comings in the currently available insurance schemes. These include the affordability of insurance premiums, the access to risk insurances for individuals and the availability of information about risks. They suggest the UNFCCC as a suitable platform to enhance risk insurance as a means of risk reduction and climate change adaptation.

Acknowledgements

The European Regional Development Fund's (ERDF) Baltic Sea Region Programme 2007–2013 part-financed the BaltCICA project and the publication of this book. The International Union of Geological Sciences' Commission on GeoEnvironment (IUGS/GEM) part-financed activities to seek for potential case studies of the outreach chapter. Professor Joy Pereira from the South East Asia Disaster Prevention Institute (SEADPRI) strongly supported the book by identifying and contacting authors for the outreach chapter. The editors would also like to thank the numerous reviewers of each article for their constructive comments.

References

Adger, W.N., Dessai, S., Goulden, M., Hulme, M., Lorenzoni, I., Nelson, D.R. and Wreford, A., 2009. Are there social limits to adaptation to climate change? *Climatic Change*, 93 (3), pp.335–354.

COM (Commission of the European Communities), 2009. *White paper – Adapting to climate change: Towards a European framework for action*. [pdf] Brussels: COM. Available at: <http://eur-lex.europa.eu/lexuriserv/lexuriserv.do?Uri=CELEX:52009DC0147:en:NOT> [Accessed 22 August 2012].

Dannevig, H., Rauken, T. And Hovelsrud, G., 2012. Implementing adaptation to climate change at the local level. *Local Environment*, 17 (6–7), pp.597–611.

Dessai, S. And Hulme, M., 2004. Does climate adaptation policy need probabilities? *Climate Policy*, 4 (2), pp.107–128.

Ekman, M., 1996. A consistent map of the postglacial uplift of Fennoscandia. *Terra Nova*, 8 (2), pp.158–165.

Ekman, M., 1999. Climate changes detected through the world's longest sea level series. *Global and Planetary Change*, 21, pp.215–224.

Fenger, J., Buch, E. And Jacobsen, P.R., 2001. Monitoring and impacts of sea level rise at Danish coasts and near shore infrastructures. In: Jørgensen, A.M.K., Fenger, J. And Halsnaes, K., eds. 2001. *Climate Change Research – Danish Contributions*. Copenhagen: Gads Forlag, pp. 237–254.

German Association of Cities, 2012. *Positionspapier Anpassung an den Klimawandel – Empfehlungen und Maßnahmen der Städte*. [online] Available at: <http://www.staedtetag.de/fachinformationen/umwelt/059004/index.html> [Accessed 05 September 2012].

Graham, L.P., Chen, D., Bøssing, O., et al., 2008. Projections of future anthropogenic climate change. In: Bacc Author Team, 2008. *Assessment of Climate Change for the Baltic Sea Basin*. Berlin, Heidelberg: Springer-Verlag, Ch.3.

Hagen, E. And Feistel, R., 2008. Baltic climate change. In: R. Feistel, G. Nausch and N. Wasmund, eds. 2008. *State and Evolution of the Baltic Sea, 1952–2005*. Hoboken, New Jersey: John Wiley & Sons, Inc., Ch.5.

Heino, R., Tuomenvirta, H., Vuglinsky, V.S. and Gustafsson, B.G., 2008. Past and current climate change. In: BACC Author Team, eds. 2008. *Assessment of Climate Change for the Baltic Sea Basin*. Berlin, Heidelberg: Springer-Verlag, pp.35–131.

HELCOM (Helsinki Commission), 2007. *Climate Change in the Baltic Sea Area – HELCOM Thematic Assessment in 2007*. Baltic Sea Environment Proceedings No. 111.

HELCOM, 2012. *The brackish nature of the Baltic Sea*. [online] Available at: <http://www.helcom.fi/environment2/nature/en_GB/nature/> [Accessed 09 July 2012].

HSY (Helsinki Region Environmental Services Authority), 2012. *Helsinki Metropolitan Area Adaptation to climate change strategy*. Helsinki: HSY.

IPCC, 2007. *Fourth assessment report (AR4): Climate Change 2007. Contribution of working groups I, II and III to the fourth assessment report of the Intergovernmental Panel on Climate Change (IPCC)*. Cambridge: Cambridge University Press.

Johansson, M., Boman, H., Kahma, K.K. and Launiainen, J., 2001. Trends in sea level variability in the Baltic Sea. *Boreal Environment Research*, 6, pp.159–179.

Johansson, M.M., Kahma, K.K., Boman, H. And Launiainen, J., 2004. Scenarios for sea level on the Finnish coast. *Boreal Environmental Research*, 9, pp.153–166.

Jones, P.D. and Moberg, A., 2003. Hemispheric and large-scale surface air temperature variations: an extensive revision and an update to 2001. *Journal of Climate*, 16, pp.206–223.

Meier, H.E.M., 2006. Baltic Sea climate in the late twenty-first century: a dynamical downscaling approach using two global models and two emission scenarios. *Climate Dynamics*, 27, pp.39–68.

Meier, H.E.M., Broman, B. And Kjellström, E., 2004. Simulated sea level in past and future climates of the Baltic Sea. *Climate Research*, 27, pp.59–75.

Neumann, T. And Friedland, R., 2011. Climate change impacts on the Baltic Sea. In: G. Schernewski, J. Hofstede and T. Neumann, eds. 2011. *Global Change and Baltic Coastal Zones*. Dordrecht, Heidelberg, London, New York: Springer.

Räisänen, J., Hansson, U., Ullerstig, A. et al., 2004. European climate in the late twenty-first century: regional simulations with two driving global models and two forcing scenarios. *Climate Dynamics*, 22, pp.13–31.

Schmidt-Thomé, P., Klein, J. And Satkunas, J., 2010. Climate change, impacts and adaptation – some examples of geoscience applications for better environmental management in the Baltic Sea. *Episodes*, 33 (2), pp.102–108.

Scotto, M.G., Barbosa, S.M. and Alonso, A.M., 2009. Model-based clustering of Baltic sea-level. *Applied Ocean Research*, 31, pp.4–11.

Siegel, H., Gerth, M. And Tschersich, G., 2006. Sea surface temperature development of the Baltic Sea in the period 1990–2004. *Oceanologia*, 48 (S), pp.119–131.

Stigebrandt, A. And Gustafsson, B.G., 2003. Response of the Baltic Sea to climate change – theory and observations. *Journal of Sea Research*, 49, pp.243–256.

Suomen Kuntaliitto, 2010. *Kuntaliiton ilmastolinjauk-set*. [pdf] Available at: <http://www.kunnat.net/fi/asiantuntijapalvelut/yty/ilmastonmuutos/ilmastolinjauk set/Documents/ilmastonmuutos_ebook.pdf> [Accessed 06 September 2012].

The Congress of Local and Regional Authorities, 2012. *Making cities resilient resolution 339*. Strasbourg: COE (Council of Europe).

Winsor, P., Rodhe, J. And Omstedt, A., 2001. Baltic Sea ocean climate: an analysis of 100 yr of hydrographic data with focus on the freshwater budget. *Climate Research*, 18, pp.5–15.

2 Participatory Climate Change Adaptation in Kalundborg, Denmark

B. Bedsted & S. Gram

The Danish Board of Technology, Copenhagen, Denmark

2.1 Introduction

The municipality of Kalundborg is situated on the west corner shoreline of Zealand, Denmark. Like many municipalities along the Danish coast, they have only recently started to consider the need to develop adaptation strategies. Municipalities on the west coast of Jutland are used to dealing with storm surges from the North Sea, but coastal areas in the rest of Denmark are better protected from such surges and have dealt with them on a much less regular basis. A change is anticipated.

The municipality of Kalundborg joined the BaltCICA project because their representative in Brussels became aware of the opportunity and because the project could contribute to the increased knowledge of and experience of local climate adaptation in the municipality. Had they not joined, it is safe to conclude that the attention to climate change adaptation among citizens, stakeholders and politicians in the municipality would have been less conspicuous. While Denmark has adopted a national strategy for climate change adaptation, this strategy does not impose any obligation on municipalities to make their own strategies, nor does it provide municipalities with much information on how to proceed with such strategies.[1]

Any obligation would have been met by municipalities with claims of financial support from the Danish state, which has so far been very limited.

Thus, while the municipality of Kalundborg considered making a climate adaptation strategy before joining BaltCICA, they did not plan to make it a very detailed one, and they joined BaltCICA with the purpose of taking a close look at an area located in the south-western part of the municipality. Flooding is already an issue in this area, as it occasionally affects farmers and summer cottage owners. BaltCICA was considered to be one of several ways to gain the required knowledge to draw up a climate adaptation strategy for that particular area, and possibly to provide inspiration for such a strategy covering the entire municipality.

The case study area was selected in cooperation with the Danish Board of Technology (DBT) and the Geological Survey of Denmark and Greenland (GEUS), the two other Danish BaltCICA partners. GEUS became part of the Danish team in order to provide geological data, whereas DBT specializes in the involvement of stakeholders and citizens in political decision-making processes with regards to technological and scientific issues such as energy, biotech, healthcare, IT, biodiversity and climate change (Vig & Paschen, 2000; DBT, 2012). The case study area was

---

[1] This situation changed somewhat, when a new centre-left government won the general election of September 2011. The new minister for the Environment announced in the autumn 2011 that all municipalities should make a climate adaptation strategy within the next two years. At the time of writing this chapter, it was still not decided how detailed such strategies should be.

Climate Change Adaptation in Practice: From Strategy Development to Implementation, First Edition.
Edited by Philipp Schmidt-Thomé and Johannes Klein.
© 2013 John Wiley & Sons, Ltd. Published 2013 by John Wiley & Sons, Ltd.

selected, in order to address future problems with flooding caused by storm surges and heavy precipitation. Moreover, it was selected in order to include different and potentially conflicting interests from local inhabitants and stakeholders. The intention was to make a climate adaptation strategy for the area, and at the outset of the project it was decided that DBT would assist the municipality in these efforts by organizing both a scenario workshop and a citizen summit. While the municipality had some preliminary understanding of how these methods work, they later got somewhat surprised by the strain they put on their decision-makers and civil servants.

This chapter presents the implementation of a participatory climate change adaptation process. It focuses on practical issues and experiences in the case study of Kalundborg. For further reading related to the broad scientific and theoretical discussion on citizen participation the literature mentioned in this chapter can serve as a starting point.

2.2 Climate data

Solid facts are not abundant in climate change research, and relating to them in a planning perspective involves a number of both practical and political choices.

At the outset of BaltCICA, GEUS made a series of calculations of the expected sea level, storm surges and precipitation patterns in 2090. This year was chosen both because of the availability of climate model figures and because the main interest of the municipality was a long-term planning horizon. The starting point for the calculations was the A2 scenario developed by the Intergovernmental Panel on Climate Change (IPCC, 2000). The A2 scenario is the most pessimistic of the two development scenarios recommended by the Danish government for planning purposes. At that time, though, new research on the ablation of the ice cap on Greenland was emerging with estimates of sea level rise between 90 cm and, worst case scenario, up to two meters for A2 (DMI, 2009). Although those estimates were subject to a great deal of uncertainty, the Danish partners agreed

to choose what we considered a conservative estimate of 80 cm sea level rise in 2090.

On this basis, the municipality itself made some modelling, using advanced 3D software.[2] This led to the production of maps, showing potential consequences of storm surges in the future, partly in combination with incidents of heavy precipitation.

Rather than working with different IPCC scenarios, the Danish partners agreed that it would be better to work only with the one chosen. One might argue that by doing so, the uncertainty of future developments reflected by the different IPCC scenarios was downplayed. The counter argument was that for practical purposes it would be too difficult to relate to multiple scenarios in the decision-making process, and that uncertainties would be a prominent fact, regardless of how many climate scenarios were addressed. Thus, throughout the study, the uncertainties were highlighted while maintaining a combination of the A2 scenario and the latest research on the expected sea level rise.

2.3 The case study area

The case study area around Reersø and Tissø is an exemplarily Danish rural area, and there are many more like it along the Danish coastline. It is dominated by farmland and to a lesser extent by protected nature areas, scattered settlements and summer cottage areas. It is inhabited by approximately 12 000 residents (out of which 321 live all year round in their summer cottages), including 6839 in the hamlets of Gørlev and Høng, two areas that are not, however, expected to be seriously affected by future floods.

The summer cottages in the low-lying areas by Ornum Strand, Bjerge Sydstrand, Bjerge Nordstrand and on the peninsula of Reersø are expected to get most seriously affected by future floods. Altogether, there are 3036 summer cottages in the area. Equally exposed are some permanent residences, large farmland areas and internationally protected nature areas with meadows, bogs, streams and lakes. The area around Flasken and Vejlen is particularly vulnerable,

[2]Encom Discover 3D – extension module for MapInfo.

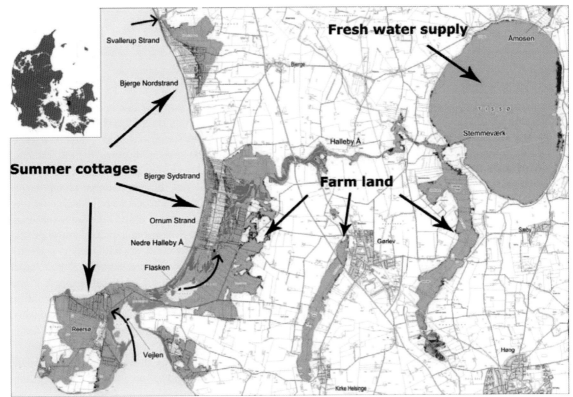

Figure 2.1 Salt water flooding in 2090.
Source: Map produced by the Municipality of Kalundborg.
Notes: [a]light blue: (80 cm above current sea level) areas expected to be permanently flooded by 2090.
[b]green: (150 cm above current sea level) areas currently flooded at 100-year incidents.
[c]yellow: (210 cm above current sea level) areas expected to be flooded at 20-year incidents in 2090.
[d]red (230 cm above current sea level) areas expected to be flooded at 100-year incidents in 2090.
[e]blue arrows: point at the locations where the water will enter first.

at the mouth of the stream called Nedre Halleby Å, currently almost unregulated and with a delta and lagoon-like character.

The infrastructure in the area holds public roads, sewage systems, electrical supply, water supply and drainage. It holds groundwater supplies for drinking water and fresh water from Tissø Lake (the source of Nedre Halleby Å) is used for industrial purposes in Kalundborg. The area is somewhat important for tourism in the municipality of Kalundborg and includes several locations of interest with regards to cultural heritage. A large part of the rain falling on the middle and western parts of Zealand flows through this area before reaching the sea.

The map above shows how the case study area is foreseen to be affected in 2090 (Figure 2.1). In particular, residences in the town of Reersø and summer cottages on the peninsula of Reersø, Ornum Strand, Bjerge Nordstrand and Bjerge Sydstrand are exposed to future floods. In a situation of flooding from the sea combined with heavy precipitation, low-lying summer cottages at Bjerge Sydstrand will be particularly exposed, because rain water from a large catchment area in the hinterland will flow in that direction and meet salt water from the flooding. This scenario is not pictured on the map below, though. The accumulated cost of damages to private properties by 2090 is estimated by a private consultancy, NIRAS,

Figure 2.2 The decision-making process from local climate modelling to a municipal adaptation strategy.

to be approximately 242 million Euro (Municipality of Kalundborg, 2011).

2.4 Methods in general – the entire process

DBT was among the first organisations in Denmark to initiate a public and political debate on climate change adaptation.[3] One of the early lessons learned was that decisions with regards to spatial planning in future flood-prone areas are often of a political nature, although they may seem only technical in nature to the people making them. Protecting an area from flooding may benefit landowners, but may not be sustainable from a societal perspective in the long run. Thus, small decisions about the introduction of protective measures increase expectations and demands for future decisions about additional protective measures and exclude a democratic debate on alternative possible futures, for example on whether one should continue protecting current land use or let nature take its course. The methods chosen for the decision-making process in Kalundborg were designed to stimulate such democratic debate.

Two specific methods were chosen in combination in order to build up a deliberative decision-making process: A scenario workshop and a citizen summit. The scenario workshop was designed to involve local stakeholders in the development of different possible future land uses and adaptation measures. The scenario workshop method was first developed in the early 1990s to find ways of developing urban ecology (Andersen & Jæger, 1999). The citizen summit was designed to consult ordinary citizens about their views on the abovementioned possible futures, adaptation measures and principles for an adaptation strategy. The methodology was developed by America Speaks

around the year 2000 and has been used and moderated by the DBT since 2005 (DBT, 2006). The idea was that while the scenario workshop involved stakeholders from the case study area, the citizen summit should involve citizens from the entire municipality. The rationale for this combination of approaches was partly that, although local stakeholders could contribute with local knowledge and innovative solutions, they may have a tendency to look for (costly) protective solutions, whereas citizens with no personal stake in the case study area may give higher priority to other adaptation options. As the project developed, it was decided that the citizen summit should also address adaptation options in other parts of the municipality, thus allowing citizens to compare and prioritize adaptation options in different parts of the municipality.

Throughout 2009, the climate modelling was made and the scenario workshop took place in autumn 2009. In 2010, the adaptation options developed at the scenario workshop were further elaborated by DBT and Kalundborg. Through dialogue with the administration and the politicians, adaptation options for other parts of the municipality vulnerable to future flooding (including the city of Kalundborg) were developed, and alternative (sometimes conflicting) guidelines for an adaptation strategy were identified. In March 2011, a citizen summit with 350 participants took place, in which citizens deliberated and voted for general adaptation guidelines and for different adaptation options for both the case study area (developed by the scenario workshop) and for other parts of the municipality. In 2011, the results were analysed, debated by the politicians, and the administration started drafting up an adaptation strategy, based partly on the results from the citizen summit, partly on further assessments of climate impacts in the municipality, and partly on fairly general guidelines from government agencies and ministries. The whole decision-making process is illustrated in Figure 2.2. The following part of the chapter describes the different project phases and the methodological approach in more detail.

[3]Read more about DBT climate change adaptation scenario workshops in 2003 at http://www.tekno.dk/subpage.php3?article=1089&toppic=kategori11&language=uk

2.5 Scenario workshop – in detail

Scenario workshops ensure a participatory involvement at a very early stage of the development of concrete adaptation measures, thus increasing the likeliness of their implementation by the stakeholders involved, although such results cannot be guaranteed no matter how well the involvement is organized. They require a limited amount of technical data and can set the stage for the subsequent development of concrete adaptation options. They can do so by developing more general developmental visions for the local society, and by identifying more concrete technical issues and conflicts of interest that need to be dealt with.

Preparations for the scenario workshop in the case study area commenced four to six months prior to the workshop and included two main tasks besides the practical preparations:

1. Development of future scenarios to raise awareness and to provoke creativity at the workshop.
2. Identification of local stakeholders to participate in the workshop.

2.5.1 Development of scenarios

The starting point of the scenarios was the preparation of maps pointing out the effects of severe floods and other climate change impacts in 2090. As described above, it involves a number of practical and political choices to choose data, but making the maps also requires a significant amount of local knowledge; for example, knowing the precise height of various infrastructural constructions (roads, sluices, dykes, etc.).

After the maps had been created, the consequences for the local community became clearer, and preparations of the scenarios, to be presented to the stakeholders at the scenario workshop, began. These preparations involved telephone interviews with local stakeholders who would presumably be affected by future floods, and they involved extensive discussions among the Danish BaltCICA partners. Local stakeholders had knowledge about previous flood events and the possible effects of future flooding, and they had experience in and ideas of how to meet the challenge. This knowledge was used to develop the scenarios (DBT, 2009).

The scenarios used at the scenario workshop are short (two to four pages each) and explain in a journalistic style the effects of having chosen different strategies to deal with future flooding. They are science fiction made to provoke debate among stakeholders and, thus serve as a starting point for the development of adaptation options at the scenario workshop. Thus, the word 'scenario' is used rather differently for this method, than it is in climate modelling and foresight studies.

The scenarios were written by a science journalist who took part in the research among local stakeholders. It was eventually decided to develop three future scenarios:

1. A basic scenario or 'laissez-faire' scenario based on the assumption that it is not desirable to do anything special in advance to alleviate the impact of future climate changes, beyond what is within immediate economic reach. Initiatives will be taken ad hoc.
2. A so-called 'protection scenario' based on an attempt to protect current land uses as much as possible, including residential areas, infrastructure, commerce and agriculture, against the consequences of future climate changes. Initiatives will be launched in order to protect existing economic interests, even if this has negative consequences for the environment and nature.
3. A so-called 'adaptation scenario' based on the need to adapt to future climate changes rather than fight against them. Current land uses, such as farming practices, will be re-evaluated and adjusted to the changing environment, and more space will be allocated to wetlands. The scenario also assumes that much consideration will be given to environment and nature.

Each of the three scenarios is an attempt to describe the pros and cons of the different adaptation options involved as well as the possible effect of these options on the local stakeholders and community at large. Approximately three weeks prior to the workshop, the scenarios were sent to the participating stakeholders.

2.5.2 Identification of stakeholders

While researching for the development of the scenarios, a simultaneous research was carried out to identify potential participants in the scenario workshop. Participants should be local stakeholders with an interest likely to be affected by climate change and with a position in the local community investing them with the power required to push for the implementation of adaptation measures, if needed. Thus, a 'stakeholder' can be more narrowly or broadly defined depending on what one would define as a relevant stake in the future development of the area in question. The DBT are strong advocates of a rather broad and inclusive definition, for both practical and principled reasons (Bedsted, 2007).

The key to a successful involvement process is to consider carefully, when and how to involve who and to communicate precisely and up front about, what kind of influence those involved will have on the decision-making process. Eventually, 28 participants[4] were invited to participate in the scenario workshop, including local politicians, local and regional officials (technicians, civil servants), farmers and representatives from home owners' associations, nature and environmental organizations, outdoor organizations, harbour authorities, youth (from secondary school), the tourist and business committee, the water supply sector, dyke and pump associations and the archaeological society. They all received written invitations and were contacted by telephone in order to explain to them the kind of process they were invited to take part in. Only a few declined, and if so, pointed to a more relevant participant. Great care was taken both before and during the workshop to explain to the participants what their input would be used for in the decision-making process. They were told explicitly that they would have no privileged say in the future decision-making process, but that most of their suggestions for adaptation measures would be discussed at a citizen summit.

2.5.3 The scenario workshop programme

The scenario workshop took place over two days with three weeks in between in the autumn of 2009. The working method alternated between group work and plenaries. The programme was structured around three consecutive phases: the *Critical analysis phase*, the *Visionary phase* and the *Implementation phase*. Groups of four to six participants were brought together representing different and presumably conflicting interests. One could also choose to make the groups including members with similar, rather than conflicting inter-

ests. One could also choose to alternate between these two group constructions throughout the workshop. There are many pros and cons of each procedure, the description of which exceeds the limits of this article.

The scenario workshop form and rules are there to ensure that everyone is heard, that all ideas are included in the debate and that participants work towards formulating an action plan.

- *Critical analysis phase* (day 1)

The task of the critical analysis phase is to criticize the scenarios – to provide both positive and negative criticism based on the views, knowledge and experiences of the participants. The scenarios represent different possible futures. They are not predictions, and the task does not involve choosing the preferred scenario or assessing which is the most probable. The scenarios are there to inspire criticism, which can lead to the development of new visions and adaptation measures.

- *Visionary phase* (day 1)

Using the knowledge gained from the critical analysis phase, the purpose of the visionary phase was to have the groups develop their own visions for the future. Participants could include elements of the pre-constructed future scenarios and make up their own. Groups were allowed to come up with more than one vision for the future.

In the time span of three weeks in between day one and day two, the results of the first workshop day were analysed and the visions developed in the different groups merged into four different visions. These four visions were sent to the participants prior to day two, and they were invited to state in advance which of the visions they would prefer to elaborate on during the second day of the workshop. Day two started with corrections and acceptance of the merged visions.

- *Implementation phase* (day 2)

The purpose of the implementation phase was an action plan for the implementation of each of the four visions. Taking into consideration how to implement a vision in real life, a number of barriers became apparent. These barriers may be economic, cultural, social, organizational, political or technical (for a discussion of barriers specific to climate change adaptation see, for example, Adger et al., 2009). Dealing with these barriers may lead to adjustments of the vision to be elaborated on. The action plans were drawn out along a timeline displaying who was responsible for what

[4]The ideal number of participants for a scenario workshop is between 25 and 40.

and when, that is, responsibilities of the landowners and the municipality, respectively.

The outcome of the scenario workshop was four different visions, arguments for and against these visions, plans for their implementation, a long list of technical issues to be clarified and a clearer sense of the political choices involved in the identification and implementation of adequate adaptation measures.[5]

The four different visions developed were:

1. Transforming the area into a nature area.
2. Phasing out vulnerable properties but allowing interim protection.
3. Establishment of onshore dykes and river dykes.
4. Construction of large offshore dykes.

Of the four visions, protection based on offshore dykes and transforming the area into a nature area seemed the most radical and controversial in the eyes of the Kalundborg administration, the former probably most so. It was fairly evident that these two visions had not been among the first considered had the administration made the decision according to regular procedures. There was agreement, though, among the Danish partners to stick to the four visions and thus fulfil the promise made to stakeholders at the scenario workshop that their ideas for the future would be put to a vote later on the citizen summit.

2.6 Transnational cooperation

The scenario workshop methodology was presented and debated at several meetings with BaltCICA partners, and it was applied in both Klaipeda City, Lithuania and Hamburg, Germany. Because the practical and political context differed in both places, so did the method, and the project partners had extensive exchanges (including study trips) in order to determine how to put the methodology to best use (for a discussion of the importance of adjusting participatory methodologies to the political context, see Klüver et al., 2004). In Klaipeda City, it was a given premise that some kind of protection should be implemented in order to prevent flooding in parts of the city from

the Smeltale River. It was just not clear how to do it exactly, and therefore the scenario workshop was able to focus more closely, than was the case in Kalundborg, on a few alternative options to choose from. In Hamburg, a much larger area than Kalundborg was the object of interest, and it therefore involved an increased multitude of interests. Not directly integrated with the ongoing decision-making process in Hamburg, the scenario workshop offered stakeholders the opportunity to explore adaptation measures in a new and innovative setting.

Other BaltCICA partners used parts of the scenario workshop methodology in their countries. A survey among the partners shows that the two most appealing elements seem to have been the use of future scenarios as a starting point for discussions and the efforts made to involve a wide variety of stakeholders in such discussions.

2.7 Developing adaptation options for the citizen summit

After the scenario workshop in Kalundborg, much effort was put into developing the different visions drawn out by the stakeholders. A consultancy firm, NIRAS, was involved in order to estimate the practical viability, the environmental consequences and economical costs of implementing the adaptation options in those visions. Assessments were also made by technical experts within the municipality.

At the same time, discussions began between the municipality and DBT about adaptation challenges and options in other parts of the municipality. Flood scenarios were modelled (Figure 2.3) for other areas in the municipality, and alternative adaptation options identified. DBT's role in this process was partly to assist the administration with a clarification of the adaptation options available, but mostly to identify the political choices involved in choosing one adaptation measure over the other. Such choices may seem obvious at the outset but could change once further examined. For example, it seemed obvious that the municipality has a responsibility to protect the city of Kalundborg from storm surges, and therefore the question seemed to be, if it should be protected one way or the other. When further examined, though, it turned out that a more pertinent question was whether or

---

[5]Rather than ending up with alternative solutions, the scenario workshop can also be designed to target a consensus-oriented output, helping participants reach an agreement on one joint vision for the future.

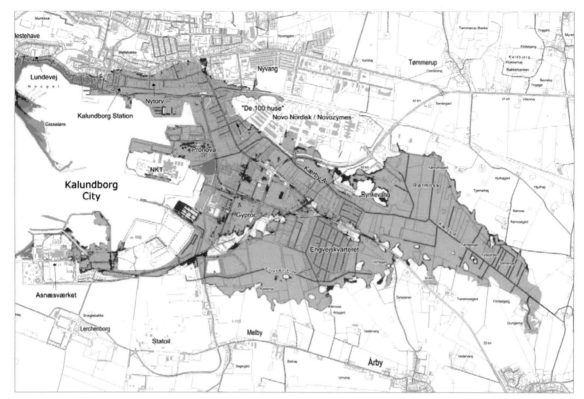

Figure 2.3 Anticipated flooding in Kalundborg City.
Source: Map produced by the Municipality of Kalundborg.
Notes: [a]blue: (80 cm above current sea level) areas expected to be permanently flooded by 2090.
[b]green: (150 cm above current sea level) areas currently flooded at 100-year incidents.

[c]yellow: (210 cm above current sea level) areas expected to be flooded at 20-year incidents in 2090.
[d]red: (230 cm above current sea level) areas expected to be flooded at 100-year incidents in 2090.

not one should make private companies on the waterfront co-responsible for protecting the city. Thus, what seemed at first sight to be a technical issue turned out to be equally relevantly dealt with as a political one, which is a point often made in literature about technical changes (e.g. Winner, 1986).

In general, it became the role of DBT to encourage the municipality to identify and make clear the most pertinent political decisions that had to be taken in order to make an adaptation strategy, even if such decisions could cause controversy. The process included discussions with the political Committee for Technical and Environmental Issues ('Udvalget for Teknik og Miljø'). The politicians were presented to the citizen summit concept and gave their input to relevant questions to address. They fully accepted the idea that citizens could be consulted, when it came to

political decisions they were responsible for. They also saw such a consultation as a possibility to examine public acceptance of political actions that were expected to be unpopular, such as implementing regulations for handling rain water on private properties.

2.8 Citizen summit – in detail

A citizen summit allows a large number of citizens to deliberate on a set of questions with fellow citizens and vote individually on alternative answers. Contrary to other methods of involving citizens in a decision-making process (such as consensus conferences and interview meetings) a citizen summit leaves little room for citizens to develop their own suggestions for dealing with a certain challenge (for an overview of different participatory methods, see Steyaert & Lisoir, 2005). It is a method often used at the stage of a decision-making process, in which clear political choices have already been identified, and decisions should ideally be made in the near future.

A citizen summit can involve from 100 to many thousand citizens, and the large number of participants contributes to making it an effective tool at the political level. Unlike conventional opinion polls, it aims at providing the participating citizens with balanced and science-based information, as well as the opportunity to deliberate for a full day with other citizens prior to rendering their judgement. It thus allows for detailed questions and well-considered responses. In spite of this, the municipality was worried about putting sensitive issues to the vote. They feared that citizens would mostly have their own private interests in mind and show less concern for the kind of priorities the administration would have to make.

2.8.1 The citizens

In Kalundborg, 350 citizens participated in the citizen summit on 5 March 2011. They were chosen to reflect the demographic distribution in the municipality with regards to age, gender and geographical residency. An invitation to the citizen summit was sent to 7000 randomly selected citizens in the municipality. Five hundred were selected out of the positive responses. Out of those, some cancelled well in advance of the summit and others cancelled a few days before, that is, due to illness. Because the statistical validity of results from citizen consultations like this one is the source of much discussion, suffice it to say that the sample of citizens is large and diverse enough to give a sense of the general trends in the opinions of the citizens of Kalundborg (for a discussion of representativeness and citizen consultations, see Agger, Jelsøe, Jæger & Phillips, 2012).

2.8.2 The questions

The questions were prepared in 2010. As described above, the questions should reflect the most pertinent political decisions that need to be taken in order to develop an adaptation strategy. Identifying such questions can be a long process, and it is closely related to the development of the information material for the citizens, because the research involved can lead to new questions.

Eventually, a set of 19 questions was developed. It was divided into six thematic subjects:
1. Personal experiences with flooding and demographic data.
2. Vulnerable rural areas (such as the case study area).
3. Kalundborg City.
4. Dividing responsibilities between citizens and authorities.
5. General climate adaptation strategy.
6. Involvement of citizens in planning for climate adaptation.

For example, citizens were asked what kind of development they prefered for the case study area:
1. Transforming it into a nature area.
2. Phasing out vulnerable properties but allowing interim protection.
3. Establishment of onshore dykes and river dykes.
4. Construction of offshore dykes.

Different financing mechanisms were discussed and voted on as well.

Another question was:

Should the municipality have the authority to alter the status of exposed summer cottage areas, so they can be transformed, in the long term, into nature areas with periodic and permanent flooding?
• Yes
• No
• Don't know

2.8.3 Information material

The information material for a citizen summit must provide the participating citizens with background information enabling them to answer the questions. It aims at providing both facts and unbiased discussions of the pros and cons of voting one way or another. It also includes ethical and political arguments for voting one way or the other, thus giving citizens a starting point for their deliberations (Gudowsky & Bechtold, 2012).

For example, the material provided information about potential economic and environmental consequences of the different adaptation options. It provided support to see such options from the perspective of different stakeholders (farmers, nature conservationists, tax payers etc.) in order to make their own assessments of what a fair and reasonable solution should look like.

Photo 2.1 Citizen Summit in Kalundborg on 5 March 2011.
Source: Photo by Jørgen Madsen, DBT.

An information booklet of 32 pages was developed by DBT in close cooperation with the Kalundborg administration (Municipality of Kalundborg, 2011). It was sent to the citizens three weeks prior to the summit, whereas the questions were not made publicly known in advance in order to avoid the fact that citizens could make up their minds prior to the deliberation. At the citizen summit, different speakers were invited to introduce the main discussions presented in the information material in order to prompt deliberation among the citizens.

2.8.4 Deliberation and voting

The 350 citizens, divided into tables of five to seven people (Photo 2.1), were led by a head facilitator and group moderators through a programme divided into six thematic sessions. During the thematic sessions, citizens voted on alternative answers to a total of 19 questions, and each thematic session was introduced by the facilitator and a short presentation (five minutes) by an invited speaker. Also, computer animations of potential future flooding, produced by the Municipality of Kalundborg, were presented.[6] The participants then engaged in moderated discussions at their tables, the purpose of which was to give all participants time to listen to other opinions and reflect prior to voting. Depending on the number of questions and the complexity of the issues, each thematic session lasted between 35 and 75 minutes. Modera-

tors were trained in advance to provide facilitation at the tables and included several local politicians. Each thematic session concluded with citizens casting their votes anonymously on one to five questions. Electronic voting equipment allowed the results to be presented instantaneously on a large screen.

Contrary to the scenario workshop method, the citizen summit methodology was not applied by other BaltCICA partners. In practical terms it is a more complex, time consuming and expensive method, which made it a difficult one to apply after the BaltCICA budget was approved. A similar method was, however, applied in a number of countries in 2009 for the first ever global citizen consultation, World Wide Views on Global Warming, leading up to the climate COP15 in Copenhagen in November 2009 (WWViews, 2009; Rask, Worthington & Lammi, 2012).

2.9 Interpretation of results – in details

Citizens left the summit with a copy of the voting results, some of which were reported the following days in both local and national media. Although the voting results were immediately obvious to the average person, an analysis of the political implications was produced by DBT (DBT, 2011a; DBT, 2011b) and subsequently discussed with the municipal Committee for Technical and Environmental Issues at a joint meeting.

To the question about the preferred development for the case study area, two thirds of citizens (Figure 2.4) voted in favour of making a decision now that will allow the coastline to move further inland,

---

[6] Computer animations of potential future flooding, produced by the Municipality of Kalundborg: http://www.kalundborg. dk/Til_borgeren/Klima/Borgertopm%C3%B8de.aspx

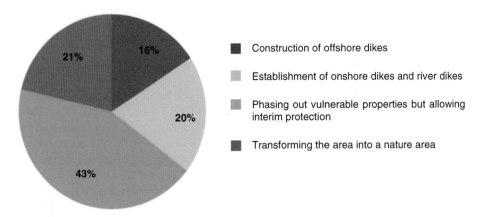

Figure 2.4 Voting results to four alternative development solutions for the case study area.

and thereby eventually discontinue current activities in these areas such as summer cottages and farming. About one third of the participants wanted a collective solution based on dykes. This result differs significantly from the results from the scenario workshop. Here, local stakeholders were more supportive of various dyke solutions. These results point to the importance of careful consideration as to when and how which citizens and stakeholders should be included in a planning process.

To the question; 'Should the municipality have the authority to alter the status of exposed summer cottage areas, so they can be transformed, in the long term, into nature areas with periodic and permanent flooding?', 85% of the participants answered in favour, 10% against, and 5% 'don't know'. Three quarters of the participants felt that private industry at the harbour area in Kalundborg City shares the responsibility of protecting it from future floods; almost all participants voted in favour of letting the municipality make heavier demands as to the handling of precipitation on private properties; three quarters voted in favour of not establishing new infrastructure (such as sewer systems) in areas prone to future flooding; eight out of ten authorised the municipality to earmark farmland for occasional flooding in the event of heavy precipitation, in order to protect other vulnerable areas such as Kalundborg City and summer cottage areas from flooding; nine out of ten wanted coastal planning to be more centralized; and nine out of ten wanted the municipality to make a long-term climate adaptation strategy based on the

current expectations of climate change, rather than acting only when problems arise or legal obligations to act are imposed.

Thus, the general trend was that citizens authorized the municipality to engage more actively in climate change adaptation and make decisions that could overrule and disregard the interests of private property owners for the sake of the common interest.

2.10 Towards a climate strategy and its implementation

At the time of finishing this article (June 2012), the Kalundborg administration was in the process of drafting a climate adaptation strategy. The strategy will include an analysis of expected climate impacts and vulnerable areas in the municipality. It will also include priorities and guidelines with regards to adaptation measures. Once the strategy is completed, it is to be agreed upon among the politicians, and it can then be implemented in the Municipality Plan 2013–24, which is the legal document for future initiatives in the municipality.

2.11 Discussion

When assessing the effects of participatory processes like the one described in this chapter, one should take several parameters into consideration (Cruz-Castro &

Snz-Menéndez, 2004). Thus, they can result in both increasing knowledge, forming attitudes and opinions and initializing actions. And they can do so, both with regards to scientific, societal and policy issues.

There are a number of both principled and practical reasons to involve stakeholders and citizens in decision-making processes (Klüver, 2002). Among the principled reasons is the search for social coherence by involving those affected by the decisions in question, as well as a general wish to strengthen the democratic legitimacy and transparency of policy making. Among the practical reasons are the inclusion of a wider range of knowledge, values and ideas; ensuring higher credibility and sustainability of decisions made; and easier implementation of those decisions by people who have participated in a fair and open process (for examples of how citizen participation leads to greater legitimacy and better implementation of decisions, see Goodin and Dryzek, 2006). In short, the involvement of stakeholders and citizens in decision-making processes gives promise of higher legitimacy and effectiveness.

2.12 Conclusions

In Kalundborg, knowledge of climate change and adaptation options has increased considerably among the politicians, stakeholders and citizens.[7] As a result of the BaltCICA project, climate adaptation is now more widely recognized in the municipality as not merely a technical issue but also a political question about future land use, sharing responsibilities and priority of interests. Citizens and stakeholders want decisions to be taken soon. Property owners want to know if their assets will be protected in the future or not, and the administration will think twice before making infrastructure investments in areas prone to future flooding if their future protection is unrealistic or unwished for.

The citizens gave the municipality a broad mandate to make political decisions about long-term strategies for climate change adaptation, even if such strategies would disregard private interests for the sake of more

important and common interests. A climate change adaptation strategy for the municipality is in the making, but it remains to be seen what kind of decisions will be made and actions initialised when the Municipality Plan is adopted in 2013. According to Head of the Committee for Engineering and Environment, Mr Bertel Stenbæk, the results from the citizen summit have provided the politicians with a better idea of what kind of climate adaptation solutions the citizens of Kalundborg prefer, and some of the results have already been used for planning purposes.[8]

Generally speaking, policy-making has become more complex and reliant on scientific theory, knowledge and models, some of which are more disputed than others. Stakeholders offer their advice, but have their own interests in mind, and opinion polls are too superficial to give politicians a sense of public opinion on complex issues. It is therefore increasingly important to make use of the many tools available for participatory decision-making both on a local, national and global scale. Climate adaptation is a good example of this, and we expect the municipality of Kalundborg to benefit from having applied such tools in their management of an uncertain future.

References

Adger, N.A., Dessai, S., Goulden, M., Hulme, M., Lorenzoni, I., Nelson, D.R., Naess, L.O., Wolf, J. and Wreford, A., 2009. Are there social limits to adaptation to climate change? *Climatic Change*, 93, pp.335–354.

Agger, A., Jelsøe, E., Jæger, B. and Phillips, L., 2012. The creation of a global voice for citizens: The case of Denmark. In: Rask, M., Worthington, R. and Lammi, M., eds. 2012. *Citizen Participation in Global Environmental Governance*. London: Routledge, pp.45–69.

Andersen, I.-E. and Jæger, B., 1999. Danish participatory models – Scenario workshops and consensus conferences: towards more democratic decision-making. *Science and Public Policy*, 26 (5), pp.331–340.

Bedsted, B. ed., 2007. *Enablers of Science-Society Dialogue*. [pdf] Available at: <http://www.tekno.dk/pdf/projekter/forSociety/p07_ForSociety_Task_2-2_Final_Report_110707.pdf> [Accessed 01 June 2012]

---

[7]The municipality also informs about the project on its webpage (in Danish only): http://www.kalundborg.dk/Til_borgeren/Klima/BaltCICA.aspx

[8]Video about the participation process in Kalundborg, available at http://www.tekno.dk/subpage.php3?article=1595&toppic=kategori11&language=uk&category=11

Cruz-Castro, L. and Snz-Menéndez, L., 2004. Shaping the impact: the institutional context of technology assessment. In: Decker, M. and Ladikas, M., eds. 2004. *Bridges between Science, Society and Policy*. Berlin, Heidelberg: Springer-Verlag GmbH, pp.101–127.

DBT (Danish Board of Technology), 2006. Citizens' Summit. [online] Available at: <http://www.tekno.dk/subpage.php3?article=1232&toppic=kategori12&language=dk> [Accessed 01 June 2012]

DBT, 2009. *Three future scenarios for the Kalundborg case study.* [pdf] Available at: <http://www.baltcica.org/documents/BaltCICA%20scenariosENG_with%20logos.pdf> [Accessed 01 June 2012]

DBT, 2011a. *Analysis of results from citizen summit on climate change adaptation, in Kalundborg on the 5th of March 2011.* [pdf] Available at: <http://www.tekno.dk/pdf/projekter/baltcica/p12%20Analysis%20of%20results%20from%20citizens%20summit%20on%20climate%20change%20adaptation%20in%20Kalundborg%20on%20the%205th%20of%20March%202011.pdf> [Accessed 01 June 2012]

DBT, 2011b. *Results from the citizen summit on climate change adaptation – Saturday, the 5th of March, 2011, Kalundborg.* [pdf] Available at: <http://www.tekno.dk/pdf/projekter/baltcica/p11_BaltCiCa_-_citizen_summit.pdf> [Accessed 01 June 2012]

DBT, 2012. *Teknologirådet.* [online] Available at: <www.tekno.dk> [Accessed 08 June 2012]

DMI (Danish Meteorological Institute), 2009. *Fremtidige havniveauændringer – Et resumé af den aktuelle viden i foråret 2009.* [pdf] Available at: <http://www.dmi.dk/dmi/dkc09-07.pdf> [Accessed 08 June 2012]

Goodin, R.E. and Dryzek, J.S., 2006. Deliberative impacts: The macro-political uptake of mini-publics. *Politics & Society*, 34 (2), pp.219–244.

Gudowsky, N. and Bechtold, U., (in press) The role of information in public participation. *Journal of Public Deliberation*. (Submitted)

IPCC, 2000. *IPCC Special Report on Emissions Scenarios: Summary for Policymakers – A Special Report of IPCC Working Group III.* [online] Available at: <http://www.grida.no/publications/other/ipcc_sr/> [Accessed 20 June 2012]

Klüver, L., 2002. Project management – a matter of ethics and robust decisions. In: Joss, S. and Bellucci, S., eds. 2002. *Participatory Technology Assessment – European Perspectives*. London, UK: Centre for the Study of Democracy, University of Westminster, pp.179–208.

Klüver, L., Bellucci, S., Berloznik, R., Bütschi, D., Carius, R., Cope, D., Cruz-Castro, L., Decker, M., Gram, S., Grunwald, A., Hennen, L., Karapiperis, T., Ladikas, M., Machleidt, P., Sanz-Menendez, L., Peeters, W., Staman, J., Stephan, S., Szapiro, T., Steyaert, S. and Van Est, R., 2004. Technology assessment in Europe: Conclusions & wider perspectives. In: Decker, M. and Ladikas, M., eds. 2004. *Bridges between Science, Society and Policy*. Berlin, Heidelberg: Springer-Verlag GmbH, pp. 88–98.

Municipality of Kalundborg, 2011. *Baggrundsmateriale til borgertopmøde om klimatilpasning.* [pdf] Available at: <http://www.tekno.dk/pdf/projekter/baltcica/p11_Kalundborg_web_udgave%20baggrundsmateriale%202011%2021%20feb.pdf> [Accessed 08 June 2012]

Rask, M., Worthington, R. and Lammi, M. eds., 2012. *Citizen Participation in Global Environmental Governance*. London: Routledge.

Steyaert, S. and Lisoir, H. eds., 2005. *Participatory Methods Toolkit – A practitioner's manual.* [pdf] Available at: <http://www.kbs-frb.be/uploadedFiles/KBS-FRB/Files/EN/PUB_1540_Participatoty_toolkit_New_edition.pdf> [Accessed 20 June 2012]

Vig, N.J. and Paschen, H., 2000. *Parliaments and Technology: The Development of Technology Assessment in Europe*. New York: State University of New York Press.

Winner, L., 1986. Do artifacts have politics? In: Winner, L., 1986. *The Whale and the Reactor: A Search for Limits in an Age of High Technology*. Chicago: University of Chicago Press, pp.19–39.

WWViews (World Wide Views), 2009. *World Wide Views on Global Warming – The Project.* [online] Available at: <http://www.wwviews.org/node/259> [Accessed 20 June 2012]

3

Adaptation to Sea Level Rise: Calculating Costs and Benefits for the Case Study Kalundborg, Denmark

Markus Boettle[1], Diego Rybski[1] & Jürgen P. Kropp[1,2]

[1]Potsdam Institute for Climate Impact Research (PIK), Potsdam, Germany
[2]University of Potsdam, Institute of Earth and Environmental Science, Potsdam, Germany

3.1 Introduction

While global warming is still the focus of climate research, severe impacts need to be anticipated and appropriate adaptation strategies to mitigate adverse effects are required. In particular, coastal regions are facing serious consequences. Due to the complex nature of impacts from sea level rise (and climate change in general), decisions on adaptation are difficult to take. Cost-benefit analysis (CBA) is one possible tool to support such decisions and to estimate the economic efficiency of adaptation investments. Unlike, for example, multi-criteria decision analysis (see Chapter 4 in this book or Triantaphyllou, 2000), where the input of subjective preferences is necessary, CBA is a purely rational and monetary approach which is probably one of the reasons for its wide acceptance.

Although sea level rise is a steadily ongoing process, the consequences are not in general likely to be from a progressing, permanent inundation of areas, but from single floodings, which are likely to become more frequent and severe in the future (IPCC, 2011). Therefore, the assessment of impacts should be based on extreme events, such as extreme sea levels due to storm surges.

This chapter is organized as follows. In the second section we introduce a stochastic framework for the assessment of annual flood damages and describe its application in the case study of Kalundborg. The third section investigates possible factors influencing flood risk and the fourth section describes the resulting cost-benefit analysis, again exemplifying the case study of Kalundborg in order to demonstrate the applicability of the approach. The final section summarizes and draws conclusions.

3.2 Risk assessment

3.2.1 Risk

Many definitions of the term risk can be found in literature. Commonly, risk is vaguely defined as probability times consequence, where the probability refers to a certain flood event and consequence stands for the corresponding monetary damage. Since we are not only interested in one specific event (e.g. a 100-year flood) but in all possible floods (i.e. events of arbitrary return level), the flood risk is obtained by summing the risks of all flood magnitudes (Merz, Hall, Disse & Schumann, 2010; Poussin et al., 2012). In our context, flood risk describes the average annual flood damage in a specific region. Thus, the term is associated with a single year and estimating the risk of a longer time period necessitates a summation of the annual risks. However, the calculation of risk requires knowledge about the frequencies of flood events on the one

Climate Change Adaptation in Practice: From Strategy Development to Implementation, First Edition.
Edited by Philipp Schmidt-Thomé and Johannes Klein.
© 2013 John Wiley & Sons, Ltd. Published 2013 by John Wiley & Sons, Ltd.

hand and information about the corresponding consequences on the other hand.

3.2.2 Extreme value theory

Extreme events such as floods are typically characterized by their exceedance probability or their recurrence time (annuality). Extreme value theory is commonly used to describe such frequencies in a mathematical framework (Leadbetter, Lindgren & Rootzen, 1983; Coles, 2001; Hawkes, Gonzales-Marco, Sanchez-Arcilla & Prinos, 2008). In particular, the Generalized Extreme Value (GEV) distribution is employed to describe the maximum annual water level at a certain gauge, using three parameters: the location μ, the scale σ, and the shape ξ. In order to estimate these parameters from historical gauge data, several algorithms are available (e.g. via L-moments (Hosking, 1990; Wang, 1990) or maximum likelihood (Phien & Fang, 1989; Hosking, 1990)). However, μ, σ, and ξ contain all the necessary information about the stochastic occurrence of extreme water levels in the current state, that is, for the present environmental conditions. In view of rising mean sea levels and changing storm intensities, these parameters should be considered as altering (Woth, Weise & von Storch, 2006; Mudersbach & Jensen, 2010). Consequently, it needs to be assumed that the risk will alter over time.

3.2.3 Damage functions

Apart from the stochastic description of flood events, their interrelation with the associated damages needs to be assessed comprising two steps. Firstly, the affected assets need to be determined. For this purpose hydrodynamic models or simplified flood algorithms (Poulter & Halpin, 2008) provide characteristics on how each asset is affected (e.g. the inundation height of each building). Secondly, the economic damage needs to be elaborated, typically by a damage function. Although the damage is influenced by several factors (such as inundation duration, contamination load, warning time, or flow velocity (Wind, Nierop, de Blois & de Kok, 1999; Kreibich et al., 2005; Thieken, Müller, Kreibich & Merz, 2005; Middelmann-Fernandes, 2010) the inundation depth is usually considered as the main factor and so called stage-damage functions are commonly used on the building level (Smith, 1994; Dutta, Herath & Musiake, 2002; Apel, Aronica, Kreibich & Thieken, 2009). Such

functions relate the inundation level of a certain asset with the corresponding damage and different types of functions can be applied to take account of the variety of assets or buildings (Blong, 2003; Kang, Su & Chang, 2005; Penning-Rowsell, 2005). The integration of all building damages within the considered area for flood events of variable magnitude then leads to a *macroscopic damage function* providing the total damage for the case study as a function of the sea level (Hallegatte et al., 2011; Boettle et al., 2011).

3.2.4 Integration

Since one does not know when a certain flood event occurs, the damage function is linked with the stochastic properties of the above mentioned extreme water levels, namely with the GEV distribution. The flood risk can then be calculated by means of these two components. Considering the maximum annual water level as a random variable with probability density p and the damage function D as a transformation, basic stochastics provide the following formula for the risk R:

$$R = \int D(x)\,p(x)\mathrm{d}x, \qquad (3.1)$$

where x takes all possible water levels. In fact, this term accounts only for one flood event per year (considering the heaviest flood) and neglects any additional floods. Still, this represents just a small shortcoming, since more than one flood event per year is rare in most areas.

3.2.5 Example

We want to show how this procedure is performed for a case study region south of the city of Kalundborg in Denmark (Figure 3.1), which comprises mostly summer cottages, located in low-lying areas near the coast. So far, no flood defences are implemented in the area but in view of rising sea levels a debate on how assets can be protected from future storm surges is ongoing. One specific protection measure is described in Section 3.3.

Basically, the whole risk assessment procedure involves three steps: (i) the estimation of GEV parameters from historical gauge data, (ii) the elaboration of a macroscopic damage function, and (iii) the actual calculation of the risk according to Equation (3.1).

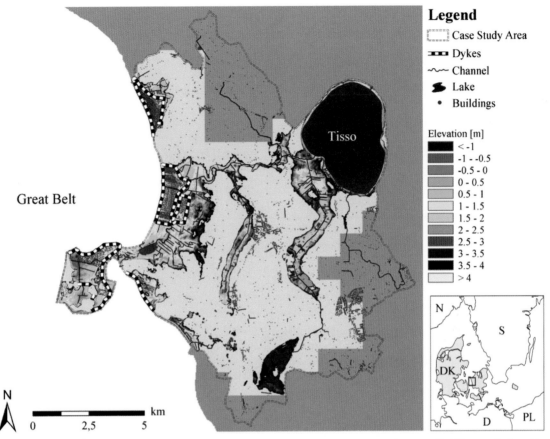

Legend
- Case Study Area
- Dykes
- Channel
- Lake
- Buildings

Elevation [m]
	< -1
	-1 - -0.5
	-0.5 - 0
	0 - 0.5
	0.5 - 1
	1 - 1.5
	1.5 - 2
	2 - 2.5
	2.5 - 3
	3 - 3.5
	3.5 - 4
	> 4

Figure 3.1 Case study area south of the City of Kalundborg with locations of the proposed protection measures (bold, black and white line). The elevation data is colour coded (light grey indicates elevations above 4 m, dark grey stands for no data). The white area represents the sea and buildings are indicated by red dots.
Source: DEM owned by Niras BlomInfo A/S Denmark.

For the Kalundborg case study, a macroscopic damage curve is presented in Boettle et al. (2011) on the basis of a high resolution elevation model and detailed cadastral information using a flood-fill algorithm (four-nearest neighbours). The function, shown in Figure 3.2, provides monetary building damages for sea levels between zero and four meters and is based on a linear building damage function. For further details we refer to Boettle et al. (2011).

Regarding the distribution of maximum annual water levels, 32 maximum annual water level measurements at the gauge in Kalundborg between 1971 and 2006 were used.[1] Although this dataset barely allows reliable estimates for the GEV parameters to be derived, we perform the analysis with this information since more extensive data were not available. In addition to the maximum sea levels, mean sea level data from a gauge close to Kalundborg (Korsør) is publicly available[2] and was used to deduce a linear upward trend in the corresponding period of approximately 0.16 cm per year. After subtracting this trend from the time series the GEV parameters $\mu \approx 91.3$, $\sigma \approx 16.96$, and $\xi \approx 0.00$ were obtained using maximum likelihood estimation for censored sample data (Phien & Fang, 1989). The parameters imply that flood events of arbitrary height are possible within the model and

---

[1] We want to thank Jacob Arpe for the provision of data.

[2] http://www.psmsl.org/

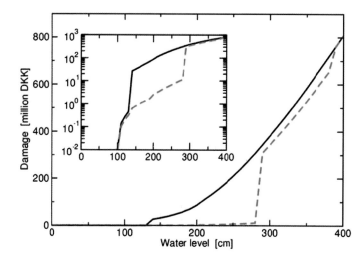

Figure 3.2 Damage function for the case study Kalundborg without flood defence (black) and with proposed dyke protection (dashed green) obtained as linear interpolation of damages at water levels with 10 cm distance. The inset shows the same curves in semi-logarithmic scale.

no upper bound for water levels exists (Coles, 2001). However, with regard to computational aspects, it is only possible to consider water levels up to a certain limit. In our case a limit of four meters above mean sea level has been chosen, which means a consideration of extreme events up to a recurrence time of approximately five billion years (given the present parameters). In practice, the integral in Equation (3.1) is approximated by the following sum

$$R \approx \Delta x \sum_{i=1}^{N} D(x_i) p(x_i), \qquad (3.2)$$

where x_i takes values between zero and four meters with equal distances Δx and $N := 4\text{m}/\Delta x$ being the number of discretization steps. Performing this summation, a total flood risk of approximately 2.8 million Danish Crowns (DKK) is obtained for the case study Kalundborg in 2010. This figure represents an average value and not a prediction of damages in 2010. In contrast, it is very likely that the real damage is far away from the expected value and it needs a much longer observation period to find a settling of the average close to the expected damage. For instance, considering a period of 100 years a total damage of around 280 million DKK is estimated (assuming constant conditions), but no information on the composition of this damage is provided. Possibly, the whole damage

stems from one single event and no other damages occur.

3.3 Risk influencing factors

The calculations above refer to one specific year and several factors, for example, environmental or anthropogenic effects, can change the underlying conditions. On the one hand, a rise in sea level or changing wind patterns can affect the occurrence probabilities of extreme water levels and therefore impact the expected damages via the GEV parameters (Mudersbach & Jensen, 2010). This interrelation is little understood so far, but adverse effects on the flood risk can be expected at least from sea level rise (IPCC, 2011). On the other hand, socio-economic development and land use changes affect the distribution of assets and therefore the damage function and the expected damages. In particular, the implementation of flood defence systems or no further settlement in flood-prone areas can effectively mitigate flood damages (Pielke Jr & Downton, 2000).

In the following we investigate the influence of two factors on the flood risk in the Kalundborg case study: (i) sea level rise and (ii) the implementation of a flood protection measure. A more general perspective on these effects can be found in Boettle, Rybski & Kropp (in preparation).

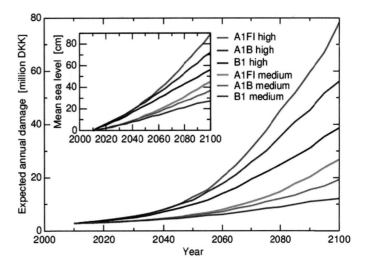

Figure 3.3 Development of the expected annual damages (undiscounted) in the case study Kalundborg for several SRES scenarios. Corresponding sea level projections relative to the current sea level for the region of Vestsjælland are obtained from the DIVA tool (Hinkel & Klein, 2003; Vafeidis et al., 2008) and are shown in the inset. All lines represent a linear interpolation of annual values.

3.3.1 Sea level rise

Since in general, the impact of rising mean sea levels on extremes is not clear (Woodworth, Menendez & Gehrels, 2011), we follow Hallegatte et al. (2011) and use the very intuitive approach of adding mean sea level rise to today's extremes. However, future mean sea levels depend on several unpredictable factors, such as global CO_2 emissions. We therefore employ SRES emissions scenarios (IPCC, 2000) in order to see how the flood risk could evolve. The corresponding local sea level projections for the region of Vestsjælland have been provided by the Dynamical Interactive Vulnerability Assessment (DIVA) tool (Hinkel & Klein, 2003; Vafeidis et al., 2008), considering medium and high climate sensitivity. Assuming all other factors (such as storm intensities, economic values, location of assets) to be constant, we consider the expected damage in the case study as a function of time. As Figure 3.3 illustrates, an increase of the flood risk by a factor between four (B1, medium) and 28 (A1FI, high) by 2100 can be found. The developments of the corresponding mean sea levels are shown in the inset. Consequently, the average annual damages in the case study will be many times higher than today, which represents an increasing threat to house owners. Additionally, the results show that the ranges between medium and high climate sensitivity are approximately of the same order as the differences between the scenarios. Hence, the uncertainties due to the unknown socio-economic future and due to a

vague understanding of the climate sensitivity are of similar relevance.

3.3.2 Flood protection

Currently, several possibilities of protecting the most flood-prone areas in the case study from future storm surges are being discussed. The construction of dykes and the removal or a retrofitting of threatened structures with stilts is among the suggestions. In our context, such protection measures would affect the shape of the macroscopic damage function since smaller damages from certain sea levels can be expected. We want to examine the effectivity of one of these adaptation options provided by the municipality of Kalundborg, namely the construction of several dykes around low-lying areas. The suggested dykes, illustrated in Figure 3.1, have a total length of approximately 18 km and a height of 2.8 m above mean sea level, whereas the dykes at the coastline have an additional metre of height to protect from wave overtopping. Regarding the costs, the total construction would entail expenses of approximately 265 million DKK.[3] However, an analogous procedure as for the no-protection scenario provides a damage function for the adaptation scenario (depicted as a dashed green line in Figure 3.2). Based on this damage function, an expected annual damage of approximately 127 000 DKK for the year

---

[3] Estimated costs provided by the Municipality.

2010 was calculated, which means a flood reduction of more than 95% compared to the current situation. Despite this impressive number, the efficiency evaluation of the described project needs to consider further aspects, such as the related implementation costs and the avoided damages over a longer time period. A detailed analysis is described in Section 3.4.

In view of this example, the question about the appropriate height of the dykes arises. In practice, this issue is not always open to discussion due to legal regulations prescribing the protection height (see e.g. Chapter 4 in this book). Nevertheless, we want to investigate the general effect of varying protection levels on the expected annual damage in Kalundborg. Therefore, we calculate the flood risk assuming different protection heights. Instead of performing a flood simulation and deriving a damage function for each height as in the example described above, we approximate the new damage functions by setting the original damage function to zero for all water levels below the given protection height, that is,

$$D_\omega(x) := \begin{cases} 0 & \text{if } x \le \omega \\ D(x) & \text{if } x > \omega \end{cases}, \qquad (3.3)$$

where D denotes the original damage function (Figure 3.2, black line), x the sea level, and ω the supposed protection height. Now we are able to derive the flood risk for arbitrary protection levels according to Equation (3.2). The resulting risks are displayed as a function of ω in Figure 3.4. It can be seen that the expected damage does not change visibly for protection levels below 100 cm. This is mainly due to the fact that it can avoid flooding only from very small extreme events, which contribute in only a minor way to the flood risk (as suggested by the damage function in Figure 3.2). This behaviour is followed by a slow reduction between 110 cm and 140 cm. For higher protection levels the expected damage decreases roughly exponentially and finally almost vanishes. For instance, protection levels above 270 cm mitigate the residual risk to an almost negligible amount of less than 10 000 DKK per year. These values differ considerably from the more specific example above, where a protection height of 280 cm at several locations was supposed and a residual risk of approximately 127 000 DKK was found. The reason for this deviation is that the suggested dykes in the example do not protect all assets and some buildings would still suffer from regular flooding. This explanation is supported by the inset of Figure 3.2, where the dashed green line exhibits damages of more than one million DKK for flood levels around 160 cm – in contrast, Equation (3.3) provides no damages for a protection height of 280 cm.

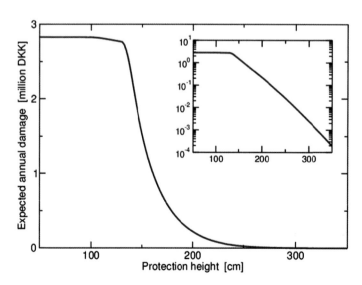

Figure 3.4 Expected annual damage in the case study Kalundborg as a function of the protection height, presuming no damages occur up to the corresponding water level. The inset shows the same curve in semi-logarithmic scale.

3.4 Cost-benefit analysis

In order to evaluate the economic efficiency of an investment, all related costs and benefits need to be taken into account. In our context, benefits emerge exclusively in terms of avoided damages and we consider the benefits only implicitly by discussing the costs. However, regarding a specific adaptation measure, the avoided damages correspond to the difference between the residual damage and the damages in the no-adaptation case. Hence, we compare all costs occurring in the adaptation scenario with those related to the no-adaptation policy. In order to take the preference of present to future consumption into account, future costs are commonly *discounted* to their present value by a certain *discount rate* (Halsnæs et al., 2007). While high discount rates (e.g. 5%) rather reflect the return rate of private-sector investments, low rates (e.g. 1%) have a social aspect representing the interests of future generations (Portney & Weyant, 1999). The discount rate plays a crucial role for the valuation of future damages. For example, a damage of 100 000 € in 50 years has a present value of 60 804 € considering a 1 % discount rate but only 8720 € at a 5 % discount rate. Thus, high discount rates depreciate upcoming damages and therefore make adaptation measures less profitable since the costs for implementation typically emerge in the near future and are therefore not discounted. Especially for long-term decisions this exponential depreciation of future costs

becomes dominant. To mitigate this effect, declining rates have been proposed (e.g. Weitzman, 2001; 2010). However, since discount rates are discussed very controversially, we will not choose one specific but will consider several constant rates.

Making an investment one wants to know, if and when it amortises, that is, when the (discounted) avoided damages exceed the expenditures and the investment becomes profitable. Hence, we need to cumulate the costs for the lifetime of the project. Figure 3.5 shows the development of costs for several SRES scenarios with (solid lines) and without (dashed lines) considering the described protection measure and disregarding discounting. In the adaptation scenarios the costs start at the construction costs of 265 million DKK followed by a less steep increase, representing lower annual damages. One can see that the point of amortisation (intersection of solid with dashed lines) depends strongly on the underlying sea level rise scenarios (inset of Figure 3.3). Table 3.1 summarises the years, when the investment is amortised considering several sea level scenarios and discount rates. The importance of the discount rate becomes apparent: While the investment amortises until 2070 in all sea level scenarios disregarding discounting, it does not become profitable until 2100 in any case for discount rates higher than 4%. The results also suggest that the uncertainty due to unknown climate sensitivity is of a similar magnitude as for the choice of the discount rate. Surprisingly, the underlying SRES scenario plays only a

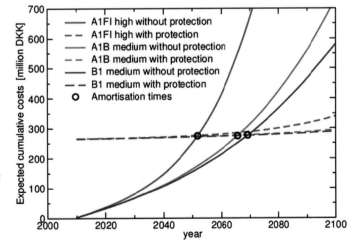

Figure 3.5 Comparison of the expected cumulative costs supposing the proposed protection measure (dashed lines) with the no-protection option (solid lines) considering several sea level scenarios and disregarding discounting. Corresponding scenario projections for the region of Vestsjælland are obtained from the DIVA tool (Hinkel & Klein, 2003; Vafeidis et al., 2008) and are shown in the inset of Figure 3.3. All lines represent linear interpolations of annual values.

Table 3.1 Amortisation years for the described protection measure and several SRES scenarios, climate sensitivities and discount rates

Scenario Climate sensitivity		B1		A1B		A1FI	
		High	Medium	High	Medium	High	Medium
Amortisation year at discount rate	0%	2055	2070	2053	2066	2052	2065
	1%	2063	2088	2059	2081	2059	2078
	2%	2076	–	2069	–	2068	2099
	3%	–	–	2087	–	2082	–
	4%	–	–	–	–	–	–

Note: A dash "–" indicates that the investment will not be amortised until 2100. The corresponding sea level scenarios are obtained from the DIVA tool (Hinkel and Klein, 2003; Vafeidis et al., 2008) and are illustrated in the inset of Figure 3.3.

minor role for the amortisation time, which is already indicated by the developments of the corresponding mean sea levels (inset of Figure 3.3).

It is worth mentioning that these results are based on purely monetary values and exclude any non-monetary aspects (e.g. social or ecological). Therefore, it might be reasonable to implement an option, although it is not efficient in a CBA sense.

3.5 Conclusions

A method to perform CBA for adaptation measures in the context of sea level rise has been described and the analysis for a specific protection system in Kalundborg was presented. It was found that in general the results depend fundamentally on assumptions regarding future mean sea levels and on the choice of the discount rate. For a time horizon until 2100, it was found that the proposed dyke system is a profitable investment in all sea level scenarios (with an earliest amortisation time of 2052) only for discount rates less than 2%. For rates above 4% no sea level scenario led to an amortisation before 2100, which shows the crucial character of the discount rate in a cost-benefit analysis. Due to this fundamental effect, the discount rates need to be discussed in detail and agreed upon by the decision makers and stakeholders.

Our approach also comprises the estimation of annual damages and thus allows a further investigation of the effects of sea level rise and protection measures on the flood risk. For instance, we have shown

that the expected annual damage strongly depends on the rate of sea level rise and an increase by a factor between four and 28 by 2100 was estimated (disregarding discounting), which implies great uncertainty due to the unknown development of future mean sea levels. It can be concluded that in any case adaptation to projected future conditions can considerably reduce expected damages in the region.

With regard to different protection levels in the Kalundborg case study, it is found that the residual annual damage decreases approximately exponentially with the height, assuming that all assets are protected up to a certain water level. Since the implementation costs for protection measures typically increase monotonically with the height, the optimal protection level can be derived. This optimum describes the minimum of the sum of all costs related to the protection measure and the residual damages.

Our results could be of particular importance for local authorities responsible for the decision making process regarding adaptation to sea level rise, because they include the dynamics of sea levels as well as possible adaptation plans. In general, we obtained insights into the interplay of protection height, sea level rise and the corresponding residual damage.

Acknowledgements

The authors would like to thank Anika Nockert, Lena Reiber and Olivia Roithmeier for creating the map. We appreciate financial support by the BaltCICA

Project (part-financed by the EU Baltic Sea Region Programme 2007–13).

References

Apel, H., Aronica, G.T., Kreibich, H. and Thieken, A.H., 2009. Flood risk analyses – how detailed do we need to be? *Natural Hazards*, 49 (1), pp.79–98.

Blong, R., 2003. A new damage index. *Natural Hazards*, 30 (1), pp.1–23.

Boettle, M., Kropp, J.P., Reiber, L., Roithmeier, O., Rybski, D. and Walther, C., 2011. About the influence of elevation model quality and small-scale damage functions on flood damage estimation. *Natural Hazards and Earth System Science*, 11 (12), pp.3327–3334.

Boettle, M., Rybski, D. and Kropp, J.P., (in preparation) How changing sea level extremes and protection measures alter coastal flood damages.

Coles, S., 2001. *An Introduction to Statistical Modeling of Extreme Values*. London: Springer.

Dutta, D., Herath, S. and Musiake, K., 2002. Direct flood damage modelling towards urban flood risk management. *Proceedings of the Joint Workshop on Urban Safety Engineering 2001*, (ICUS Report), pp.127–143.

Hallegatte, S., Ranger, N., Mestre, O., Dumas, P., Corfee-Morlat, J., Herweijer, C. and Wood, R.M., 2011. Assessing climate change impacts, sea level rise and storm surge risk in port cities: a case study on Copenhagen. *Climatic Change*, 104 (1), pp.113–137.

Halsnæs, K., Shukla, P., Ahuja, G. et al., 2007. Framing Issues. In: B. Metz, O.R. Davidson, P.R. Bosch, R. Dave and L.A. Meyer, eds. 2007. *Climate Change 2007: Impacts, Mitigation. Contribution of Working Group III to the Fourth Assessment Report of the Intergovernmental Panel on Climate Change*. Cambridge, UK: Cambridge University Press, pp.117–167.

Hawkes, P.J., Gonzales-Marco, D., Sanchez-Arcilla, A. and Prinos, P., 2008. Best practice for the estimation of extremes: A review. *Journal of Hydraulic Research*, 46 (2), pp.323–332.

Hinkel, J. and Klein, R.J.T., 2003. DINAS-COAST: Developing a method and a tool for dynamic and interactive vulnerability assessment. *LOICZ Newsletter*, 27, pp.1–4.

Hosking, J.R.M., 1990. L-moments: Analysis and estimation of distributions using linear combinations of order statistics. *Journal of the Royal Statistical Society*, 52 (1), pp.105–124.

IPCC, 2000. *Special Report on Emissions Scenarios (SRES): A Special Report of Working Group III of the Intergovernmental Panel on Climate Change*. Cambridge: Cambridge University Press.

IPCC, 2011. *Managing the Risks of Extreme Events and Disasters to Advance Climate Change Adaptation (SREX): A Special Report of Working Group I and Working Group II of the Intergovernmental Panel on Climate Change*. Cambridge: Cambridge University Press.

Kang, J.-L., Su, M.-D. and Chang, L.-F., 2005. Loss functions and framework for regional flood damage estimation in residential area. *Journal of Marine Science and Technology*, 13 (3), pp.193–199.

Kreibich, H., Thieken, A.H., Petrow, T., Müller, M. and Merz, B., 2005. Flood loss reduction of private households due to building precautionary measures – lessons learned from the Elbe flood in August 2002. *Natural Hazards and Earth System Science*, 5 (1), pp.117–126.

Leadbetter, M.R., Lindgren, G. and Rootzen, H., 1983. *Extremes and Related Properties of Random Sequences and Processes*. New York: Springer.

Merz, B., Hall, J., Disse, M. and Schumann, A., 2010. Fluvial flood risk management in a changing world. *Natural Hazards and Earth System Science*, 10 (3), pp.509–527.

Middelmann-Fernandes, M., 2010. Flood damage estimation beyond stage–damage functions: an Australian example. *Journal of Flood Risk Management*, 3 (1), pp.88–96.

Mudersbach, C. and Jensen, J., 2010. Nonstationary extreme value analysis of annual maximum water levels for designing coastal structures on the German North Sea coastline. *Journal of Flood Risk Management*, 3 (1), pp.52–62.

Penning-Rowsell, E., 2005. *The Benefits of Flood and Coastal Risk Management: A Manual of Assessment Techniques*. London: Middlesex University Press.

Phien, H.N. and Fang, T.-S.E., 1989. Maximum likelihood estimation of the parameters and quantiles of the general extreme-value distribution from censored samples. *Journal of Hydrology*, 105 (1–2), pp.139–155.

Pielke Jr, R.A. and Downton, M.W., 2000. Precipitation and damaging floods: Trends in the United States, 1932–97. *Journal of Climate*, 13, pp.3625–3637.

Portney, P.R. and Weyant, J.P., 1999. *Discounting and Intergenerational Equity*. Washington DC: RFF Press.

Poulter, B. and Halpin, P.N., 2008. Raster modelling of coastal flooding from sea-level rise. *International Journal of Geographical Information Science*, 22 (2), pp.167–182.

Poussin, J.K., Ward, P.J., Bubeck, P., Gaslikova, L., Schwerzmann, A. and Raible, C.C., 2012. Flood risk modelling. In: J. Aerts, W. Botzen and M.J. Bowman, eds. 2011. *Climate Adaptation and Flood Risk in Coastal Cities*. London, New York: Earthscan, pp.93–121.

Smith, D.I., 1994. Flood damage estimation – a review of urban stage-damage curves and loss functions. *Water SA*, 20 (3), pp.231–238.

Thieken, A.H., Müller, M., Kreibich, H. and Merz, B., 2005. Flood damage and influencing factors: New insights from the August 2002 flood in Germany. *Water Resources Research*, 41 (12), pp.1–17.

Triantaphyllou, E., 2000. *Multi-Criteria Decision Making Methodologies: A Comparative Study*. Applied Optimization, 44, Dordrecht: Kluwer Academic.

Vafeidis, A.T., Nicholls, R.J., McFadden, L. et al., 2008. A new global coastal database for impact and vulnerability analysis to sea-level rise. *Journal of Coastal Research*, 24 (4), pp.917–924.

Wang, Q.J., 1990. Estimation of the GEV distribution from censored samples by method of partial probability weighted moments. *Journal of Hydrology*, 120, pp.103–114.

Weitzman, M.L., 2001. Gamma discounting. *American Economic Review*, 91 (1), pp.260–271.

Weitzman, M.L., 2010. Risk-adjusted gamma discounting. *Journal of Environmental Economics and Management*, 60 (1), pp.1–13.

Wind, H.G., Nierop, T.M., de Blois, C.J. and de Kok, J.L., 1999. Analysis of flood damages from the 1993 and 1995 Meuse floods. *Water Resources Research*, 35 (11), pp.3459–3465.

Woodworth, P.L., Menendez, M. and Gehrels, W.R., 2011. Evidence for century-timescale acceleration in mean sea levels and for recent changes in extreme sea levels. *Surveys in Geophysics*, 32 (4–5), pp.603–618.

Woth, K., Weisse, R. and von Storch, H., 2006. Climate change and North Sea storm surge extremes: an ensemble study of storm surge extremes expected in a changed climate projected by four different regional climate models. *Ocean Dynamics*, 56 (1), pp.3–15.

4 Coastal Protection and Multi-Criteria Decision Analysis: Didactically Processed Examples

Markus Boettle[1], Philipp Schmidt-Thomé[1,2] & Diego Rybski[1]

[1]Potsdam Institute for Climate Impact Research (PIK), Potsdam, Germany
[2]Geological Survey of Finland (GTK), Espoo, Finland

4.1 Introduction

Sea level rise represents an unprecedented challenge for coastal protection authorities and regional planners. With growing consciousness about climate change, decision makers become aware of the problems and risks emerging from sea level rise. Adaptation plans can help to reduce these threats but require information about the efficiency of certain measures to develop such plans. One approach to quantify efficiency is cost-benefit analysis (CBA), where costs are compared to the monetary benefits (see Chapter 3 in this book). The disadvantage is that it is purely based on monetary values and neglects other aspects, such as social, ecological or cultural. Thus, multi-criteria decision analysis (MCDA) has been proposed in order to also consider values, which are intangible, that is, for which no market price exists and which, therefore, cannot be easily measured in monetary terms. Even though approaches for monetization exist (e.g. Farber, Costanza & Wilson, 2002), their application is rather complex and laborious. Moreover, MCDA avoids monetary measurements of all aspects and allows the decision maker to set her/his personal priorities. Recently, MCDA has been applied in several studies in the context of environmental management (Kiker et al., 2005; Rogers, Seager & Gardner, 2004; Sparrevik et al., 2011; Yatsalo et al.,

2007; de Bruin et al., 2009). However, standard CBA and MCDA approaches might only be applicable in planning concepts if the methods are applied in such a way that the holistic concept plays a stronger role than the underlying mathematical concepts. In addition, MCDA applications in planning should clearly state both the potentials and limitations of applying mathematical methods in decision making.

Against this backgrounds, a BaltCICA workshop was organized by PIK and GTK and has been held in Potsdam between 29 June and 1 July 2011. The intention of the workshop was to carve out the above mentioned potentials and limitations. In order to design the workshop to be as realistic as possible two case studies were chosen in which complex decision-making processes had just been finalised. It was not intended to develop MCDA into a tool that could replace current decision-making processes, but to support these. Assuming that it is possible to reproduce real decision making, MCDA methods could operate as an additional support tool for further decision-making processes on climate change adaptation. The MCDA methods were therefore selected, presented and applied in such a way that the main focus of their application lay in an intensive communication among the involved participants. Furthermore, a strong focus lay on keeping the discussion and weighting process transparent to enable traceability of the decisions taken.

Climate Change Adaptation in Practice: From Strategy Development to Implementation, First Edition.
Edited by Philipp Schmidt-Thomé and Johannes Klein.
© 2013 John Wiley & Sons, Ltd. Published 2013 by John Wiley & Sons, Ltd.

The input from stakeholders of the State Agency for Agriculture and Nature Mittleres Mecklenburg in Rostock (StALU MM) who had actually participated in the original decision making processes of the case studies was used to develop the didactics and to stay as close as possible to the 'real-life' conditions. The examples presented here might support the further development of MCDA applications for decision-maker communication processes in the context of climate change adaptation.

The article is organized as follows. Before introducing several MCDA techniques in Section 4.3, we present background information on the two case studies in Section 4.2 in order to get an understanding of the task. The application of the methods is described in Sections 4.4 and 4.5 and finally, the results are discussed in Section 4.6.

4.2 Background of the case studies

The coast of Mecklenburg-Western Pomerania consists of unconsolidated pleistocene and holocene glacial, glaciofluvial, fluvial and eolic sediments. The coastline is very dynamic: 65% of the outer coast is facing erosion (up to 35 m/100 years) and 13% of the coast is experiencing sedimentation.

Two locations in the federal state of Mecklenburg-Western Pomerania at the German Baltic Sea coast were selected to apply the method: Markgrafenheide and Ostzingst. The two decision problems in the case studies differ greatly from each other, as in the first case the objective of coastal protection concentrates on a ring protection for a single settled area, whereas in the second case a larger hinterland is to be protected. While in Markgrafenheide the coastal protection measures have already been constructed, in Ostzingst the decision is taken but not implemented so far.

4.2.1 Coastal protection in Mecklenburg-Western Pomerania

This section briefly describes the legal framework of coastal protection in the considered case studies. This information was vital for the workshop participants because it set the 'boundary conditions' under which the coastal protection options should be developed.

The responsibility for coastal protection in Germany is allocated under the federal states leading to different coastal protection regulations in each state. In the case of Mecklenburg-Western Pomerania coastal protection is centralized under the Coastal Department at the State Agency for Environment and Nature in Rostock (StAUN). The Coastal Department is responsible for basic research and geodatabases, technical planning and construction as well as legal regulations and technical approval. The first master plan for coastal and flood protection of Mecklenburg-Western Pomerania was published in 1993, which acknowledged that the coast 'is a mosaic in space and time, its character is variation' (MBLU, 1993). This means that coastal dynamics, including coastal erosion and retreat, are accepted as natural processes and that the present coastline may change over time. Therefore, the protection of humans and their assets against erosion and flooding must respect nature as well as its processes and act 'as close as possible' to natural coastal dynamics. Because of that, all materials used for coastal protection shall be naturally degradable (sands, gravel, boulders, wood). Artificial materials (e.g. concrete, steel) may be used only in places where it can be proven that natural materials would lead to no effect (MBLU, 1993).

The legislative regulation states that continuous settlements may be considered for coastal protection under public funding, in contrast to single houses or estates.

The 2007 'Action Plan on Climate Protection for Mecklenburg-Western Pomerania' formulated recommendations on how to further develop coastal protection (MWAT, 2007). New buildings in flood-prone areas shall be avoided by subsequent planning regulations and current coastal protection installations shall be adapted to the effects of climate change and sea level rise, including retreat options.

The master plan from 1993 is currently being replaced by the 'Set of Rules for Coastal Protection' ('Regelwerk Küstenschutz Mecklenburg-Vorpommern') (MLUV, 2009c). The intention of the Set of Rules is to move from the static, rather inflexible master plan towards a system of rules that covers all relevant sectors of coastal protection. These rules have four main topics, each of which is divided into several subtopics. Currently, the Set of Rules is in preparation (MLUV, 2009b; 2009c; 2010a;

2010b; 2011a) and it is especially acknowledged that adjustments in the regulations are unavoidable under the ongoing climatic changes (MLUV, 2009c).

4.2.2 Boundary conditions of the case studies

In our context, decision making is not only influenced by the interests of different stakeholder groups, but is also limited by certain rules as introduced above. In Mecklenburg-Western Pomerania the philosophy of coastal protection provides *socio-cultural boundary conditions*, which can be summarised as follows:

(i) Coastal dynamics, including coastal erosion and retreat, are accepted as natural processes and the present coastline may change over time.

(ii) The protection of humans and their assets against erosion and flooding must respect nature and its processes and act 'as close as possible' to natural coastal dynamics.

In addition to this rather informal philosophy, a legal framework describes binding *legal boundary conditions* of coastal protection. These are:

(iii) Coastal protection is only possible for urban settlement areas.

(iv) Natural protection measures are to be employed as far as possible, e.g. low coasts have a dune protection.

(v) The materials used for coastal protection shall be natural (e.g. sands, gravel, boulders, wood - degradable by natural processes).

(vi) Off the water line non-natural material may be used in case it proves to be better than other material.

(vii) The level of protection (i.e. protection height) is set by a design flood plus sea level rise.

The decision-making process must respect all legal aspects (e.g. building codes, environmental protection) and include governance approaches, that is, the integration of local stakeholder and decision-maker groups. Final decisions on coastal protection are taken by integrating all factors. In case of conflicts, appraisal methods (Abschätzverfahren) involving relevant public bodies and stakeholders are applied.

The coastal protection components in Mecklenburg-Western Pomerania are based on the concept that dunes form a natural protection. Dykes, groynes and boulders have been used since early settlements in the area and thus comprise the *traditional protection*. In contrast, *engineering protection* measures are mod-

ern technical approaches including man-made material, that is, pile walls, concrete walls and sea locks.

As a result of the boundary conditions, the coastal protection in Mecklenburg-Western Pomerania is made up of the following components:
- physical conditions (coast types: e.g. flat, cliff; dynamics: e.g. abrasion, sedimentation, steady);
- philosophy (socio-cultural boundary conditions);
- legal framework (legal boundary conditions);
- design flood (set by the local flood heights reached in 1873, to date the highest recorded flood);
- sea level rise of 50 cm by the end of the 21st century (set by an agreement between the two federal states Mecklenburg-Western Pomerania and Schleswig Holstein) (MLUV, 2011a).

These conditions provide the framework within which all decisions have to be taken. This means that for the MCDA only legally approved options should be used. It is not possible to assume higher sea level rise rates, to use a different design flood, or to use other materials. These boundary conditions are extremely important in the MCDA as they limit the set of options and can also make criteria dispensable (e.g. the protection level is prescribed in our cases and is therefore not suitable as criterion). If no such boundary conditions are considered, MCDA would not be practicable because unrealistic options would be compared. Therefore, it is vital to set the boundary conditions before considering potential criteria and options of MCDA.

4.3 Introduction and methods of multi-criteria decision analysis

Multi-criteria decision analysis is a tool to choose between different competing options. In the context of adaptation to sea level rise, it enables a valuation and ranking of different adaptation strategies and can therefore support decision making.

Before a decision process starts, the decision to be taken has to be specified, i.e. it has to be made clear from which *options* one wants to choose. Thus, a precise definition of the options is inevitable. Options could be the construction of a protection measure (such as a dyke) or the retreat from flood-prone areas. In our context it is very vital to consider only options that satisfy certain requirements (e.g.

legal regulations) and are therefore implementable in practice.

In order to decide, it is also essential to describe the goals that should be achieved by the implementation of the chosen option. Goals can have monetary, social, cultural, or ecological character. Typically, goals are competing (e.g. price and quality) and there is no option that can fully achieve all of them. Within the MCDA process goals are represented through *criteria* with respect to which options are evaluated. It is essential that all possible goals and interests are represented by the set of criteria to obtain acceptance for the method results among all stakeholders. On the other hand, redundancies between the criteria should be minimized since this could lead to an overrating of certain goals. However, each criterion needs to be quantifiable (on an arbitrary scale). Considering a specific one, we call the proxy that is used to measure the performances of the options an *indicator*. For non-quantifiable criteria such as *aesthetics* a ranking of the options or an expert estimation could serve as indicator. At this stage, raw scores of different criteria are not comparable since they use different units. Therefore, a normalization comprising two steps is performed. First, it has to be ensured that high values indicate desirable results. This is for example not the case if we consider costs, which we aim to minimize. In such cases we replace the performance value (e.g. 1000 €) by 1 divided by the value (e.g. 1/1000 €). In the next step, each value is divided by the sum over all values in the corresponding criterion. Please note that other types of normalization are possible (e.g. dividing by the maximum score instead of the sum of all scores). In any case, it follows that for each criterion the sum over all scores equals 1 and that all scores lie between 0 and 1. Thus, the scores of all criteria use the same scale and are consequently comparable.

We would like to mention that, in fact, a normalization is only indispensable for the MAUT approach (Section 3.2). The PROMETHEE method (Section 3.3), for instance, typically normalizes the values implicitly by the preference function, whereas the ELECTRE method (Section 3.5) does not need any normalized numbers. However, a normalization at this stage makes the application of different methods more comprehensible and does not cause any undesired effects. A table containing all scores is

called *decision matrix* and is commonly of the following form:

$$M = \begin{pmatrix} a_{1,1} & \cdots & a_{1,n} \\ \vdots & \ddots & \vdots \\ a_{m,1} & \cdots & a_{m,n} \end{pmatrix}, \qquad (4.1)$$

where the i-th row represents the performances of the i-th option A_i with respect to the criteria $1, \ldots n$.

Since the choice of criteria should not be decisive for the analysis, *weights* are used to distinguish the different importances of criteria. These weights need to be assigned, for example, by the decision maker, and are naturally very subjective. However, the weights play a crucial role in the entire process and their elaboration should therefore involve stakeholders in order to achieve a better acceptance of the results.

The workflow of the whole MCDA process is illustrated in Figure 4.1 and shows where stakeholder involvement is desirable (blue background). In the following, we want to give a short introduction to the MCDA methods that have been applied to the case studies and assume that the decision matrix is given in the form of Equation (4.1).

4.3.1 AHP

In our context, the Analytic Hierarchy Process (AHP) method (Saaty, 1980) is used as an auxiliary tool for the determination of weights. The direct assignment of weights to the criteria can be a challenging task and an indirect approach using AHP might be helpful. Here, the criteria are compared pairwise and the decision maker is asked to make judgements on the degree of preference of one criterion over another by a value between one (= no preference, equal importance) and five (=strong preference). In the literature, a range between one and nine can often be found, which leads to more widespread weights.

The idea of the approach is to define the weights in a way that they give a best fit to these judgements. AHP, which provides such fits, is a mathematical method and is based on the *eigenvalue approach* (Triantaphyllou, 2000). We use a simple approximation of the eigenvector (Teknomo, 2006).

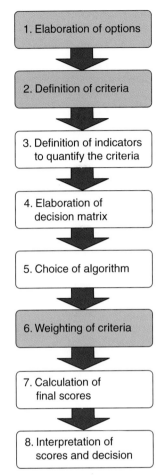

1. Elaboration of options

2. Definition of criteria

3. Definition of indicators to quantify the criteria

4. Elaboration of decision matrix

5. Choice of algorithm

6. Weighting of criteria

7. Calculation of final scores

8. Interpretation of scores and decision

Figure 4.1 Workflow of the MCDA process. Steps 1, 2 and 6 can involve stakeholder groups (indicated by blue colour).

4.3.2 MAUT

The Multi-Attribute Utility Theory (MAUT) approach is based on the economic concept of utility functions and provides a complete ranking of the options (Keeney & Raiffa, 1993). A utility function transforms the raw performance value of an option (e.g. costs of 1000 €) to a comparable dimensionless scale, usually in the interval [0,1]. Assuming a decision matrix as described in Equation (4.1), this is already given, so the performance values are comparable among the different criteria. The MAUT method then calculates the score for each option as the sum of its performances, weighted by the weights of the corresponding crite-

rion. Hence, the score of the alternative A_i can be written as

$$S(A_i) = \sum_k w_k a_{i,k}, \qquad (4.2)$$

where w_k denotes the weight of the k-th criterion and $a_{i,k}$ the performance of alternative A_i with regard to the k-th criterion. For further details, please refer to (Triantaphyllou, 2000).

4.3.3 PROMETHEE I

The Preference Ranking Organization Method for Enrichment Evaluation (PROMETHEE) is a so-called outranking method introduced by Brans and Vincke (Brans & Vincke, 1985; Brans, Vincke & Mareschal, 1986). Outranking methods do not assign an absolute score to each option (as in e.g. MAUT) but try to outrank options by pairwise comparisons, where outranking means: alternative A_i outranks A_j, if there are sufficiently strong indications that A_i is at least as good as A_j from the decision maker's point of view (i.e. supposing the decision maker's criteria weights).

Given a normalized decision matrix (4.1), we use for each criterion k the *preference function*

$$p_k(i, j) := \max(a_{i,k} - a_{j,k}, 0), \qquad (4.3)$$

where $a_{i,k}$ and $a_{j,k}$ denote the performances of the options A_i and A_j with respect to criterion k as described in Equation (4.1). This is only one possible preference function and a different choice might be advisable. A set of possible types of functions can be found in Geldermann and Rentz (2007). However, a preference function provides a value between 0 and 1 and indicates the strength of preference of option A_i over option A_j with respect to criterion k. In order to consider all criteria, we follow Geldermann and Rentz (2007) and define

$$\pi(i, j) := \sum_k w_k p_k(i, j) \qquad (4.4)$$

with criteria weights w_k. The function π provides information about the preference of one option over another considering all criteria.

Now, we follow again Geldermann and Rentz (2007) and calculate the so-called *leaving* and *entering flow* Φ^+ and Φ^- of an option A_i, respectively, as follows

$$\Phi^+(A_i) := \sum_j \pi(i, j) \qquad (4.5)$$

$$\Phi^-(A_i) := \sum_j \pi(j, i) \qquad (4.6)$$

The leaving flow Φ^+ of an option is a number that measures the *outranking character* of the option. On the other hand, the entering flow Φ^- measures the *outranked character* of an option. Basically, the higher the leaving flow and the lower the entering flow, the better the alternative. The PROMETHEE I method prefers an option A_i to an option A_j, if A_i has a higher leaving flow *and* a lower entering flow than A_j.

It is worth mentioning that the method is not always able to compare each pair of options. Accordingly, it can happen that the method yields no preference order for a chosen pair of options. The calculation of the flows and additional details can be found in Brans, Vincke and Mareschal (1986) or Geldermann and Rentz (2007).

4.3.4 PROMETHEE II

The PROMETHEE II approach uses the same basis as the PROMETHEE I, but tries to overcome the shortcoming, that it possibly cannot compare all alternatives. For this reason, the *net flow* is calculated from the leaving and the entering flow for each option A_i:

$$\Phi^{net}(A_i) := \Phi^+(A_i) - \Phi^-(A_i) \qquad (4.7)$$

The net flow comprises information about the outranking as well as the outranked character of an alternative. Now, the alternative with the highest net flow provides the preferred option and a complete ranking of alternatives is obtained. For further details, please refer to Fülöp (2005) or Geldermann and Rentz (2007).

4.3.5 ELECTRE

Another outranking approach developed by Roy (Roy 1968) is ELECTRE (ELimination Et Choix Traduisant la REalité) which we implemented by strictly following Fülöp (2005). Basically, it tries to judge the assertion that alternative A_i outranks alternative A_j by

using a *concordance* and a *discordance index*. The concordance index $c_{i,j}$ is defined as the sum of all weights of those criteria where the score of A_i is at least as high as the score of A_j and measures the advantages of A_i over A_j, neglecting the disadvantages. The discordance index $d_{i,j}$ is somewhat more complex. It equals 0, if A_i performs better than A_j in all criteria and has a high value, if A_i is clearly inferior to A_j in at least one criterion. However, both indices lie between 0 and 1 and measure the concordance and the discordance with this assertion, respectively. The calculation methods of the two indices can be found in Fülöp (2005). However, an option A_i outranks an option A_j, if the concordance index $C_{i,j}$ is above a threshold c^* and the discordance index $d_{i,j}$ is below a threshold d^*. These thresholds have to be chosen by the decision maker ($0 \le d^* < c^* \le 1$) and are apparently of major relevance for the outcome. Choosing the concordance threshold c^* that is too high results in very few or even no outranking relations. In contrast, a low concordance threshold can lead to a vast number of outrankings, which provide only little information. The reverse effect can be found for the discordance threshold d^*.

As in the case of a PROMETHEE I approach, this method does not necessarily yield a full ranking of all options. Depending on the choice of thresholds, the method possibly does not provide any outranking statements.

4.4 The case study Markgrafenheide

Markgrafenheide is a village located roughly five kilometres east of Warnemünde. The entire village is flood-prone, not only attributable to the vicinity of the coast, but also to flood waters entering from the low-lying hinterland. It was concluded that a ring-shaped protection is the only way to guarantee flood protection for the entire village. In contrast, the type of protection measures (i.e. the choice of building materials) was left open. From a coastal protection point of view the favoured solution would be a dune with groynes on the coastline and a ring-dyke around the village. The groynes do not improve protection against storm surges but diminish coastal erosion, that is, beach and dune nourishment frequencies decrease. This represents the traditional way of protection and avoids the use of non-natural materials. However, there are several reasons for deviating from this solution as different types of protection are possible within the boundary conditions.

Because of an area close to the coast which is considered to be of high ecological value (see Figure 4.2, upper left panel), the construction of a protection measure in this area causes concerns since a part of the protected area would be used for its implementation. Although it is not legally protected, the area taken for the protection measure should be as small as possible. Thus, less space taking solutions should be preferred to wide, spacious options.

Another problem crops up in the East of Markgrafenheide where the space between buildings and

the river is very small. The construction of a dyke would imply that a couple of houses need to be removed, which would cause considerable costs for the compensation of property owners.

Basically, the construction of dunes and groynes at the shore line and dunes, dykes and pile walls in the hinterland is possible. The task is to find the optimal combination of these materials for the segments of the ring protection.

4.4.1 Options

For the workshop three options were elaborated beforehand (see Figure 4.2) and presented to the participants. One of them is today's implemented solution.

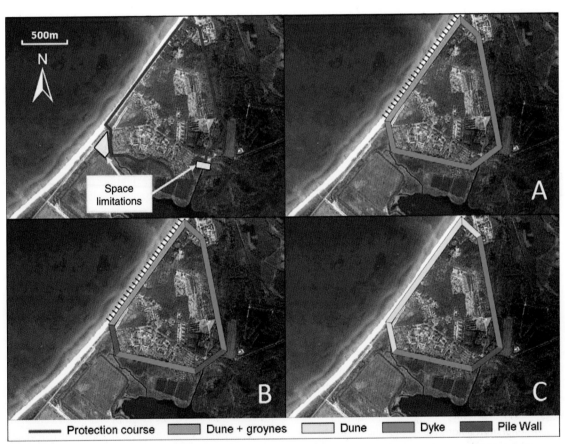

Figure 4.2 The upper left panel shows the case study of Markgrafenheide with the proposed protection course (red). The light blue regions indicate areas that should not be taken by one of the protection measures. Panels A, B and C depict schematically the suggested materials for the corresponding segments.

Source: Aerial photograph from StAUN, Coastal Department, 1998

Option A: Favourite from coastal protection perspective

As illustrated in Figure 4.2A, this solution consists of a dune with groynes at the shore line and a ring-dyke in the hinterland. Arguments in favour of option A are:

- Dunes allow sandy beaches.
- Groynes reduce abrasion.
- Dykes in the hinterland avoid mixture of components.
- Easy decommission of all components if necessary.

Option B: Mixture of protection options

Option B, shown in Figure 4.2B, represents a modification of Option A. It consists of several materials and replaces the dyke using less space by taking the pile wall at two segments of the ring protection. Thus, the ecologically valuable area is preserved as best as possible and no buildings have to be removed (in the east) for its implementation. Arguments in favour of option B are:

- Dunes allow sandy beaches.
- Groynes reduce abrasion.
- Pile wall close to the shore allows best possible preservation of the ecologically valuable area.
- Pile wall along the canal avoids the deconstruction of houses.

Option C: Tourism-oriented option

The third alternative is a tourist-friendly option (Figure 4.2C). The segments close to the coast are dunes, so that they can be used for recreational purposes. Principally, it is not permitted in Mecklenburg-Western Pomerania to enter any coastal protection installations, including dunes. This does not hold for the two additional dune segments since they are not directly at the coast. Also, the dunes along the coastline do not have groynes, which avoid a possible interference with people swimming. Furthermore, in order to avoid problems with the owners of houses that need to be removed (in the east), a pile wall is proposed in this area. Arguments in favour of option C are:

- Dunes and beach nicer without groynes.
- Additional dunes alongside accessible for tourists.
- Pile wall along the canal to avoid deconstruction of houses.

Apparently, each solution has its advantages, which represent conversely the disadvantages of the others.

So, at first glance, no option is obviously superior which suggests the application of an MCDA.

4.4.2 Criteria and Indicators

All described options provide the same protection level for Markgrafenheide. Therefore, criteria like flood safety or damage reduction are not suitable in this context and others need to be defined. Within the workshop possible criteria were discussed and due to the heterogeneity of the participants a wide range of perspectives was considered, which simulated the discussion between different stakeholder and interest groups in real decisions. Finally, it was agreed that the following set of criteria describes all important aspects:

- construction costs;
- operation & maintenance costs (O&M);
- aesthetics;
- decommission;
- loss of ecologically valuable area.

Construction costs are the sum of all one-time payments that are necessary for the implementation of an option and are naturally expressed in Euros [€]. They include the allocation of land, labour and material. Additional costs occur after the completion of the option in terms of *operation and maintenance*. Those emerge more or less regularly and are therefore measured in costs per year [€/a]. In our example, they represent the costs for beach nourishments and painting of pile walls. A less tangible criterion is *aesthetics*. It considers optical aspects as well as additional functionalities which can lead to preferences of inhabitants and tourists. Since there is no obvious way to measure the performance of an option regarding this criterion, a valuation by experts from StALU MM has been chosen, which rates each option on a certain scale. Future climate conditions cannot be predicted exactly and finding an ideal option today does not preclude that a change of adaptation strategy is necessary in the future. Hence, an adaptation option which can hardly be adjusted to changing conditions possibly needs to be dismantled at some point. Costs for *decommissioning* an implemented measure is therefore considered as potential costs [€] in the future. As described above, at one location of the protection ring the construction of a protection measure will take some of the protected natural space. The size of the *lost ecologically valuable area* [m^2] due to the construction therefore serves as criterion to measure ecological compatibility.

Table 4.1 (Normalized) decision matrix for the case study Markgrafenheide with high values indicating good performances

	Construction costs	O&M	Aesthetics	Decommission	Loss of ecologically valuable area
Option A	0.32	0.43	0.19	0.28	0.08
Option B	0.31	0.43	0.18	0.27	0.89
Option C	0.37	0.13	0.64	0.44	0.04

4.4.3 Application and results

Within the workshop, the application of the MCDA methods described in Section 4.3 was done with spreadsheets that were prepared beforehand. The sheets included the proposed protection options as well as the decision matrix (Table 4.1) and the MCDA algorithms. The participants were asked to enter their personal preferences in terms of criteria weights. They were allowed to choose between (i) a direct assignment of weights and (ii) an indirect approach where they compared each pair of criteria on a range between one (= equal importance) and five (= strong preference). In the latter, AHP was used to derive absolute weights for all criteria. Then, the results from the MCDA methods MAUT, PROMETHEE I, PROMETHEE II and ELECTRE were calculated in the background and shown in the same sheet. That way the participants were able to realize changes in the results from modified weightings immediately. Also, a comparison of results from different participants was facilitated.

Here we want to exemplify some results. The calculations were based on the (normalized) decision matrix in Table 4.1, which was calculated beforehand considering all information obtained from local stakeholders from StALU MM.

We want to illustrate how this matrix is derived by exemplifying the last column. The amount of lost eco-logically valuable area was estimated to be 4600 m^2 for option A, 400 m^2 for option B and 10 000 m^2 for option C. Since we want large numbers to indicate good performance, these values are inverted and we obtain 1/4600, 1/400 and 1/10 000, respectively (note that there are also other methods of inversion). Now, the highest value (1/400) stands for the option for which the smallest part has to be taken from the ecologically valuable area. This operation is not necessary for criteria whose indicators are designed such that high values indicate good performances. As a next step, the values need to be normalized in order to ensure the comparability between different criteria. For this purpose, the sum of all scores (1/4600 + 1/400 + 1/10 000) is calculated and the inverted scores are divided by this sum. This yields the scores 0.08 (option A), 0.89 (option B) and 0.04 (option C) as in Table 4.1. So far, all calculations are based on an objective evaluation of the options and were accepted by all participants.

The application of the MCDA algorithms from Section 4.3 requires criteria weights. This is the gist where personal preferences come into play and each participant obtains his own result. We want to illustrate the outcome of setting all weights equal as several participants did that in the workshop (see Table 4.2). Clearly, this is only exemplary, but gives an idea of how the results are presented.

Table 4.2 Results from MCDA methods applied to the case study Markgrafenheide with equal weight

MAUT		PROMETHEE I	PROMETHEE II		ELECTRE (c* = 0.7, d* = 0.3)
Ranking	Score	Outrankings	Ranking	Score	Outrankings
1. Option B	0.232	B outranks A and C	1. Option B	0.247	—
2. Option C	0.231		2. Option C	−0.029	
3. Option A	0.193		3. Option A	−0.218	

As can be seen, there are no disagreements among the methods. This can change, when different weights are chosen (see also Section 4.5). In general, giving more weight to *aesthetics* leads to option C. Giving low weight to *loss of ecologically valuable area* and *aesthetics* leads to option A. The fact that the outranking methods PROMETHEE I and ELECTRE provide only a few statements has two reasons. Firstly, the accordance and discordance indices c^* and d^* are chosen too strictly in the ELECTRE method and secondly, as can be seen from the decision matrix, each option has its own strength which makes it difficult for those methods to state outranking relations.

Despite small variations, option B achieved the best results in the participants' trials and is also the solution that has been implemented in reality. This shows that MCDA is able to reproduce 'real-life' decisions. Moreover, participants with different rankings could retrace each other's outcomes and were able to comprehend choices in favour of other options more easily. In order to test the methodology once more, a second case study with a different set up was presented to the participants of the workshop.

4.5 The case study Ostzingst

The case study Ostzingst is close to the island of Rügen and larger than the first case study (see Figure 4.3). The task here is not to protect one contiguous settlement but to protect several villages at the Bodden coast from flooding. At the same time, parts of the peninsula shall be renaturated by converting them from agricultural land to salt marshes.

The salt marshes are to be installed on the outer peninsula (Ostzingst) that lies in front of a Bodden (brackish lake). The Ostzingst encompasses altitudes between +0.40 m and −0.25 m relative to mean sea level. In case of a storm surge the sea water can inundate the entire Ostzingst which is wanted for the renaturation process. However, the flood waters will then quickly reach the Bodden coast in the south where several settlements are located, which in turn need to be protected.

Currently, three dyke systems exist, displayed in Figure 4.3: (i) an outer and (ii) an inner coast dyke on the Ostzingst protect the agricultural area on the

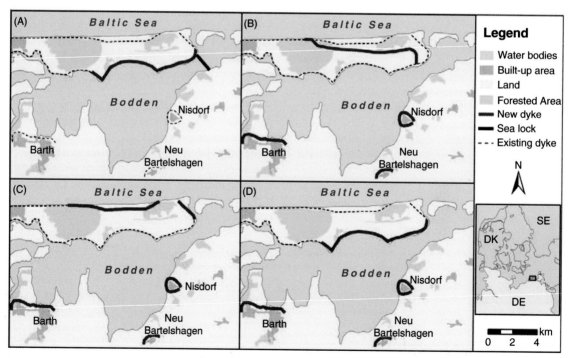

Figure 4.3 The case study area of Ostzingst with the proposed dyke locations of the options A – D (red) in the corresponding panel. The existing dykes (dashed black line) will be removed or reinforced in all cases.

peninsula from flooding and reduce the water inflow to the Bodden, such that (iii) small dykes around the settlements are sufficient to protect the area. In order to reach the goal of renaturation, at least one of the two dykes on the Ostzingst needs to be removed. However, exclusive protection by dykes around the settlements at the Bodden coast is not possible because structures are built very close to the waterline and the construction of such high dykes is not feasible. Therefore, there must be one dyke system on the Ostzingst and a second protective measure has to be reinforced or newly installed around the settlements.

4.5.1 Options

Contrary to the first case study there are several possibilities of where to build the protection measures. For the workshop we elaborated three options in addition to the recently chosen solution and asked the participants to perform an MCDA in order to see if the same decision as the real-life example is proposed.

Basically, each option consists of two parts: An outer protection at the sea coast to reduce the inflow to the Bodden and an inner part to protect the settlements against high Bodden water levels. For the latter, a sea lock (option A) or small dykes close to the settlements (the same for options B, C and D) are under consideration.

Option A

This option consists of the following components:
• Removal of the outer and reinforcement of the inner coast dyke. This enables salt water to flood the Ostzingst from the coast side and the meadows can develop quickly.
• Construction of a sea lock at the inflow of the Bodden (as done, e.g. in Greifswald, Rotterdam, and London) to protect the settlements.

The advantage of option A is that the removal of the outer dyke will enable a quick flooding of the entire Ostzingst to reestablish the salt marshes. The Bodden would be protected by the dyke on the inner coast and the barrage. This would mean that the smaller dykes on the Bodden coast would not have to be reinforced. One of the disadvantages are the inadequate subsoil properties of the inner coast of the Ostzingst that would lead to high building costs and loss of land as the dyke toe has to be widened. From an environ-

mental point of view a barrage is questionable because it hinders the exchange of water and would lead to further degradation of the Bodden water quality. Also, it would be located just adjacent to the preserve area.

Option B

This scenario proposes the following:
• Removal of the outer as well as the inner coast dyke and construction of a new dyke in the centre of the Ostzingst. This enables salt water to flood the peninsula from both sides. This central dyke would protect the Bodden from the high waves during a storm surge. Flood waters would enter the Bodden only through the open passage in the east and lead to a moderate and slow rise of the water table. This solution thus enables salt water to flood the peninsula from both sides.
• Reinforcement and construction of small dykes to protect the settlements.

The advantages of option B are that the dyke in the middle of the Ostzingst would partly lead through a forest area and would not be visible, that is, suggesting a natural landscape. The dyke would run on an existing dirt road which already has compacted ground so that no ground reinforcement for a higher dyke on the south coast would be necessary. The disadvantages are the removal of two existing, functioning dyke systems (although they are insufficient and old) as well as extra costs to reinforce small dykes on the Bodden coast. The latter disadvantage is also valid for options C and D.

Option C

This proposal includes:
• Reinforcement of the outer and removal of the inner coast dyke. This enables salt water to flood the Ostzingst from the Bodden side and the meadows can develop quickly.
• Reinforcement and construction of small dykes to protect the settlements.

The advantage of option C is the maintenance of a functioning outer dyke system that will also prevent erosion of the salt marshes. The disadvantage is that natural flooding from the southern coast will be seldom and the reestablishment of the salt marshes would take longer than with the removal of the outer dyke. For other disadvantages see option B.

Option D

The last option comprises:

• Removal of the outer and reinforcement of the inner coast dyke. This enables salt water to flood the Ostzingst from the coast side and the meadows can develop quickly (as in option A).

• Reinforcement and construction of small dykes to protect the settlements.

The advantage of option D is that the salt marshes will be reestablished rather quickly (see option A). The disadvantages are a combination of those of option A (high building costs and loss of land for inner dyke) and B (extra costs for dyke reinforcement on the Bodden coast).

Options B, C and D differ only in the location of the dyke on the Ostzingst, that is, they all propose the same dyke protection for the settlements. Option A proposes the same dyke on the Ostzingst as option D but suggests a sea lock instead of the small dykes.

4.5.2 Criteria and indicators

Similar to the Markgrafenheide case, a brainstorming session followed by a group discussion led to a set of criteria:

• construction costs;
• operation and maintenance (O&M);
• aesthetics;
• ecology.

Again it is essential that all possible interests are reflected by the set of criteria. The first three coincide with the Markgrafenheide case. In addition, the criterion *ecology* was defined in order to present possible ecological problems arising from the sea lock and the associated reduction of water exchange between the sea and the Bodden during closure periods. Furthermore, the overflow of the Ostzingst is considered in this criterion, which is essential to reestablish the salt marshes. It was agreed that the criterion *Decommission* from the first case study is not relevant in this case since all options would cause similar costs. Also, the protection level is not considered here since all options provide the same flood protection.

4.5.3 Application and Results

As in the first case study, a spreadsheet with all necessary information and calculations was prepared such that the participants were able to see the effects of

Table 4.3 (Normalized) decision matrix for the case study Ostzingst

	Construction costs	O&M	Aesthetics	Ecology
Option A	0.38	0.47	0.07	0.12
Option B	0.20	0.24	0.53	0.48
Option C	0.18	0.07	0.13	0.16
Option D	0.25	0.22	0.27	0.24

their weightings on the results for several MCDA methods.

Again, a decision matrix was elaborated in close collaboration with local stakeholders from StALU MM which represents an evaluation of the options as objectively as possible. All further calculations were based on this decision matrix (Table 4.3).

Naturally, the participants chose quite different weights for the criteria reflecting their personal preferences. We want to illustrate the results from the MCDA methods for the following weights, which were chosen by one participant and found a wide basis of consensus:

• Construction costs: 0.5.
• Operation & Maintenance: 0.2.
• Aesthetics: 0.05.
• Ecology: 0.25.

The results in Table 4.4 show an agreement of all methods regarding option C, which has the lowest scores and is outranked by all other options. In contrast, there is disagreement about the best option. Whereas in the MAUT approach option B performs best, option A is the best option following the PROMETHEE II method. Both outranking methods PROMETHEE I and ELECTRE are not able to distinguish between option A and B but rank option C unambiguously as the less favourable option. This example shows that different methods can provide different outcomes using the same input information. However, it can also be seen from the final scores of option A and B in the MAUT and the PROMETHEE II approach that they are very similar and, consequently, the resulting ranking might be blur. Similar results have been obtained with other participants' weights. It could be concluded that although

Table 4.4 Results from MCDA methods applied to the case study Ostzingst using personal weights for Construction costs (0.5), Operation & Maintenance (0.2), Aesthetics (0.05) and Ecology (0.25)

MAUT		PROMETHEE I	PROMETHEE II		ELECTRE (c* = 0.7, d* = 0.3)
Ranking	Score	Outrankings	Ranking	Score	Outrankings
1. Option B	0.308	A outranks C and D	1. Option A	0.270	A outranks C
2. Option A	0.303	B outranks C and D	2. Option B	0.251	B outranks C
3. Option D	0.245	D outranks C	3. Option D	−0.020	D outranks C
4. Option C	0.144		4. Option C	−0.323	

the scores are very sensitive to varying weights, the rankings are quite similar for a reasonable choice of weights. Recently, Mecklenburg-Western Pomerania has decided to implement option B which again shows that the real decision can be reproduced by the majority of participants by means of an MCDA.

4.6 Discussion

MCDA alone cannot be used to reach decisions – it may only be used in the long and iterative argumentation process. For certain, since the tool has a mathematical and scientifically sound background, the results of MCDA may pose strong arguments in case some involved parties question the soundness of proposed investments or solutions.

The biggest challenge of MCDA is obviously the choice of weights. But it is precisely these weightings which are also one of the greatest advantages of MCDA since they can reflect the personal preferences of the decision maker and can therefore lead to a high acceptance of the results if their choice is justified comprehensibly. On the one hand, the weighting opens MCDA for criticism as these choices are not objective. Critics of certain options might thus always be able to use the weighting as a counter argument. Furthermore, the weights could be tuned in order to obtain a desired result. However, the weighting process in MCDA can be designed transparently and can thus enable all parties, also in review processes, to pinpoint the reasons that favoured one or the other option. This enables all involved parties to openly discuss the importance of the criteria at stake.

The two examples that were discussed and analysed in the workshop have shown that real-life decisions are replicable by MCDA. It could thus also be considered to employ MCDA methods into the decision-making processes on climate change adaptation. As mentioned above, the aim was not to replace or change current decision-making procedures but to offer the involved parties an additional tool. The methodology chosen for the workshop has proven to be successful. It was of particular importance to introduce the legal boundary conditions of the case studies as these are different in every country. Those, and the philosophy behind the legal framework gave the participants the possibility to understand the potential options and to stick as close as possible to the reality. It is thus strongly recommended to give these frameworks equally strong attention as is given to the MCDA methods. Otherwise the results obtained by such a workshop might be too distant from a realistic implementation.

In a sense, MCDA combines several steps that are often handled separately in current decision-making processes. These steps should certainly not be abandoned, but if one tool can serve as an integrative agent some gaps between steps might disappear and the big picture might become more obvious. The methodologies described in this chapter were in fact applied in decision making on future flood protection in Klaipėda city (Lithuania), see Chapter 9 in this book.

Acknowledgements

The authors would like to thank the stakeholders Dr. Lars Tiepolt and Knut Sommermeier from the State

Agency for Agriculture and Nature Mittleres Mecklenburg (StALU MM) for their lectures on coastal protection in Mecklenburg-Western Pomerania and for their support in developing the case studies for this workshop. Furthermore, thanks to Anika Nockert and Anne Holsten for the preparation of maps. We appreciate financial support from the BaltCICA Project (part-financed by the EU Baltic Sea Region Programme 2007-13).

References

Brans, J.P. and Vincke, P.H., 1985. A preference ranking organization method - (The PROMETHEE method for multiple criteria decision-making). *Management Science*, 31 (6), pp.647–656.

Brans, J.P., Vincke, P. and Mareschal, B., 1986. How to select and how to rank projects: The PROMETHEE method. *European Journal of Operational Research*, 24, pp.228–238.

Bruin de, K. et al., 2009. Adapting to climate change in the Netherlands: an inventory of climate adaptation options and ranking of alternatives. *Climatic Change*, 95 (1-2), pp.23–45.

Farber, S.C., Costanza, R. and Wilson, M.A., 2002. Economic and ecological concepts of valuing ecosystem services. *Ecological Economics*, 41 (3), pp.375–392.

Fülöp, J., 2005. *Introduction to Decision Making Methods*. [pdf] Available at: <http://academic.evergreen.edu/projects/bdei/documents/decisionmakingmethods/pdf/> [Accessed 1 June 2012].

Geldermann, J. and Rentz, O., 2007. Multi-criteria decision support for integrated technique assessment. In: J.P. Kropp and J. Scheffran, eds 2007. *Advanced Methods for Decision Making and Risk Management in Sustainability Science*. New York: Nova Science, pp.257–273.

Keeney, R.L. and Raiffa, H., 1993. *Decisions With Multiple Objectives: Preferences and Value Tradeoffs*. Cambridge: Cambridge University Press.

Kiker, G.A. et al., 2005. Application of multicriteria decision analysis in environmental decision making. *Integrated Environmental Assessment and Management*, 1 (2), pp.95–108.

MBLU, 1993. *Generalplan Küsten- und Hochwasserschutz Mecklenburg-Vorpommern*. Schwerin: Ministerium für Bau, Landesentwicklung und Umwelt Mecklenburg-Vorpommern.

MLUV, 2009a. *Regelwerk Küstenschutz Mecklenburg-Vorpommern – Übersichtsheft: Grundlagen, Grundsätze, Standortbestimmung und Ausblick*. Schwerin: Ministerium für Landwirtschaft, Umwelt und Verbraucherschutz Mecklenburg-Vorpommern.

MLUV, 2009b. *Regelwerk Küstenschutz Mecklenburg-Vorpommern. 1 – 4/2009: Internes Messnetz Küste und hydrographische Datenbanken*. Schwerin: Ministerium für Landwirtschaft, Umwelt und Verbraucherschutz Mecklenburg-Vorpommern.

MLUV, 2009c. *Regelwerk Küstenschutz Mecklenburg-Vorpommern. 3 – 7/2009: Vermessungsrichtlinie und digitale Datenformate*. Schwerin: Ministerium für Landwirtschaft, Umwelt und Verbraucherschutz Mecklenburg-Vorpommern.

MLUV, 2010a. *Regelwerk Küstenschutz Mecklenburg-Vorpommern. 2 – 1/2010*. Schwerin: Ministerium für Landwirtschaft, Umwelt und Verbraucherschutz Mecklenburg-Vorpommern.

MLUV, 2010b. *Regelwerk Küstenschutz Mecklenburg-Vorpommern. 2 – 2/2010*. Schwerin: Ministerium für Landwirtschaft, Umwelt und Verbraucherschutz Mecklenburg-Vorpommern.

MLUV, 2011a. *Regelwerk Küstenschutz Mecklenburg-Vorpommern. 1 – 5/2011*. Schwerin: Ministerium für Landwirtschaft, Umwelt und Verbraucherschutz Mecklenburg-Vorpommern.

MLUV, 2011b. *Einheitlicher Zuschlag von 50 Zentimetern bei Küstenschutzanlagen an der Ostsee*. Press release, 16 August 2011.

MWAT, 2007. *Aktionsplan Klimaschutz Mecklenburg-Vorpommern, Abschnitt Klimafolgenforschung. Bericht: Arbeitsgruppe ''Ostsee/Küste'*. Schwerin: Ministerium für Wirtschaft, Arbeit und Tourismus Mecklenburg-Vorpommern.

Rogers, S.H., Seager, T.P. and Gardner, K.H., 2004. Combining expert judgment and stakeholder values with PROMETHEE: A case study in contaminated sediments management. In: A. Linkov and A. Bakr Ramadan, eds. 2004. *Comparative Risk Assessment and Environmental Decision Making*. Dordrecht: Kluwer Academic Publishers, pp.305–322.

Roy, B., 1968. Ranking and choice in pace of multiple points of view (ELECTRE Method). *Revue Francaise D'Informatique De Recherche Operationnelle*, 2 (8), p.57.

Saaty, T.L., 1980. *The Analytic Hierarchy Process*. New York: McGraw-Hill.

Sparrevik, M., Barton, D.N., Oen, A.M.P., Sehkar, N.U. and Linkov, I., 2011. Use of multicriteria involvement processes to enhance transparency and stakeholder participation at Bergen Harbor, Norway. *Integrated Environmental Assessment and Management*, 7 (3), pp.414–425.

Staun, 1998. *Markgrafenheide*. [Aerial photograph] (Staun, Coastal Department).

Teknomo, K., 2006. *Analytic Hierarchy Process (AHP) Tutorial*. [online] Available at: <http://people.revoledu.com/kardi/tutorial/ahp/> [Accessed 1 June 2012].

Triantaphyllou, E., 2000. *Multi-Criteria Decision Making Methodologies: A Comparative Study* (Applied Op.). Dordrecht: Kluwer Academic Publishers.

Yatsalo, B.I. et al., 2007. Application of multicriteria decision analysis tools to two contaminated sediment case studies. *Integrated Environmental Assessment and Management*, 3 (2), pp.223–233.

5 Preparing for Climate Change: Planning Adaptation to Climate Change in the Helsinki Metropolitan Area, Finland

Ruusu Tuusa[1], Susanna Kankaanpää[2], Jari Viinanen[3], Tiia Yrjölä[3] & Sirkku Juhola[1,4]

[1]Department of Real Estate, Planning and Geoinformatics, Aalto University
[2]Helsinki Region Environmental Services Authority (HSY)
[3]Environment Centre, City of Helsinki, Finland
[4]Department of Environmental Sciences, University of Helsinki

5.1 Introduction

Cities are at the forefront of climate change with currently more than half of the world's population living within 60 kilometres of the sea and approximately three quarters of large cities are located on the coast (UN-Habitat, 2008). Cities and the process of urbanization are the key drivers of environmental change by both contributing to the change as well as facing the consequences of the change. The vulnerability of cities to climate change is location-dependent, as urban areas have particular traits, such as location, structure, assets and density that make their residents and assets vulnerable to climate change (Gasper, Blohm & Ruth, 2011).

Urban climate governance refers to how different actors, public, private and the civil society actors articulate climate goals, exercise influence and authority, and manage urban climate planning and implementation process (Anguelovski & Carmen, 2011). This new policy issue is contributing to the emergence of new institutional arrangements and strategies. There are several cities within Europe that have now produced an adaptation strategy (Ribeiro et al., 2009), and many cities in the Baltic Sea Region are also following suit.

This chapter discusses the case study of the adaptation strategy process within the Helsinki Metropolitan Area in Finland[1] and examines the main barriers and challenges for developing an adaptation strategy. This case study explores strategic climate change adaptation from the regional and local authorities' point of view and elaborates on the process that will lead to the publication of the adaptation strategy. In this strategy process, the actors within the Helsinki Metropolitan Area have identified climate change impacts and vulnerabilities in the area and outlined strategic climate change adaptation options.[2] This article describes both

---

[1]Some parts of the regional adaptation strategy process were funded by other sources than the BaltCICA project (see Figure 5.2).

[2]BaltCICA project partners in the Helsinki Metropolitan Area are the Helsinki Region Environmental Services Authority HSY, Centre for Urban and Regional Studies YTK at Aalto University, and the city of Helsinki Environment Centre. Within the BaltCICA case study HSY outlined adaptation options for the Helsinki Metropolitan Area; the Environmental Centre focused on the city of Helsinki, and YTK has

---

Climate Change Adaptation in Practice: From Strategy Development to Implementation, First Edition.
Edited by Philipp Schmidt-Thomé and Johannes Klein.
© 2013 John Wiley & Sons, Ltd. Published 2013 by John Wiley & Sons, Ltd.

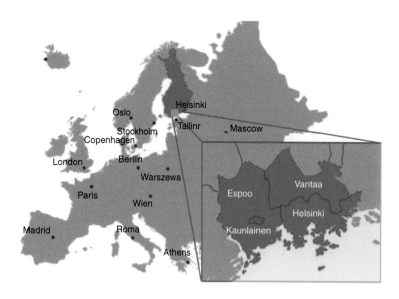

Figure 5.1 The Helsinki Metropolitan Area and its four cities.
Source: Kankaanpää, 2010.

the strategic process and its findings of the climate change impacts, most crucial adaptation needs and the current state of adaptation in the Helsinki Metropolitan Area cities. After describing the process we compare some of the findings to research literature from other countries to identify similarities and differences in adaptation barrieres and enabling factors.

The Helsinki Metropolitan Area is located on the south coast of Finland, facing the Gulf of Finland (Figure 5.1). The Area consists of four cities: Helsinki, the capital city of Finland, and the cities of Espoo, Vantaa and Kauniainen. Helsinki and Espoo are the two largest cities in Finland – Helsinki with over half a million inhabitants – and Vantaa, the fourth largest. Together the Metropolitan Area municipalities constitute the most densely populated and largest urbanized area in Finland with about one million inhabitants (Statistics Finland, 2011).

The Metropolitan Area cities have 40 years' of experience of cooperation in public transportation, waste management, air protection and regional development, more recently also in water services. The Helsinki Region Environmental Services (HSY) is a

joint body founded by the four cities. YTV, as HSY was previously called, coordinated the regional mitigation strategy process that was completed in 2007 and is now coordinating the adaptation strategy process. The sub-national level climate change adaptation strategies, including the regional adaptation strategy in the Helsinki Metropolitan Area are voluntary commitments of municipalities and the regional bodies, such as the HSY, can have a mandate for coordinating climate policy, but not control over local decision making. Hence, the member cities are mostly responsible for implementing the strategies.

This chapter is structured as follows: the next section introduces the concept of adaptation and current adaptation policy in Finland. It also provides background information on the climate policy in the Helsinki Metropolitan Area. Section 5.3 discusses the adaptation strategy process. Fourthly, the chapter describes case study outcomes and next adaptation steps and discusses the process, tools used and interaction with authorities and other stakeholders. The final section discusses the case study findings in relation to some international research results and to the general adaptation situation in Finland.

provided the case study with scientific insight. In the beginning of the BaltCICA project the regional joint authority was called the Helsinki Metropolitan Area Council YTV. In 2010 YTV was split into two authorities: Helsinki Region Environmental Services Authority (HSY) and Helsinki Regional Transport Authority (HSL).

5.2 Planned adaptation policy in Finland

According to the Intergovernmental Panel on Climate Change (IPCC) adaptation to climate change means 'the adjustment in natural or human systems in response to actual or expected climatic stimuli or their effects, which moderates harm or exploits beneficial opportunities' (Parry et al., 2007, p.6). In recent years, the majority of developed countries has produced national adaptation strategies, outlining their planned adaptation policies (Swart et al., 2009).

As the first EU country, Finland published a national adaptation strategy (Marttila et al., 2005) in 2005. The Finnish national adaptation strategy (NAS) includes cross-cutting measures and measures for 15 sectors of society, focus being on planned and anticipatory adaptation within the public sector. The NAS follows the principle of mainstreaming climate policy and to this end emphasises the importance of including adaptation into key policy areas instead of treating adaptation as a separate policy (Juhola, 2010). The NAS is to be revised in 2012–13.

The Ministry of Agriculture and Forestry conducted an evaluation on the implementation of the NAS in 2009. The evaluation concluded that the level of adaptation was highest in the water and flood management sector, and second highest in agriculture, forestry, transportation and road management, as well as spatial and urban planning. At the national level, two ministries have published an adaptation action plan for their administrative field: the Ministry of the Environment (2008) with an update in 2011 (Ministry of the Environment, 2011), and the Ministry of Agriculture and Forestry (2011). Other ministries have not yet published comprehensive adaptation strategies or action plans but some ministries have included adaptation into other plans and programmes. For example, climate change impacts are included in the Internal Security Programme (Ministry of the Interior, 2008) and in the climate policy programme of the Ministry of Transport and Communications (Liikenne- ja viestintäministeriö, 2009). There are also preliminary studies on climate change impacts on the transportation sector (Saarelainen & Makkonen, 2007, 2008; Finnish Road Administration, 2009).

The NAS does not explicitly oblige local authorities to prepare their own adaptation strategies or plans but some Finnish local authorities have begun to develop strategic adaptation plans for their city or in cooperation with other cities. Municipalities and regions have so far concentrated on climate change mitigation, but

some mitigation strategies address adaptation to some extent.

5.2.1 Climate policy in the Helsinki Metropolitan Area

The Helsinki Metropolitan Area has been among the front runners in local climate initiatives in Finland: the Helsinki Metropolitan Area Council YTV and its member cities joined the ICLEI's Urban CO_2 project in the early 1990s. Since then, YTV and its member cities have participated in several campaigns and have undertaken voluntary commitments such as the local authorities' energy efficiency agreements. The Helsinki Metropolitan Area cities and YTV began to prepare a comprehensive regional mitigation strategy in 2003 and the strategy was published in 2007 (YTV, 2007; Kaasinen, 2010).

The regional mitigation strategy did not include any adaptation measures and therefore the Metropolitan Area began to prepare the regional adaptation strategy in 2009. The adaptation strategy is a voluntary joint strategy for the area's four municipalities and the process is coordinated by HSY. The individual cities in the Metropolitan area have not yet prepared local adaptation strategies, but some climate change impact assessments have been made and some adaptation initiatives have been undertaken. For example, in 2007, the city of Espoo made an investigation on general climate change impacts and adaptation needs in Espoo and included some possible adaptation measures (Soini, 2007).

5.3 Preparation of the adaptation strategy

The regional strategy process began in 2009 and it consisted of two processes: one at regional level[3] and the other at local level. The processes were intertwined but have focused on slightly different things.

---

[3]Officially Helsinki Metropolitan Area is a sub-regional authority. In Finland regions are larger administrative units, which are divided into several sub-regions. The Helsinki Metropolitan Area sub-region is part of Uusimaa region. However, in this study the word region is used in a more general sense referring simply to something between national and local.

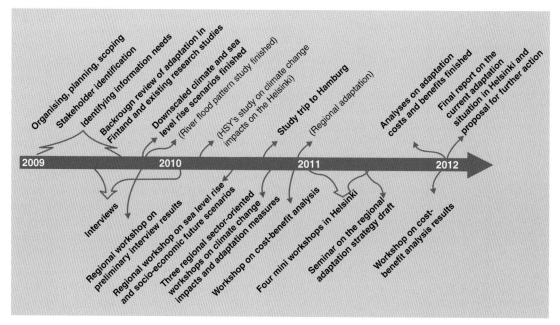

Figure 5.2 Adaptation process in the Helsinki Metropolitan Area.
Note: non-BaltCICA activities in grey.

Firstly, the Helsinki Metropolitan Area has focused on producing region-wide, comprehensive strategic adaptation measures. Strategic regional level cooperation provides an arena for discussion on adaptation issues and measures, which require, or could benefit from, regional cooperation. The focus of the regional process was on managing climate change related risks in built environment and coastal areas. Secondly, the city of Helsinki examined the existing local adaptation plans and adaptation measures, and consequently identified sectors where such plans and measures are still lacking. Based on this information, the city adaptation working group outlined a proposition on further adaptation steps the city could take in the future. Figure 5.2 illustrates the adaptation process in the Helsinki Metropolitan Area.

5.3.1 Climate change information for the adaptation strategy

Information about the changing climate, its impacts, stakeholders' views on adaptation and what adaptation measures they have undertaken are needed when outlining adaptation options and action plans. The adaptation strategy process in the Helsinki Metropolitan Area began with a background study on adaptation-related plans in local, regional and national levels, adaptation-related legislation and research, and some international adaptation examples. Also material and experiences from the previous regional mitigation strategy process and strategy implementation contributed to the planning of the regional adaptation strategy process.

In 2009, the Finnish Meteorological Institute prepared downscaled climate change and sea level rise scenarios for the Helsinki Metropolitan Area (Venäläinen et al., 2010). Although many cities around the Baltic are facing similar challenges, it is necessary to identify particular climate risks for Helsinki. According to the study, it is likely that because of climate change, high sea levels in the Helsinki area become more probable. Extreme short-term fluctuations and long-term changes in sea level have several impacts on the urban environment and ought to be taken into account when new areas and buildings are planned. Helsinki has already experienced unexpected short-term sea level rise and consequent flooding in the downtown area in January 2005 (Venäläinen et al., 2010).

Other major climate change impacts in the Metropolitan Area are related to temperatures and snow conditions. In particular, mean winter temperatures are rising and thermal winter is becoming shorter. This means reduction in snow cover and thinning of ground frost. Also, the ice on the Baltic Sea is projected to be thinner and the period of ice cover shorter. Moreover, summer heat waves are likely to be longer and there will be more warm days in the summer. Overall variability in temperatures is likely to increase. It is predicted that annual precipitation will grow and the probability of heavy rainfall may increase (Venäläinen et al., 2010).

The downscaled climate scenarios were complemented by a study investigating changes in flood patterns in two major river basins in the Metropolitan Region by the Finnish Environment Institute (Veijalainen, Sippel & Vehviläinen, 2009). The study concludes that in some areas flooding is likely to diminish because of reductions in snow cover. In other areas the probability of flood grows because of increasing heavy rainfall and winter flooding probability, but uncertainties in predicting heavy rainfalls make the flood risk assessment uncertain as well. Current flood risk maps, however, are estimated to still be current. The problem is, though, that the maps and flood instructions are not always complied with.

The studies above indicate that flooding is one of the most prominent effects of climate change in the Helsinki Metropolitan Area, hence we take a brief glance at the current framework of flood protection in Finland. Traditionally, the most important flood protection measures in Finland have been water level regulation through hydropower plants, dams and reservoirs built mostly for electricity generation. Also dredging and small-scale floodbanks have been used and in some cases rivers have been rechanneled. However, permanent flood protection constructions are mostly small-scale (Maa- ja metsätalousministeriö, 2009).

In recent years, Finland has refined legislation regarding anticipatory flood risk assessment, flood mapping and flood protection to implement the EU's Floods Directive (2007/60/EC) and to adapt to the impacts of climate change. Previously most attention was put to inland flooding which has caused the largest economic losses in Finland, but the new legislation also emphasises coastal and urban flooding caused by storm or melting water. Secondly, the new legislation elaborates responsibilities and systematizes comprehensive flood risk management and the cooperation of local and regional authorities. In brief, the state's regional Centres for Economic Development, Transport and the Environment conducted a mapping of flood risks related to inland and coastal flooding and local and regional authorities together prepare flood risk management plans. However, municipalities are responsible for storm water management. Land-use and construction planning are important means of anticipative flood protection and flood risk management available to municipalities (Kuntaliitto, 2002; Maa- ja metsätalousministeriö 2009; Flood Risk Management Act 620/2010).

The storm water management plans and flood strategies the Helsinki Metropolitan cities have prepared indicate that the cities have begun to integrate climate change adaptation needs into land use and construction but changes in practices are slow (Nurmi et al., 2008; Valkeapää, Nyman & Vaittinen, 2008; Vantaa, 2009; Espoo, 2011).

5.3.2 Stakeholders in the strategy process

The case study stakeholders were both information sources and objects: their contribution and knowledge was considered useful in the strategic adaptation process and at the same time they were the subjects of new information and ideas. The regional adaptation process involved stakeholders from public administration, private sector and among non-governmental organizations, though the focus was on public bodies. Identification of relevant stakeholders was based on two things: previous participation, for example in the regional mitigation strategy process that had taken place a couple of years before, and secondly, sectors' or stakeholders' relevance to adaptation specifically at a regional level. In the local level adaptation process in Helsinki, stakeholders were mainly city departments as the aim was to investigate the city's own adaptation measures.

The aim at both process levels was to involve as large a variety of administrative sectors as possible, especially from municipalities' administration, since one of the most important aspects of the process was to be comprehensive and cross-sectoral: to collect information on the current state of climate change adaptation on every sector and to involve the sectors at the definition of regional and local strategic goals. Both

Table 5.1 Stakeholder groups in the regional and local adaptation processes

Local administrative sectors	Regional and state administration	Others
1. City planning, construction and infrastructure **2.** General administration and finances **3.** Health and social services, well-being, culture **4.** Environment **5.** Countryside **6.** Energy supply **7.** Water supply **8.** Transportation **9.** Local rescue services	**1.** Regional joint bodies **2.** Research institutes **3.** Central state administration (ministries) **4.** Regional state administration **5.** Other cities (city of Turku) **6.** Regional rescue services	**1.** Interest groups and lobbying organisations **2.** Commerce and enterprises **3.** Non-profit associations (Finnish Red Cross)

local and regional stakeholder groups are presented in Table 5.1.

The largest stakeholder group in the regional strategy process was municipal and city administration. About 15 local administrative institutions from the Metropolitan Area cities were involved.[4] In Table 5.1 the institutions are divided into groups according to their field of activity. The group 'city planning, construction and infrastructure' was the largest one, and, together with local environmental administration, also the one most actively involved in the regional process. Secondly, state administration (both central and regional), state research institutes and regional joint authorities similar to HSY were involved. Nearly all stakeholders were involved in the regional process, whereas the local process involved mainly local administrative sectors in Helsinki and some regional joint bodies.

Thirdly, some private enterprises in sectors relevant to climate change adaptation (e.g. construction and insurance) as well as interest groups and lobbying organizations were integrated mainly in the interview phase, except the Association of Finnish Local and Regional Authorities which was involved in workshops and the adaptation strategy working group. Not many NGOs were involved in the strategy process as active partners, although many commented on the

draft strategy. Citizens were not directly involved, but individuals could comment on the regional strategy draft on the internet after the draft was published in 2010.

With regards to the working methods, workshops and group discussion are commonly used in local and regional strategy or development projects in Finland.

Workshops bring different stakeholders and different sectors together with the general goal of sharing ideas and knowledge, supporting learning and understanding and creating joint perspectives, to solving contradictions or increasing mutual understanding. The workshops were mostly organized around expert presentations and consequent group work, except the mini workshops in Helsinki, which were more like group discussions altogether because the number of participants was small. One of the regional workshops was based on future socio-economic scenarios. These case study workshops had two deliberate aims: to gather information from stakeholders and get their contribution to adaptation planning, and secondly, to inform them about climate change impacts and adaptation requirements and also implicitly assist learning.

The case study illustrates that it is useful to bring together stakeholders from different fields to interact in groups. However, it was challenging to get stakeholders involved. The aim of involving a variety of administrative fields was met to some extent. Authorities which are most often related to environmental issues or whose field of expertise is predominantly

---

[4]Not every sector had representatives from all four cities. Helsinki and its administrative sectors were the most active ones in the regional adaptation process.

climate change issues were the most active, for example, the local environmental and land-use planning agencies as well as the state's regional Centre for Economic Development, Transport and the Environment which steers, monitors and guides municipalities' land use, flood protection and so on.

5.3.3 Interviews based on extreme weather events

Interviews were one of the main instruments of gathering information from stakeholders and the case study partners carried out approximately 30 interviews between summer 2009 and January 2010. The aim was to gather information on how current weather events affect different sectors and what measures the stakeholders have undertaken. The stakeholders and sectors interviewed were selected because they were seen as the most relevant to climate change adaptation at the local and regional level.

The case study partners chose to use an interview method that was inspired by LCLIP (Local Climate Impacts Profile) developed by the UK Climate Impacts Programme UKCIP (2011). LCLIP is a toolkit to identify local climate change impacts and carry out vulnerability analysis, using extreme weather events as a basis for stakeholder interviews. The project partners searched in newspapers for a variety of extreme weather events that occurred in the region during the past five to six years. Widely reported weather events were chosen including the heat waves in the summers of 2003 and 2008, heavy rainfall in summer 2004, rapid change of conditions in winter 2005, winter flooding in 2005, drought in summer 2006, and snow storm in winter 2008. Table 5.2 illustrates an example of interview results within sectors of city planning, construction, technical sector and rescue services authorities.

The interviews revealed that the general awareness of climate change is good among the administrative sectors interviewed. Extreme weather events had already caused some damage and some organizations had consequently improved their preparedness level. Also, changes in practices had taken place. However, the awareness and preparedness level varied among organizations and positions. Persons in leading positions may have different answers than for example technical managers who have practical experience in the causes of weather events.

The interview results also indicate that climate change is seldom the only cause of action. Adaptation measures can be part of wider risk assessments or readiness planning. External drivers such as new legislation and regulation have an effect on how organizations adapt. Further, climate change is only one factor in the organisations' operational environment.

Interviews provide an opportunity to approach stakeholders who otherwise would not participate, for example in workshops. On the other hand, unlike workshops, interviews do not allow much communication between stakeholders unless group interviews are used. In general, the public authorities interviewed regarded climate change as a complicated issue which requires interaction and cooperation between authorities and administrative sectors. The interview results and especially local authorities' views are quite consistent with studies in other countries (see for example Næss, Bang, Eriksen & Vevatne, 2005; Wilson, 2006; Storbjörk, 2007; Amundsen, Berglund & Westskog, 2010).

Regarding the interview method, the case study partners considered the LCLIP-inspired interview approach useful. The interview method was a good way to collect hands-on information which would probably not have been possible, if the interviews had approached adaptation and climate change in abstract terms. The terminology related to climate change research was partly unfamiliar to the interviewees and thus could have been an obstacle in accessing the local knowledge. However, using extreme weather events as a discussion basis and talking in practical terms produced an amount of concrete information about the consequences of extreme weather events and the vulnerabilities among different organizations and sectors. Based on the interview results, it was possible to draw conclusions on existing adaptation measures and the need to develop new ones, and on the stakeholders' operational framework and adaptation barriers and costs. On the other hand, the interview approach used in the case study can be criticized for producing mostly detailed technical information and less information on the societal and institutional framework within which the adaptation decisions take place.

5.3.4 Stakeholder interaction

The strategy process included one-way communication and two-way interaction with stakeholders. The

Table 5.2 Interview results regarding floods, within sectors of city planning, construction, technical sector and rescue sector

Sector/ orgasation	What has happened/could happen?	What was done then/could be done?	How the organisation has been preparing itself for similar events?	Changes in practices?
City planning			• Flood areas have been identified • Flood risk assessments are used in land use planning • Using pilot case studies as examples • Additional assessment in flood risk areas	• Flood risk assessments and examples case studies are used in land use planning
Construction			• Recommended minimum height for buildings in local building codes have been revised in some cities • Construction ban in flood risk areas • Revised building instructions	• Run-off water drains have been improved in some places • Plans of using a flood tunnel to storage floodwater
Technical department	• River flood caused waste water to leak into river, winter flood in 2005			• An initial review has been made of urban flood risk areas in Helsinki between +1,1 and +2,0 meters • Plans of permanent flood protection constructions such as dykes
Rescue departments	• Coastal flooding in 2005 in Helsinki downtown, resque and protection work	• Sandbags were put to prevent the sea water overflow to downtown • Possibility to use pumps	• Contingency plans • Emergency duty • Meteralacial Institute will provide information about unusual weather conditions • Cooperation with local technical departments and oil destruction authorities	• Changes in communications in extraordinary situations: real time information and Instructions on rescue department's or city's web pages

interviews served both as information gathering and distribution. Many of the stakeholders involved are in regular interaction and cooperation with each other as well as with HSY and the city of Helsinki Environmental Centre – hence, information was shared. For example, the background studies on climate change impacts described in Section 5.3.1 were published in 2010 for public use. Two magazine articles were published about climate change impacts, adaptation needs and policy and strategic adaptation work in the Helsinki Metropolitan Area. In addition, HSY published a flyer in English about the regional adaptation strategy pro-cess and the city of Helsinki published a brochure about mitigation and adaptation in Helsinki city and what citizens could do.

Workshops were the principal form of two-way interaction with and between stakeholders. After most of the interviews and background studies were carried out, a series of workshops and a seminar took place. The aim was also to identify adaptation measures with stakeholders. All regional workshops utilized the general method of group discussion except the scenario workshop on sea level rise. The purpose of this workshop was, with the help of four socio-economic future scenarios, to identify consequences of different policies and to discuss possible and desired futures and how desired visions of the future could be put into reality.

The workshop results and other material were put together and a regional strategy draft was published in December 2010. In March of the following year, HSY organized a seminar to present the draft and regional adaptation work for authorities and other stakeholders and to collect feedback for the final regional adaptation strategy. In addition, local-oriented workshops in Helsinki included four mini workshops for city personnel and two stakeholder workshops on adaptation cost-benefit analysis, which is described in the next chapter. Each mini workshop focused on adaptation issues in one administrative field, the city government. The aim was to identify adaptation measures, outline adaptation options and interact with a variety of city departments.

All in all, a satisfying amount of stakeholders contributed to the case study. However, the process was not spared from the problems participation and differing levels of interests. As noticed above, often those organizations whose main tasks are related to climate change or who experience climate change impacts heavily are those who are also willing to take part in climate change related processes. In other words, it is difficult to activate organizations who do not rate adaptation high in their to-do list or who are not required by law to take it into account. The actors may also have interests that are in contradiction to adaptation or there can be incompatible preferences between policy areas. In the Metropolitan Area process, one challenge in activating administrative institutions and officials to participate in workshops was the stakeholders' lack of time and resources. It also may be that the strategic work at regional level was not compelling because the strategy does not bind the local authorities and ultimately they have to elaborate and implement the strategic adaptation goals at city level in any case.

The attempt was to maximize stakeholder interest to participate in workshops by bringing interesting experts to give presentations with new information. Also interviews were a good way to gather information from those organizations that did not attend the workshops. City planning, construction, infrastructure and the environmental administration were the most active in terms of attendance. Also rescue services departments were happy to get involved in adaptation planning. The results are complementary with the findings in 2009 regarding the state of implementing the NAS (Ministry of Agriculture and Forestry, 2009).

It is not difficult to understand why these sectors are interested in adaptation: climate change impacts on spatial and urban planning, construction, buildings, roads and networks are quite obvious, and the rescue and security authorities answer for contingency planning and operations at unusual conditions such as floods. Further, the regulations concerning these administrative fields have been revised lately. Less attention has been paid, for example, to climate change's social and health impacts.

5.3.5 Adaptation cost-benefit analysis

Information on the financial aspects of adaptation is useful in decision-making. During 2011 two studies on adaptation's costs and benefits were carried out, one at the level of the whole Helsinki Metropolitan Area and one at Helsinki city level. The regional-oriented study included two parts. The first addressed the costs and impacts of the adaptation measures outlined in the regional adaptation strategy draft. The second part consisted of an analysis of the costs of coastal flooding in two different areas in the Helsinki Metropolitan Area. The analysis identified direct and indirect impacts on six sectors and their costs. The calculations illustrated that the monetary costs of flooding varied among the type and age of buildings. The calculations included mainly direct damages and costs, for example the need to repair constructions and buildings. It was found out that it is difficult to identify and value indirect damages of floods and its impact on humans, or to define financial values for damages on cultural monuments, for example. The conclusion is that the analysis produced interesting information but the user of the analysis method and the data it produces must be sensitive to the underlying presumptions and the limits of the method.

The city of Helsinki Environment Centre carried out the local-oriented adaptation cost-benefit analysis. The focus of the analysis was on costs, that is, what causes the costs and how they are distributed among authorities or other actors. The analysis consisted of general adaptation assessment in Helsinki and eight case studies. The case studies were general examples of specific situations, not based on one real-life example. A full-length cost-benefit analysis was carried out for some of the case study situations, others were analysed in more general level. The time scale was 100 years, addressing direct and indirect monetary costs. The case studies match with some of the main sectors in the regional adaptation strategy. In each case study different possible policies and adaptation options were investigated (e.g. the different policies could be: a) not taking flood risk into consideration when building new areas, b) raising the minimum building height in new areas, or c) not building in the flood risk areas at all).

The general conclusion is that most of the local strategies in Helsinki do not include information on the financial costs of climate change impacts or adaptation measures. Secondly, often the focus is on costs but adaptation can also cause benefits and savings. Similar to the regional flood costs analysis, it was found that it might be challenging to define the financial value of indirect impacts such as those on health or quality of life.

Overall, the general assessment of the costs of adaptation measures outlined in the regional adaptation strategy draft may help to elaborate and compare the measures in the final strategy. The coastal flooding case studies and the case studies of the city of Helsinki were examples of the costs and benefits climate change and adaptation measures could have in different situations. One of the ideas behind the case study analyses was to demonstrate the use of a cost-benefit analysis and to produce information in the form relevant to decision making. The target was also to identify the benefits and trade-offs related to anticipatory adaptation.

The analyses illustrate that there are advantages and disadvantages in the use of the cost-benefit analysis method in adaptation cases. The analysis made indicates that there are several factors that must be acknowledged when using the method. Firstly, the underlying assumptions and the choices made during the analysis affect the results. Secondly, it is challenging to define financial costs or benefits of climate change impacts or adaptation options related to health, well-being or immaterial values. It is also challenging to identify and quantify indirect impacts of climate change or adaptation measures. It has to be

considered when it is realistic to use this method and when perhaps to use other methods. Thirdly, it has to be noticed that society is not static and inhabitants react to changes in their environment and also, for example, to information in the media.

5.4 Implementing adaptation measures

A draft of the regional adaptation strategy was published in December 2010 and the year 2011 was devoted to gathering feedback for the final strategy, elaborating strategic objectives and measures and updating research information considering climate change. The final regional adaptation strategy will be published in 2012 and its aim is to integrate climate change adaptation into all sectors of the local public administration.

The adaptation measures outlined in the strategy are not very detailed and need to be elaborated within administrative sectors, mostly at local level because the cities account for most of the public services provision, land-use and building planning and building control, maintaining and building technical and transport networks, and many other things relevant to adaptation. The role of the regional body HSY is to provide the member cities and other stakeholders with up-to-date information on climate change and adaptation. After the adaptation strategy is published the HSY will promote its implementation in its own functions and among member cities and other stakeholders. It is likely that HSY will launch adaptation-related projects in the future. Currently HSY, the city of Helsinki and a couple of other partners are planning a project related to developing adaptation tools.

The objective of the city of Helsinki was to investigate what the current local adaptation situation is and set the foundations for further action. The city has drawn all information gathered together into a report including the results of the background studies, the interviews, the workshops and the cost-benefit analysis. Based on this information, the report presents what kind of impacts climate change causes in different sectors of the city administration and how vulnerable the sectors are. Secondly, the report presents what kind of climate change adaptation measures and ways to cope with climate variability and extremes the administrative sectors in Helsinki have undertaken. Finally, the report proposes next adaptation steps within the city administration. The aim is to mainstream adaptation and integrate it into other policies (Yrjölä & Viinanen, 2012).

The report concludes that the city has addressed the importance of adaptation in strategic level but it lacks a coherent adaptation policy and there is limited cooperation between administrative sectors in adaptation issues. Adaptation is integrated especially into local flood and storm water management. There is a need to develop the cooperation, knowledge sharing and interaction between sectors and develop adaptation policies together, not only within each administrative sector. Secondly, adaptation should be integrated into every policy sector. The report concludes also that the private sector and citizens have an important role in adaptation, which should be studied further. The city should also integrate adaptation into the current climate communications, which mainly addresses mitigation (Yrjölä & Viinanen, 2012).

The most important adaptation priority in the city of Helsinki is related to constructions and infrastructure due to the age and location of the buildings, compact urban structure and the high investment and repair costs. Also, transport systems are vulnerable to climate change. The city can prepare for climate change in these sectors with land-use and building planning, in which the local autonomy is relatively high. The report recommends the use of green infrastructure, for example parks and green zones, roof tops and unbuilt areas where storm and flood waters can be saturated or channelled. Integrating mitigation and adaptation objectives can be challenging but there are measures, which benefit both (Yrjölä & Viinanen, 2012). The report will be presented to the city of Helsinki Environment Committee. How the proposed adaptation measures will be implemented in the future depends on the municipal council and municipal executive board.

5.5 Discussion: barriers and incentives for adaptation at local level

Finland brought out the NAS relatively early. However, there is not yet much national level regulation concerning adaptation at regional and local levels, and no additional state funding for implementation. Some general instructions and regulations are included in existing steering mechanisms, such as the national land use guidelines. Research on, for example, climate change's social impacts and concrete sector and local-oriented adaptation is still conducted to a minor degree. At the moment, with the lack of consistent national regulation on adaptation, many of the local adaptation measures are voluntary and depend on the local authorities (Juhola, 2010). Similarly, the upcoming regional adaptation strategy in the Helsinki Metropolitan Area does not legally bind the municipalities. Hence, it is interesting to study what the incentives and barriers for local adaptation are.

One of the barriers for adaptation the local authorities express in research literature is the lack of information and knowledge related to climate change and adaptation, and the uncertainties in the information, including its relevancy (Wilson, 2006; Storbjörk, 2007; Urwin & Jordan, 2008; Amundsen, Berglund & Westskog, 2010; Glaas, Jonsson, Hjerpe & Andersson-Sköld, 2010). The case study interviews and the study in the city of Helsinki indicate that some of the local authorities in the Helsinki Metropolitan Area find the lack of knowledge and the changing and uncertain climate data challenging. The information and climate change scenarios also become outdated quickly. In this case, a variety of regionally specific information on climate change impacts and on the consequences the climate-related events have within different fields of administration and stakeholders. Both research data and local context-specific data from stakeholders were gathered. The adaptation cost-benefit analysis was aimed at producing information in the form, which is relevant in decision making. The case study was among the first studies in Finland to collect such context-specific data on adaptation. Further and elaborated studies are needed, especially regarding adaptation implementation and the adaptation costs and benefits.

According to literature the problem may also be the way information is utilized and communicated to different groups rather than the lack of information itself (Frommer, 2011). The ability to use knowledge from other knowledge producers is also important (Glaas et al., 2010). Furthermore, Amundsen, Berglund and Westskog (2010) and Glaas et al. (2010) found that the local departments, which are used to cope with climate variability account for most of the adaptation in municipalities as a matter of routine. This makes adaptation reactive, technical and short-sighted. Information and experience sharing and interaction between sectors and different kinds of stakeholders promote learning and increase local adaptive capacity and coherence in adaptation policy (Wilson, 2006; Pahl-Wostl, 2009; Glaas, Jonsson, Hjerpe & Andersson-Sköld, 2010; Frommer, 2011).

The local and regional authorities interviewed in the case study regarded that there is a need for cooperation in adaptation issues between sectors and authorities. This case study attempted to bring stakeholders together, specifically different fields of local and regional public administration, but one project has only limited time and resources. However, financial constraints, differing interests and prioritization of more urgent issues are challenges to adaptation and cross-sectoral cooperation. The amount of different city departments in large cities, such as the city of Helsinki, also makes it difficult to cooperate and create cross-sectoral adaptation policies. The commitment and ownership of adaptation issues is important as well.

Besides horizontal level interaction, the relationship and interplay between local and national levels has a major impact on local level adaptation. According to the literature, one of the most important barriers for local level adaptation is the lack of cohesive adaptation policy at national level, the ill-defined roles of different levels of administration, and the lack of instructions, regulation and resources from national level to regional and local levels. Regulation from the state has an effect on the local room of political manoeuvre: it can support local adaptation, or hinder it. The lack of clear national adaptation policy may make it difficult for the local authorities to know what to adapt to, and how (Bulkeley & Betsill, 2005; Næss, Bang, Eriksen & Vevatne, 2005; Wilson, 2006; Storbjörk, 2007; Urwin & Jordan, 2008; Amundsen, Berglund & Westskog, 2010).

5.6 Conclusion

Adaptation is a relatively new concept to local and regional authorities in Finland, and in the Helsinki Metropolitan Area. Some sectors of public administration have reacted to changes in weather events and extremes and some fields of local administration have assessed climate change impacts, vulnerabilities and adaptation measures and included them in their plans. The problem is that often the strategies and plans address only one or two climate-related issues, but there is no connection to other issues. The objective of this case study was to approach adaptation comprehensively, to relate it to all major sectors of administration, and to bring different sectors and organizations together to discuss adaptation.

Finland has strategic adaptation targets but there is not yet much detailed guidance and instructions at the local level. Context-sensitive analysis and experiences on how adaptation targets can be successfully implemented at local level and integrated into other policy sectors are needed. Local and regional level adaptation could perhaps be promoted by emphasizing positive synergies between adaptation and other policy targets, finding no-regrets solutions to adaptation which serve to meet other policy targets as well, and highlighting the long-term benefits which adaptation brings. Nonetheless, comprehensive local and regional adaptation requires not only coherent adaptation policy at national level but also political leadership and support at regional and local level, especially if adaptation is based on voluntary action.

References

Amundsen, H., Berglund, F. and Westskog, H., 2010. Overcoming barriers to climate change adaptation – a question of multilevel governance? *Environment and Planning C: Government and Policy*, 28 (2), pp.276–289.

Anguelovski, I. and Carmin, J., 2011. Something borrowed, everything new: Innovation and institutionalization in urban climate governance. *Current Opinion in Environmental Sustainability*, 3 (3), pp.169–175.

Bulkeley, H. and Betsill, M.M., 2005. Rethinking sustainable cities: Multilevel governance and the 'urban' politics of climate change. *Environmental Politics*, 14 (1), pp.42–63.

Espoo, 2011. *Espoon hulevesiohjelma*. [pdf] Available at: <http://www.espoo.fi/download/noname/%7BDCC506 15-8F19-4C42-BA5E-5A62AD272664%7D/11123> [Accessed 5 July 2012].

Finnish Road Administration, 2009. *The Effect of Climate Change on the Routine and Periodic Maintenance of Roads.*

Helsinki: Finnish Road Administration, Central Administration. Finnra reports 8/2009.

Flood Risk Management Act, 2010. *FINLEX*. 620/2010. [online] Available at <http://www.finlex.fi/fi/laki/alkup/2010/20100620> [Accessed 9 July 2012].

Frommer, B., 2011. Climate change and the resilient society: utopia or realistic option for German regions? *Natural Hazards*, 58 (1), pp.85–101.

Gasper, R., Blohm, A. and Ruth, M., 2011. Social and economic impacts of climate change on the urban environment. *Current Opinion in Environmental Sustainability*, 3 (3), pp.150–157.

Glaas, E., Jonsson, A., Hjerpe, M. and Andersson-Sköld, Y., 2010. Managing climate vulnerabilities: formal institutions and knowledge use as determinants of adaptive capacity at the local level in Sweden. *Local Environment*, 15 (6), pp.525–537.

HSY Helsingin seudun ympäristöpalvelut – kuntayhtymä, 2010. *Pääkaupunkiseudun ilmasto muuttuu – Sopeutumisstrategian taustaselvityksiä*. Helsinki: HSY Helsingin seudun ympäristöpalvelut – kuntayhtymä.

Järvinen, S., Kankaanpää, S., Lounaisheimo, J. and Aarnio, P., 2010. Ilmastonmuutoksen vaikutukset pääkaupunkiseudulla. In: HSY Helsingin seudun ympäristöpalvelut – kuntayhtymä, 2010. *Pääkaupunkiseudun ilmasto muuttuu – Sopeutumisstrategian taustaselvityksiä*. Helsinki: HSY Helsingin seudun ympäristöpalvelut – kuntayhtymä, pp.1–47.

Juhola, S., 2010. Mainstreaming climate change adaptation: The case of multilevel governance in Finland. In: Keskitalo, E.C.H., ed. 2010. *Developing Adaptation Policy and Practice in Europe: Multi-level Governance of Climate Change*. Dordrecht: Springer Verlag, pp.149–183.

Kaasinen, S., 2010. *Helsinki Metropolitan Area Council YTV – working for climate protection*. Helsinki: HSY Helsingin seudun ympäristöpalvelut – kuntayhtymä.

Kankaanpää, S., 2011. *The Helsinki Metropolitan Area and its four cities*. Scale, Helsinki: HSY.

Kuntaliitto, 2002. *Hulevesiopas*. Helsinki: Suomen Kuntaliitto.

Liikenne- ja viestintäministeriö, 2009. *Liikenne- ja viestintäministeriön hallinnonalan ilmastopoliittinen ohjelma 2009–2020*. Helsinki: Liikenne- ja viestintäministeriö.

Maa-ja metsätalousministeriö, 2009. *Tulvariskityöryhmän raportti*. Helsinki: Maa- ja metsätalousministeriö.

Marttila, V., Granholm, H., Laanikari, J., Yrjölä, T., Aalto, A., Heikinheimo, P., Honkatukia, J., Järvinen, H., Liski, J., Merivirta, R. and Paunio, M. eds., 2005. *Finland's National Strategy for Adaptation to Climate Change*. Publications of the Ministry of Agriculture and Forestry 1a/2005. Vammala: Vammalan kirjapaino Oy.

Ministry of Agriculture and Forestry, 2009. *Evaluation of the Implementation of Finland's National Strategy for Adaptation to Climate Change 2009*. Ministry of Agriculture and Forestry 4a/2009. Vammala: Vammalan kirjapaino Oy.

Ministry of Agriculture and Forestry, 2011. *Action Plan for the Adaptation to Climate Change of the Ministry of Agriculture and Forestry 2011–2015 – Security of Supply, Sustainable Competitiveness and Risk Management*. Vammala: Ministry of Agriculture and Forestry.

Ministry of the Environment, 2008. *Adaptation to Climate Change in the Administrative Sector of the Ministry of the Environment. An Action Plan to Implement the National Strategy for Adaptation to Climate Change*. Reports of the Ministry of the Environment 20en/2008. Helsinki: Ministry of the Environment, Department of the Built Environment.

Ministry of the Environment, 2011. *Adaptation to Climate Change in the Administrative Sector of the Ministry of the Environment. Action Plan Update for 2011–2012*. Reports of the Ministry of the Environment 18en/2011. Helsinki: Ministry of the Environment, Department of the Built Environment.

Ministry of the Interior, 2008. *Safety first – Internal Security Programme*. Government resolution, Helsinki: Ministry of the Interior.

Næss, L.O., Bang, G., Eriksen, S. and Vevatne, J., 2005. Institutional adaptation to climate change: Flood responses at the municipal level in Norway. *Global Environmental Change*, 15 (2), pp.125–138.

Nurmi, P., Heinonen, T., Jylhänlehto, M., Kilpinen, J. and Nyberg, R., 2008. *Helsingin kaupungin hulevesistrategia*. Helsinki: Helsingin kaupungin rakennusvirasto, katu- ja puisto-osasto.

Pahl-Wostl, C., 2009. A conceptual framework for analysing adaptive capacity and multi-level learning processes in resource governance regimes. *Global Environmental Change*, 19 (3), pp.354–365.

Parry, M., Canziani, O., Palutikof, J., van der Linden, P. and Hanson, C. eds., 2007. *Climate Change 2007 – Impacts, Adaptation and Vulnerability. Contribution of Working Group II to the Fourth Assessment Report of the IPCC*. Cambridge: Cambridge University Press.

Ribeiro, M., Losenno, C., Dworak, T., Massey, E., Swart, R., Benzie, M. and Laaser, C., 2009. *Design of guidelines for the elaboration of Regional Climate Change Adaptation Strategies*. Vienna: Ecologic Institute.

Saarelainen, S. and Makkonen, L., 2007. *Adaptation to climate change in the road management*. Prestudy. Helsinki: Finnish Road Administration.

Saarelainen, S. and Makkonen, L., 2008. Adaptation of railway management to climate change. Preliminary study. Publications of the Finnish Rail Administration A 16/2008. Kuopio, Helsinki: Kopijyvä Oy.

Soini, S., 2007. *Ilmastonmuutos ja siihen varautuminen Espoossa*. Espoon ympäristökeskuksen monistesarja 2/2007. Espoo: Espoon ympäristökeskus.

Statistics Finland, 2011. *Population*. [online] Available at: <http://www.stat.fi/tup/suoluk/suoluk_vaesto_en.html> [Accessed 9 July 2012].

Storbjörk, S., 2007. Governing climate adaptation in the local arena: Challenges of risk management and planning in Sweden. Local Environment, 12 (5), pp.457–469.

Swart, R., Biesbroek, R., Binnerup, S., Carter, T.R., Cowan, C., Henrichs, T., Loquen, S., Mela, H., Morecroft, M., Reese, M. and Rey, D., 2009. *Europe adapts to climate change – Comparing national adaptation strategies*. PEER Report No 1. Helsinki: Partnership for European Environmental Research.

UKCIP (UK climate impacts programme), 2011. *LCLIP: Local Climate Impacts Profile*. [online] Available at: <http://www.ukcip.org.uk/lclip/> [Accessed 9 July 2012].

UN-Habitat, 2008. *Cities and Climate Change Adaptation*. Nairobi: UN-Habitat.

Urwin, K. and Jordan, A., 2008. Does public policy support or undermine climate change adaptation? Exploring policy interplay across different scales of governance. *Global Environmental Change*, 18 (1), pp.180–191.

Valkeapää, R., Nyman, T. and Vaittinen, M., 2008. *Helsingin kaupungin tulvastrategia – Tulviin varautuminen Helsingin kaupungissa*. Helsinki: Ramboll.

Vantaa, 2009. *Hulevesiohjelma*. Kuntatek 2/2009. Vantaa: Vantaan kaupunki, Kuntatekniikan keskus.

Veijalainen, N., Sippel, K. and Vehviläinen, B., 2009. Tulvien muuttuminen Vantaanjoella ja Espoonjoella. In: HSY Helsingin seudun ympäristöpalvelut – kuntayhtymä, 2010. *Pääkaupunkiseudun ilmasto muuttuu – Sopeutumisstrategian taustaselvityksiä*. Helsinki: HSY.

Venäläinen, A., Johansson, M., Kersalo, J., Gregow, H., Jylhä, K., Ruosteenoja, K., Neitiniemi-Upola, L., Tietäväinen, H. and Pimenoff, N., 2010. *Pääkaupunkiseudun ilmastotietoa ja skenaarioita*. In: HSY Helsingin seudun ympäristöpalvelut – kuntayhtymä, 2010. *Pääkaupunkiseudun ilmasto muuttuu – Sopeutumisstrategian taustaselvityksiä*. Helsinki: HSY.

Wilson, E., 2006. Adapting to climate change at the local level: The spatial planning response. *Local Environment*, 11 (6), pp.609–625.

Yrjölä, T. and Viinanen, J., 2012. *Keinoja ilmastonmuutokseen sopeutumiseksi Helsingin kaupungissa*. Helsingin kaupungin ympäristökeskuksen julkaisuja 2/2012. Helsinki: Helsingin kaupungin ympäristökeskus.

YTV Helsinki Metropolitan Area Council, 2007. *Climate Strategy for the Helsinki Metropolitan Area to 2030*. Helsinki: YTV Helsinki Metropolitan Area Council.

6

Adaptation to Floods in Riga, Latvia: Historical Experience and Change of Approaches

Laila Kūle, Agrita Briede, Māris Kļaviņš, Guntis Eberhards & AndrisLočmanis[1]

University of Latvia, Faculty of Geography and Earth Sciences, Riga, Latvia
[1] Riga City Council

6.1 Introduction

It is widely acknowledged that coastal areas are among the most vulnerable to climate change impacts and, at the same time, are considerably affected by anthropogenic impacts. Therefore, these areas need particular attention (EEA, 2006). Studies at European level (EEA, 2010) list the following climate change impacts affecting coastal areas: the impacts of sea level rise, increased flooding, more frequent storm surges, changes in temperature, precipitation and ice regime, changes due to increased coastal erosion, increased salt water intrusion in groundwater layers, and changes in coastal ecosystems. In response to extreme weather events and other coastal natural and man-made hazards, inhabitants of coastal areas – urbanised communities in particular – have developed community-based adaptation strategies over the centuries, derived from various types of knowledge (Ensor & Berger, 2009). Floods have been recognized amongst the major natural hazards, causing immense losses every year. Coastal urban territories are particularly vulnerable to flood risks, as is the case with Riga. In Latvia, which is characterized by an excessive moisture regime, rivers can flood due to snow melting and increased precipitation while coastal areas are at risk of storm surges. The 15 km long coastline of Riga City

and about 60% of its urban waterline are vulnerable to sea level rise (see Figure 6.1). In the context of climate change, storm surges and flash floods due to intensive precipitation are expected to increase (Avotniece et al., 2010). Outdated technical infrastructure of the urban water supply system is an additional cause of flash floods. The Riga City dominates in many fields in the context of Latvia's development and associated with that, significant human and man-made resources are concentrated in the capital.

The Riga municipality is the largest in the country, with 700 100 inhabitants or 31.4% of the total population of Latvia living in Riga in 2011 and producing 53% of the total GDP in 2009 (CSB, 2011). The average population density is 2303 persons per km^2 (CSB, 2011), but it can range between nine and 15 981 persons per km^2 among its 58 spatial analytical units (Riga City Council, 2012), while especially flood-prone and coastal areas have lower densities. Additionally, 17% of the city is covered by nature reserves which are mainly situated in the coastal and flood-prone areas (Riga City Council, 2012). Daily, many people commute between the suburbs and Riga City so that the population runs up to approximately one million people in the daytime. It is situated in the delta area of the three large rivers: Daugava, Lielupe and Gauja. All three rivers are treated separately in terms of the EU Water Framework Directive and thus three

Climate Change Adaptation in Practice: From Strategy Development to Implementation, First Edition.
Edited by Philipp Schmidt-Thomé and Johannes Klein.
© 2013 John Wiley & Sons, Ltd. Published 2013 by John Wiley & Sons, Ltd.

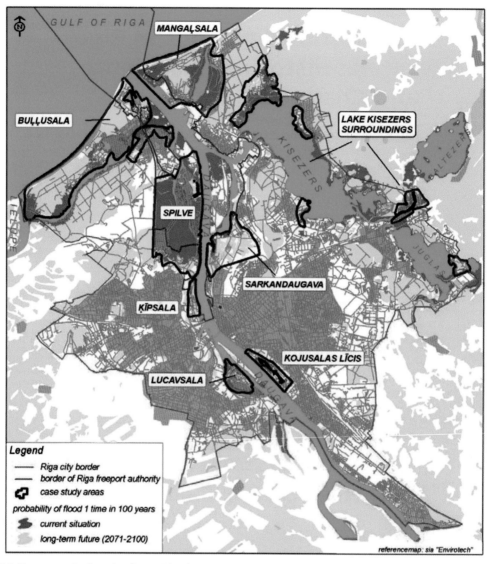

Figure 6.1 Hot spots under discussion that are identified as sites (including technical structures) for climate change adaptation measures, with focus on flood risk management.
Source: Map prepared by Andris Locmanis, and reproduced with permission from Riga City Council.

separate integrated river basin management plans are being prepared in 2009. Beside current water quality and quantity safeguarding, issues of flood prevention and climate change adaptation should be incorporated in these plans in 2015 (European Commission, 2012). As there is a need to coordinate prevention measures against coastal floods in the lower part of three large rivers the case of Riga is particularly challenging with respect to city development and spatial governance.

Directive EC 2007/60 of the European Parliament and of the Council, which specifies the structure and objectives of flood risk management plans and the recommended flood mitigation measures, acknowledges that specific local aspects should be considered in each particular case. Studies at European level do not acknowledge that Latvia in general would be highly vulnerable to flood risks in comparison with other countries (Lugeri et al., 2010) and they also do not record any major flood disasters in the period between 1950 and 2005. Studies at a more detailed level indicate that Riga and other local areas can sustain flood damages and thus flood prevention measures have to be proposed and implemented.

The aim of this study is to look at the historical experience and approaches used to reduce flooding in Riga. This chapter will describe the basic concepts of flood risk management approaches, knowledge types and stakeholder involvement, as well as providing insights into the historical context and current practices of flood prevention management in Riga City. It will also review historical extreme flood events in Riga since the 13th century. To encompass the various types of knowledge relevant for flood risk management, different information sources were examined.

6.2 Relevant aspects for flood risk management

6.2.1 Stakeholder involvement in flood risk management

Adaptation measures to reduce flood risks (Table 6.1) are basically regulated by Directive EC 2007/60. However, the Directive gives little information about the development of flood prevention strategies and implementation of flood management plans on the local level. Nevertheless, the need to find a good governance concept, supporting the implementation process and leading to acceptance and proper application of a flood risk management plan, is obvious. It is of importance to ensure the necessary multi-stakeholder participation in its preparation and decision-making process. A higher quality of the decision-making process and its outcomes can be achieved by considering international success stories and historical experiences with flooding.

Adaptation to flood risks should be a combination of top-down (represented by the EU Flood Directive EC 2007/60 and national governments) and bottom-up processes initiated by interest groups (public and private land developers, entrepreneurs, housing management organizations and citizen groups) or municipalities with flood-prone areas. This concept is stated in Article 10 (2) of the Flood Directive as follows: 'Member States shall encourage active involvement of interested parties in the production, review and updating of the flood risk management plans'. An integrated flood management approach by the World Meteorological Organization (WMO, 2009, p.19) acknowledges that it is important to make use of the strengths of the top-down and bottom-up approaches by determining the appropriate combination of elements from these approaches. It also states that members of as many sectors as possible have to be involved, as they represent different types of knowledge. Coordination and cooperation among institutions affected by flood risks or involved in their management are part of the strategy of how to overcome geographical and functional boundaries and achieve synergy for all the institutions involved (WMO, 2009).

Another key problem in flood risk management is that the Flood Directive replaces traditional flood defence strategies with a risk-based management concept (Samuels et al., 2009). Flood risk management is a part of integrated water resources management, and its systematic actions are divided into groups by the cycle of preparedness for, response to and recovery from a flood event (WMO, 2009). Other aspects of flood risk management are listed in Table 6.1. Again, it is essential for flood risk management to consider stakeholder involvement as manner of a participatory approach and to make sure that stakeholders are well represented in the discussion and decision-making process in relation to flood prevention (WMO, 2009).

Professional and public stakeholders need to build up their capacity of understanding and application of flood risk management (WMO, 2006), which is not a fixed set of tangible measures, but an evolving process of transition to more adaptive flood risk management in order to cope with the emerging uncertainties due to climate change/variability (Tippett & Griffiths, 2007).

The term 'stakeholder', initially used in business management (Freeman, 1984); stakeholder involvement is now broadly applied to different aspects of governance, including environment and resource management (Grimble & Wellard, 1997), coastal management and flood prevention (Heitza, Spaeter, Auzut & Glatrona, 2009; Werff, 2004). This approach is related to the paradigm shift in flood risk management, requiring expert information supply,

Table 6.1 Three dimensions of flood risk management

Context	Process	Content
External: political; legal; social and economical; spatial; and locational. *Internal*: social; cultural; political; institutional; available resources; capabilities; and physical conditions.	*Stages of process:* understanding context; flood-related data collection and monitoring; flood trends and analysis; scenario selection; model creation and assessment; criteria selection; identification of risks, problem areas and hot spots; prioritization; selection of alternative measures; communication and approval of selected measures; creation of organizational and financing structure; implementation; and evaluation. *Planning principles*: sustainability; legitimacy; procedural equality; justice; people first principle; social equity; resources targeted to the most vulnerable; maximisation of utility (greatest risk reduction per unit of resource input); a long-term and visionary approach; proactive, strategic, precautionary approaches; scientific data and evidence-based, ecosystem approach; water-basin approach; the water cycle management approach; multi-hazard approach; risk management approach; community-based approach, multi-scalarity; cost-effectiveness; integration; transdisciplinarity; responsive and participatory processes; openness; public participation; and empowerment. *Governance types*: bottom-up; top-down; subsidiarity; cross-sectoral; short-term; medium-term; long-term; multi-scale; stakeholder involvement. *Organizing strategic planning:* project-based planning; ongoing planning. *Strategic planning mode:* programming; portfolio planning; scenario-based planning. *Learning:* knowledge creation; review and assessment; formal and informal learning; policy transfer; institutional networking; knowledge distribution channels, targeted to specific audiences.	*Goals and specific targets* *System analysis*: controllable; not controllable variables. *Strategic alternatives* as combination of measures. *Structural measures* (flood hazard reduction): barriers; barrages; dams; river regulation and channel improvements; diversion channel creation; dykes, levees and embankments; improved drainage, stormwater and rainwater. networks; flood abatement through forestation; wetland creation and landscaping; improved evacuation network; flood proofing. *Non-structural measures* (flood vulnerability reduction): flood proofing; flood risk identification and assessment; flood forecasting and warning; preparedness, evacuation and post-disaster planning; integration of flood management aspect in regulations of economical activities; regulations of existing property management, land use, and building; flood-sensitive land use; flood prevention integration into sectorial, spatial and development planning; flood warnings and raising public awareness; resilience-building; resistance capacity-building; strengthening local institutions; property purchase and relocation; home owner adaptation; flood aware targeting of public investments, services and infrastructure; flood insurance and other risk sharing mechanisms; compensation.

Source: Modified from Hutter & Schanze, 2008; Parker, 2007; Johnson et al., 2007; Tran et al., 2008; Neuvel & van den Brink, 2009; Glavovic, 2008; WMO, 2009; Harries & Penning-Rowsell, 2011.

co-thinking, co-design, co-management, consultation, participation and action, and consequently, involving a broad range of stakeholders and practitioners (Werritty, 2006; Tippett & Griffiths, 2007; Mostert et al., 2007). Flood risk management can be implemented if adaptive capacity development, knowledge and adaptive management (Hillman, 2009) are a part of the process. The involvement of stakeholders – that is, all persons, institutions or organizations with an interest in involvement in the issue, either because they will be affected (positively or negatively), or because they can influence (positively and/or negatively) the outcome – is closely linked to social learning (Ridder, Mostert & Wolters, 2005). Stakeholders can be directly or indirectly involved in various stages of different kinds of events and other activities relating to flood risk management and the related decision-making process. But it is always important to inform all stakeholders about the process and the outcomes at each stage. Identification and involvement of a large number of individuals and groups with interest in the issue may encumber the ability to take a decision that would solve the problem (Harrison & Qureshi, 2000). Therefore, stakeholders' assignment in focus groups, as well as the conduct of discussions, expert interviews and workshops, and other means of stakeholder identification and selection are recommended respectively. Studies advise the assignment of stakeholders in terms of the ratio between their interest in the issue and the ability to change the situation, taking into account aspects like authority, finances, knowledge, capabilities, vulnerability and others. The diverse patterns of stakeholder structures will appear if these aspects are referred to flood risk management – for instance, stakeholders who are most vulnerable often lack resources and authority to cope with or prevent flood impacts. Various thematic areas, categories and typologies can be used in order to structure a vast and diverse pattern of stakeholders and their discourses in relation to the issue.

Stakeholder assignment is linked with 'the notion of "interactional field"', a social situation/space defined by its contextualities, a cluster of actors and processes with geographically, socially, economical [*sic*], and politically defined boundaries', and all these factors imply 'a look at social spaces in time/dynamics' (Aligica, 2006, p.85). Harrison and Qureshi (2000) stress that stakeholder identification is a process that

needs to be repeated, as discussions will reveal groups and individuals that have not been identified before. Stakeholders can provide information and knowledge not only about the issue, but also on other stakeholders involved. Stakeholder participation process is greatly dependent on institutions responsible for the issue concerned. It is highly important to provide a policy framework and capacity in terms of human, financial and knowledge resources, and time- and place-related aspects for the stakeholder involvement process and its management. Stakeholders have to have the ability and capacity to participate (Weber & Christopherson, 2002). Early involvement of stakeholders is important for a decision-making process with a successful outcome (Reed, 2008).

6.2.2 Social learning as a tool to diminish uncertainties in flood risk management

Social learning is based on a dialogue in which stakeholder independence, the need for interaction, openness, mutual trust and cultural tolerance, common visions, critical self-reflection and strong leadership are recognized. Stakeholder participation ensures that different perspectives on the problem are taken into account whereby ambiguities and uncertainties related to multiple framing of a problem, including the multi-disciplinarity of knowledge (relating to natural, technical and social systems), can be diminished (Raadgever et al., 2011). Face-to-face interaction, communication among stakeholders (knowledge creators, mediators and users), dialogical learning and negotiation are among the uncertainty management strategies proposed for the water management needs (Raadgever et al., 2011). 'Uncertainties about the seriousness of flooding problems, cause-effect relations, or the effects of policy options' can also serve as stimulus for more extended public consultations and other stakeholder involvement activities including 'seeking help from epistemic communities' (Meijerink, 2005, p.1063). Stakeholders can exchange and correct existing knowledge, thus providing additional details related to local specifics. Stakeholder participation adjusts planned decisions to the real situation and to the will of society and its segments, thus contributing to justice. Early stakeholder involvement facilitates the implementation of planned actions, and therefore, is a crucial element of governance aspects. Political commitment, expected and planned organizational changes and the increase of institutional capacity are aspects that also need to be considered in the development of the stakeholder participation process. The other two types of uncertainties beside ambiguity – that is, epistemic (lack of knowledge, information and theoretical understanding) and ontological (unpredictability) uncertainties (Raadgever et al., 2011; Merz & Thieken, 2005) – can be reduced by conducting research and investing in science.

6.2.3 Flood risk management and the relevance of different types of knowledge

Participatory and adaptive flood management strategies are closely linked to knowledge management and integration. As knowledge constitutes a capacity for action and provides tools to comprehend the situation by structuring it and controlling contingent circumstances (Stehr & Ufer, 2009), it is important to be aware of and use all types of knowledge. For the purposes of understanding, there are various approaches to structuring knowledge. Cooperation among natural, technical and social knowledge, implemented through partnerships and coordination across disciplinary boundaries, is crucial for integrated flood risk management (McFadden, Penning-Rowsell & Tapsell, 2009), also indicating a paradigm shift in flood prevention that traditionally has been the responsibility of technical sciences alone (WMO, 2009). The new approach in flood prevention that recognizes the multi-scalarity of water cycle processes does also recognize the importance of embracing local knowledge, community knowledge and place-based approaches, ensuring that these aspects are integrated in the strategy processes. Often, local inhabitants have lived with floods for a long time and consequently, have developed coping strategies which are not always known by the higher levels of administration, experts and academia. It is important not only to access different types of knowledge but also 'to engage diverse ways of knowing within and between scientific and local communities and constituencies of interest' (German, 2010, p.118). When proposing a flood risk management strategy, it has to be communicated effectively to and accepted by local entrepreneurs, employees, resi-

dents or visitors to such a degree that it could be successfully implemented as a flood prevention, disaster emergency and recovery tool.

If different types of knowledge management are to be integrated into a flood risk management then not only explicit or codified knowledge that is materialized in scientific literature, factual information, mathematical formulas, databases and documents; but also tacit knowledge (Smith, 2001) that is usually conveyed through face-to-face interaction, conversation, storytelling, observation, imitation, shared experience and practice, should be considered as the latter is relevant for process management and the formation of attitudes and values. Although important, tacit knowledge is difficult to interpret and transfer to other contexts, as well as to capture and to integrate tacit knowledge into formal types of knowledge. Proximity, mutual trust of the people involved, as well as commitment and leadership are crucial factors for the development of tacit knowledge (Holste & Fields, 2010). Another classification of knowledge is used in policymaking. The first type, according to this classification, is traditional 'academic' knowledge which is rooted in past research is based on peer review and is independent; the other two types are 'fiducial' and 'bureaucratic' knowledge (Hunt & Shackley, 1999). Fiducial knowledge is the basis for policymaking. Bureaucratic knowledge is produced jointly by users. Since it is a synthesis for a specific context/use and also often for a specific political situation, it is filtered and judged for particular needs. Both the fiducial and bureaucratic types of knowledge are produced on the basis of contracts and are often validated through the status of authors. All types of knowledge are interlinked and the distribution channels that have a particular bearing on how different knowledge types are interchanged and integrated among involved stakeholders. Haas (2004, p.574) maintains that 'usable knowledge is accurate information that is of use to politicians and policy-makers'. The observations indicate that scientific knowledge is seldom directly transferred to policy documents and their implementation, even if scientific consensus has been accomplished, including flood risk management (Meijerink, 2005). Haas believes that through the process of communicating scientific knowledge to public authorities, knowledge obtains such characteristics as credibility, legitimacy and saliency. 'In practice credibility and legitimacy are mutually reinforcing, as a procedural approach to developing consensual knowledge is likely to generate both accurate and acceptable knowledge' (Haas, 2004, p.574). Four criteria of usable knowledge are identified: adequacy, value, legitimacy and effectiveness (Clark & Majone, 1985). The bridging between knowledge and knowing implies organizational learning and dynamic capabilities of involved institutions. This has not only to focus on internal and external knowledge transfer but also on knowledge integration processes in which the development of understanding and the creation of new knowledge occur through individual interactions and are affected by social contexts (Eisenhardt & Santos, 2001). Equally important are knowledge integration processes as well as interpretation and institutionalization of knowledge. As Albrechts (2001, p.738) notes 'institutionalisation is a process by which ideas and practices become durable reference points for social action. This institution-building (the design of arenas) requires a certain degree of consensus about underlying values' and a commitment to administrative and financial agreements between different levels of government, sectors and private institutions.

Interdisciplinary scientific evidence is applied by downscaling projections to the local level, also included in the area of flood prevention and climate adaptation. Still, there are many pending questions such as the one of how to transfer the results of studies into legitimized and operational activities of local governments. The understanding of and responding to flood processes cover complex issues, different spatial and time scales and, thus, need various academic disciplines and policy fields to be involved in a coordinated manner. Due to uncertainties caused by the irregularity of flood events and insufficient knowledge of flood event prediction, action is often required before a complete understanding of the problem. Flood prevention includes information gathering, knowledge creation, communication and implementation as well as reasonable community actions, taking into account the existing incompatibility between nature and society and the need to overcome physical, administrative, social and political boundaries. All these stages are interlinked in a non-linear manner. They include social learning and new knowledge production processes.

6.3 Historical context of flood risk management approaches in Riga

Various types of knowledge have been used to comprehend the past flood experiences in the Riga City as well as to prepare a cartographic representation of contemporary flood risk management hot spots. The City of Riga has been confronted with the risk of floods since the dawn of settlement. However, the perception of the hazard and actions taken has been quite different. The following account is based on both scientific and historical records of various origins. It also provides an example of how various types of knowledge are utilized for the needs of stakeholders in the current and future flood management setting. The aim of the historical integration is to review information on historical weather events, particularly floods and their characteristics. The frequency of extreme weather events may increase due to climate change. Knowledge of historical weather events can play a critical role in communicating possible future damages caused by climate change. A foundation of trust can be built especially by addressing the people's experiences with or place-based knowledge of such events in former times that are rooted in the natural and cultural context. Social interaction at a closer distance and personal and social memories are particularly important for forming trust (Korczynski, 2000; Swain & Tait, 2007) needed also for action planning, as in the case of flood prevention. The Riga City has got both natural (dunes, beaches and wetlands) and man-made coastal protection structures. Traditionally, Latvian human settlements were not situated in the dunes or flood-prone areas. However, the situation changed when the importance of Riga for trade between East and West as well as its military significance and position as an outpost on the Eastern coast of the Baltic Sea determined a rapid growth of the city. There is evidence that Riga had a protection system on the Daugava River as early as the end of the 13th century (Biedriņš & Ļakmunds, 1990). Two types of constructions were used: (1) coastal wooden stilts to protect the city's fortification wall against ice and foundations against washing-out; and (2) compacted gates for flood protection, although the latter did not function well for the purpose for which they had been built.

Before the construction of hydro power plants (HPP), the causes that determine or facilitate the occurrence of floods in the Daugava River basin were basically of natural origin. The severity of flood depends on rainfall intensity and amount, wind direction and strength, snow melting intensity along with water inflow into the river basin during spring flood, ice and sludge congestions, air temperature in combination with humidity, topography of the area, hydrogeological circumstances as well as morphometric and hydraulic characteristics of the river bed.

Historical records of floods in Riga had already started in the 14th century (Moskovkina, 1960). Historical records of floods in Riga reflected in the following paragraphs are taken from various types of chronicles (including parish chronicles), municipal and state registers, newspapers as well as scientific literature. Several sources reported a catastrophic flood on the Daugava River in 1358. On the basis of annals, the newspaper *Rigasche Stadt-Blätter* wrote that water stood above man's head in the Riga Dome Cathedral's aisle. To keep the memory of this event alive, an iron cross was mounted to the cathedral building's wall, marking the water level of the Daugava River in the year 1358, which is estimated to have risen around 5.5 to 6 metres above the mean summer water level. The spring flood levels in the Daugava basin were also catastrophic in 1578, when vast areas around Riga were submerged and the water level could have possibly risen by 5 or 6 metres, causing huge damage. Great spring floods also occurred in the years 1589, 1597 and 1615 – when water levels possibly rose by 5.5 to 6 metres. In 1615, a huge ice dam formed by the former Bisenieki Isle caused a rapid rise in water levels up the river.

Some major floods were also triggered by storm surges, as was the case after a fierce storm on 30 May 1626. Large masses of sea water were pressed in the Daugava River (presumably with northwest winds) from the Gulf of Riga. These waters, together with the spring flood waters of the Daugava, caused an unusually high water stage. The entire city and surrounding pastures were inundated, many buildings were ruined, the wind downed lots of trees and a lot of people and livestock perished. This natural disaster was caused by the concurrence of two elements: water and storm. Disastrous spring floods with large piles of ice

on the Daugava River occurred again in 1649. Then, there were heavy storms on the sea in November and December 1704. Furthermore, in late June 1708, there was heavy rain and subsequent flooding in the Daugava River. The water level in Riga rose by 4.5 metres. In the vicinity of the city, fields and gardens were submerged for four weeks. All plantations and sowings perished.

A particularly harsh winter in 1708/1709 led to a frozen over Baltic Sea and the newspaper *Rigasche Stadt-Blätter* wrote that there was an ice thickness on the Daugava River that reached 1.7 m. On 6 April 1709 the ice started to drift. Since the Gulf of Riga was still covered with thick ice, the ice drifts carried by the river flood waters were piled on the isles and shores of the Daugava River. The water level rose catastrophically, reaching the absolute height mark of 4.68 m on 16 April. Compiling the historical data, Ludvigs (1967, p.231) wrote that

in November [1708] an exceptionally strong storm raged ... The storm-blown water flooded the Daugava River banks and isles, washing away houses, livestock and people. Several ships were smashed and cast ashore. The storm was followed by severe frosts, which persisted almost continuously throughout the winter. The ice cover on the Daugava River reached the thickness of 1.5 m. 22 ships were stranded in the ice. ... When the spring thaw began, the stream brought the ice from the Daugava upriver downwards, while the ice at the downriver did not break, remaining where it was. Consequently, a huge ice dam was formed. The water then broke two new outlets to the sea, flooding Pārdaugava [the left side of the Daugava] and isles of the Daugava. The ice-bound ships could not be salvaged ... The Zaķusala Island of the Daugava River alone lost 52 houses. .... The masses of ice and flood waters broke through the Riga city gates, flooding the streets, buildings and cellars. Water in the Dome Cathedral rose up to the altar.

A cross in the Riga Dome Cathedral's wall is said to act as a reminder of this disastrous flood. The water in the city reached levels similar to human height. All isles and the valley on the left side of the Daugava were flooded, whereas the water on the right side of the river reached Kube Hill and the Citadel.

There were great flood damages in Riga also in the spring of 1783, when dams were washed out and broken on 11 spots. In the same year, the Lakagígar eruption in Iceland occurred and similar flooding events were reported in Europe (Brázdil et al., 2010). Another catastrophic flooding with ice drifts in Riga arose after the harsh winter of 1795. The catastrophic flooding on 12 April 1814 was caused by a thick cover of sludge and ice. Further, a severe flood occurred two years later (in 1816), when large ice jams were deployed opposite Catherine's dam in Riga. In 1829, after a harsh winter without thaws, ice drifts started on 9 April on the Daugava River in Riga. Again, ice piled up on the many low isles and banks. The river bed was obstructed, and the water level rose rapidly upstream of the jam. The low-lying areas of the Daugava valley in Pārdaugava up to Māra's Pond Mills, Catherine's dam, Sarkandaugava and St Petersburg's suburbs were submerged. The old town was saved after much effort.

In 1837, catastrophic spring floods with ice drifts took place on rivers throughout Latvia. The end of 1856 also came with major floods in Latvia. Sharp frost started early in September and continued throughout November, followed by a heavy thaw and major floods at the end of the every month. In Riga many streets and the isles of the Daugava River were flooded, since several dams had been destroyed. Bridges had been carried away. Unprecedented floods struck the city of Riga at the beginning of March 1871. Strong winds from the sea forced large amounts of ice from the Gulf of Riga into the Daugava River mouth, making high ice piles. Their height reached 70 feet (21.35 m) and a width of around 20–30 m. The ice was spread over a length of two to three versts (2.12–3.18 km). The ice piles reached the river bottom more than 5 m deep. However, they did not remain for long in the Daugava River, since intense ice drifting started on 17 April. The piles were carried into the sea. Therefore, major flooding failed to appear except for in the low Pārdaugava (Stakle, 1941). At the beginning of the last century, the largest flooding in Riga related to ice jams occurred in 1917. Serious flooding also occurred in the years 1924, 1929, 1932, 1936 and 1937. To sum up, over a period of almost 600 years, from the 14th century until the early 20th century, Riga and its inhabitants endured devastation due to

catastrophic flooding caused by ice jams more than 20 times.

Learning from identified and analysed past events can be crucial for public and institutional awareness and for the creation of the most appropriate measures for future flood prevention, taking into account the successes and failures of public responses to historical flood events. The vast majority of the rise of catastrophically high water levels in the period of 1600 to 1700 can be explained by rapid deforestation, land cultivation and reclamation related to population increase, development of agriculture, construction of buildings in towns and countryside, building of ships, also exports of timber, production of coal and extraction of tar. As commonly accepted, snow melts faster and water drains quicker to the rivers in woodless than in wooded areas.

Protective dams were built along the Daugava River bank starting as early as the 17th century. In the middle of that century, the city fortification system was improved and the embankments were also used for flood protection. However, the flood protection systems could not completely protect the city from water inflow. Ice and spring floods also contributed to the risk, changing the water flows which, in turn, transformed the river bed almost every year, so that shipping routes had to be adapted frequently. The main cause of disastrous floods on the Daugava River in Riga over a period of more than 750 years, from the founding of the city until the Ķegums HPP was built in 1939, was the wide and shallow bed of the Daugava River with many larger and smaller isles and sandbars, often changing their location and size. The river washed away some isles during flood and some were formed elsewhere. The sandy, gritty and pebbly material carried from the river sections with rapids from Pļaviņas to the downstream end of the Dole Island accumulated in the river within the city limits of Riga and was partly washed into the sea. During spring floods, drift ice piled up on sandbars and isles, completely damming up the river with huge ice jams. Upstream of the congestion, water levels were rising fast causing flooding of vast areas of the Daugava valley and the City of Riga. In some years, strong north-western storms pressed large masses of ice from the Gulf of Riga into the river mouth, forming large ice piles which also caused flooding in the city. As a consequence of these periodical processes, some 500 to 600 years ago, especially the present-day city centre with its multi-storey dwellings, lay about 2–3 m lower. During the last centuries, especially beginning with the demolition of the city ramparts, the territory was banked and elevated up to 5–5.5 m above sea level to avoid flooding. The 5 to 8 meters thick cultural layer serves as evidence for that. The present-day topographic maps of Riga which show those parts of the city that were often inundated during the last centuries, demonstrate effectively the extent of floods (Figure 6.1). If, in some distant future, the sea level in the Gulf of Riga and the coastal areas rises by 4 to 5 metres due to storm surges, the scene would be similar to those 400 to 500 years ago.

In the 1880s, the residents started to canalize the Daugava by deepening the riverbed, straightening the watercourse, removing certain isles and sandbars and building dams. As a result, the risk of ice jams and devastating floods in the city have decreased. Moreover, the situation was improved considerably by building ramparts along the city, shutters at the ends of the streets leading to the river, watergates at both ends of the city canal, as well as by elevating the compound in low-lying built-up areas, using debris as material for banking. In the 19th and 20th centuries, the city started to use icebreakers in the Daugava River before the spring flood season. The construction of Ķegums HPP prevented the formation of ice jams in the city. Nevertheless, the occurrence of disadvantageous wind directions that can lead to severe storms in autumns and winters still threaten the city's low-lying built-up areas situated in the Daugava valley to be flooded in present-day Riga. Such weather conditions are causing water level rises in the Daugava, the Lielupe and the Gauja downstream of about 1.5–1.8 m. During the devastating storms of 1969 and 2005, the water level at the Riga hydrological measurement site at Daugavgriva reached a height mark of 2.1–2.15 m above sea level so that large, vacant and built-up city territories were flooded. As the climate becomes warmer and storms increase in power, water levels rise caused by wind-driven surge into the Gulf of Riga and in such extreme weather events the levels in the City of Riga might reach heights of about 2.5 to 3 metres in the near future. The experiences from previous centuries constitute a valuable source of information on how to adapt to flood risks, while the ongoing global climate change processes bring forward new challenges, requiring new strategies and approaches to reduce flood risks.

6.4 Initiatives of flood risk management in Riga

6.4.1 Flood modelling taking account of climate change

Within the project 'Riga against flood!' (2012), extensive flood modelling works were carried out to give a well-reasoned picture of the current situation as well as possible future consequences of climate change. The time frame was set with three periods: the present, near future (2021–2050) and distant future (2071–2100). According to these, six scenarios were modelled: floods with return periods of one in two, five, 10, 20, 100 and 200 years. All these scenarios were modelled for the main factors that cause flood in Latvia: autumn/winter storm surges, spring meltwater and heavy rain. In total, more than 50 flood scenarios and maps were produced, providing for urban planners with detailed information about flood threats in present times and in the future.

Currently, the national flood prevention legislation contains only one reference to one of the scenarios, that is, the present-day scenario with a probability of occurrence of 1% in 10 years. It is mentioned in the context of a building ban (with the exception of road infrastructure) in flood-prone areas. Such a practice is inherited from the Soviet-era. Nonetheless, this regulation has become a keystone for planners and policy makers, as well as in negotiations between inhabitants and the representatives of the economic sector. Riga city planners have amassed experience through various knowledge exchange trips and interdisciplinary projects (ASTRA, BaltCICA, Riga Against Flood!) by becoming acquainted with other cities' flood prevention practices.

6.4.2 Knowledge transfer in the fields of flood prevention and climate change adaptation

Various types of knowledge and channels have been used to acquire new knowledge in the preparation of flood risk management. The current situation in

Riga is challenging city planners as the urban area is characterized by a great diversity of residential and economic activities, population densities, settlement patterns and land-use types. With the support of the BaltCICA project, the Riga municipality's urban planners and a multidisciplinary group of researchers from the University of Latvia closely cooperated in transferring and integrating the specialized climate change knowledge into regulatory and operational activities of the local government, particularly into the field of spatial planning regulations. Knowledge transfer of new approaches in flood prevention to the municipality of Riga City occurred mainly pioneered by spatial planning as a project team and experts of the project Riga Against Flood! were located in the planning office, supervised by politicians and staff members of the city's Development Department. Community politicians' and experts' project-financed study visits were organized in order to become acquainted with flood prevention, climate change adaptation and spatial planning policies with the aid of concrete example cases, that is, the cities of Antwerp (Belgium), the Hague (the Netherlands) and Hamburg (Germany) in 2010 as well as Rotterdam (the Netherlands) in 2011 (Riga Against Flood!, 2012).

The epistemic community of planners – particularly those with a background in the natural, environmental or geography sciences – plays an important role in transferring scientific knowledge to the applicable policy documents of the Riga municipality. In 2010, the Development Department of the Riga City Council was composed of 80 employees, 12 of whom had a background in the natural sciences. Scientific knowledge transfer is crucial, for instance, for developing the city's adaptive capacity to flood events by determining the appropriate use of the municipality's human, technical and financial resources. Further information sources are research and spatial development projects. The INTERREG III B project 'Developing Policies and Adaptation Strategies to Climate Change in the Baltic Sea Region ASTRA, 2005–2007' initially identified the affected areas in Riga City in case of future sea level rise. Moreover, the Latvian National Research Programme 'Climate Change and Waters KALME, 2006–2009' provided a wide range of scientific evidence and financed public awareness activities.

The application of new knowledge obtained resulted in the preparation of cartographic representation of flood risk management hot spots. Mapping was based on a scientific model that took into account the existing data and the predicted future environmental changes up to the year 2100. This knowledge was corrected with regard to other environmental, social and political factors relevant at the local level and through communication with stakeholders who have an interest in these local areas or have tacit knowledge of flood prevention or flood-prone places.

6.4.3 Flood Risk Management Plan – mapping of vulnerable areas of various time frames and flood probabilities

The Flood Risk Management Plan draft for Riga City has been prepared by local planners in 2011 and the strategic environmental impact assessment has been produced in 2012 (Riga Against Flood!, 2012). Acquired knowledge on flood prevention approaches with reference to climate change adaptation is included in the plan draft. It includes the analysis of 22 types of vulnerable areas – recreational areas; natural environments; inner-city areas; residentially, industrially and commercially used areas; areas of mixed use; roads with and without pavement; areas of technical infrastructure and various types of the harbour area – and four types of particular objects – social institutions, cultural heritage sites, natural reserves and industrial sites with permits of integrated pollution prevention and control – in relation to flood risks in the current situation, in the near (2021–50) and distant (2071–2100) future. 0.5%, 1%, 5%, 10%, 20% and 50% probabilities have been used in the spatial analysis. However, in the descriptive part of the Flood Risk Management Plan draft emphasis was given to objects exposed to a current flood probability of 0.5% and 1%, while in the future there are objects which will be exposed to higher flood risks (Table 6.2). The implementation of the Flood Risk Management Plan will include investments in flood prevention structures as well as amendments to the existing long-term development strategies, development programmes, spatial planning documents, building regulations and strategic documents of affected sectors, inclusive of civil defence, public health, transport, culture and social sectors and nature conservation.

Table 6.2 The size of vulnerable areas according to various conditions of probability and climate in Riga city as identified by the draft Flood Risk Management Plan

Climate	Size of vulnerable areas according to various conditions of probability (in 1000 m²)					
	0.5%	1%	5%	10%	20%	50%
Current	27 005	22 107	10 418	7477	4959	0
Near future (2021–2050)	31 472	26 460	13 833	9742	7489	2880
Distant future (2071–2100)	41 230	34 411	15 004	14 059	11 118	6976

Source: Riga Against Flood!, 2012.

6.4.4 Process of the preparation of the Flood Risk Management Plan

Prior to envisioning the objectives and strategies for flood risk management in the domain of planning, the involved persons should become familiar with the 'Concept of Risk Management' in flood risk mitigation. This phase is particularly relevant as stakeholders have to change their traditional ways of dealing with flood risk issues and have to develop new skills and understanding. The process of acquisition of new knowledge and its implementation in practice needs time and continuous support by stakeholders, researchers in particular. Active cooperation between academics and practitioners in considering international experiences is of utmost importance seen from this perspective. At the end of this process, stakeholders need to be able to demonstrate their acquired knowledge by developing their vision on how to deal with future floods. Therefore, they should consider the current situation (hazard and impact identification and analysis), analyse uncertainties (hydrological, social, economic and political) as well as drivers of future development (risk changes), such as climate change and urbanization. After agreeing on a consensus on flood risk management objectives, the Riga community planners can start the concrete planning phase. At the end of this stage, a set of various flood risk management options should be prepared. This point is the milestone in the management process. In the final phase of flood risk adaptation planning one set of mitigation and adaptation measures has to be agreed on. The assessment of the various options of flood risk management is based on relevant criteria which were defined beforehand (see Chapter 4). Again, it would

be favourable if experts coached this process by introducing the approach of Multi-Criteria Decision Analysis (MCDA), by elaborating a decision matrix of the relevant criteria and by providing a decision support tool which all stakeholders could use for weighting the criteria in order to choose the best option that, finally, would serve as the decision (see Chapter 4). Good guidance is necessary in exploring the possibilities to minimize the remaining conflicts of interest between different stakeholder groups. Probably, not all conflicts can be avoided, with the result that the stakeholders have to agree on an 'acceptable level of conflict' by defining priorities. In urban areas like Riga, many stakeholder groups are affected by the actions taken within a flood risk management plan. The stakeholders' analysis should provide the existing political, social and institutional structures with special reference to the organizational structure of flood risk management within the area of interest (Table 6.3) (EC, 2003).

Stakeholder workshops and meetings have been organised by the University of Latvia to broaden the spread of information and consultations outside Riga City's municipal institutions. In the course of this, the University of Latvia transferred the climate change adaptation knowledge of other BaltCICA project partners.

6.4.5 Application of Multi-Criteria Decision Analysis for assessing flood-prone areas in Riga City

Within the BaltCICA project, interdisciplinary seminars on using multi-criteria decision analysis methods for assessing flood-prone areas were organized. In the

Table 6.3 Parameters characterizing stakeholders involved in flood risk adaptation planning and their interrelations

Parameter	Description	Involved stakeholders
Level of impact	Shows to what extent stakeholders are affected by the (non-) implementation of a flood risk management plan	From those being impacted to a higher degree (vulnerable) – residents, employees and visitors at low coastal areas – to ones with low vulnerability
Level of influence	Explains to what extent stakeholders can influence the flood risk adaptation planning and outcomes	From those with high influence, as government and municipal institutions, to residents, particularly vulnerable and with low level of influence
Level of interest	Explains the extent to which stakeholders express their interest to be involved in flood risk adaptation planning	From stakeholders with high interest in environmental, safety and insurance aspects to those with little interest, like many indoor businesses and residents of apartment and rented housing
Level of understanding	Indicates the level of knowledge and awareness of flood risk management (information, methods, legislative framework, including the EU Floods Directive and local practises) and climate change adaptation among various stakeholders	From climate researchers and experts, having a high degree of understanding, to lay persons, local politicians and local mass media that in general have lower understanding
Level of capabilities	Explains physical (technical), financial, social, cultural and political capabilities needed for coping with flood risk and the related social changes.	From state government bodies with high capabilities to lay persons or small businesses with low capabilities
Diverging interests, conflicts and overlapping responsibilities	Explains conflicting interests or ambitions among institutions	Conflicts between organizations acting in one sector, but at different spatial levels or geographical units; conflicts between stakeholders acting in one geographical unit/sector, but at various policy levels

case of Riga City, its specific geographical location in the vicinity of three large river deltas was brought into focus. Although the flood-prone areas are not always directly connected, each of them can affect the development of the respective district in the event of a flood which, in turn, can result in possible impacts on the entire city of Riga and its surroundings. In order to prevent the possibility of flooding, different kinds of flood prevention measures, particularly investments, should be taken in the near and distant future. The city planners should develop a sequential plan on how to protect these areas against flooding. Flood-prone areas in question are very diverse: areas with multi-storey residential buildings (Bolderāja), single-family house estates built at different periods (Mangaļsala, Bukulti,

Trīsciems), small river delta areas (Trīsciems by the Langas River and the built-up area around the southeast end of Juglas Lake by the Juglas River), areas located in and around the territory of the Freeport of Riga – subject to special tax and entrepreneurship-related legislation (Spilve, Sarkandaugava), industrial areas (Sarkandaugava), areas within the protected territory of the historic city centre of Riga which is part of the UNESCO World Heritage List (Ķīpsala), as well as areas which are not in active commercial use at the moment but will have potentially developed in the future (Spilve, Lucavsala).

With respect to MCDA, the Multi-Attribute Utility Theory was selected as the most appropriate method. The interdisciplinary group formed within BaltCICA

Table 6.4 Twelve criteria selected for the initial MCDA of the areas that either are flood-prone at present, or may become so in future

Criteria suggested for MCDA	Reference period	
	Now	Changes in the future
Size of the flood-affected areas	X	X
Population number in flood-affected areas	X	X
Proportion of senior citizens in flood-affected areas	X	
Number of work places (employed) in flood-affected areas	X	X
Economic losses in flood-affected areas	X	X
Number (existence) of high risk objects and critical infrastructure, public infrastructure	X	
Number (existence) of public infrastructure and cultural heritage objects	X	
Existence/non-existence of evacuation routes	X	

evaluated information about the available criteria for analysis. Twelve criteria were selected for the initial analysis that comprises the socio-economic spheres and takes into account the available indicators of the areas that either are flood-prone at present, or may become so in future (Table 6.4).

After a detailed evaluation of the criteria from qualitative and quantitative data aspects, only three criteria were selected for the final analysis: (1) population density in the flood-prone areas, (2) economic losses in the flood-prone areas, (3) changes in economic losses in the future. According to the near-future scenario development prospects for Latvia and Riga (Riga Against Flood!, 2012), the population figure of the flood-prone areas will not change considerably. At the same time, economic losses will significantly increase in many areas. Stakeholder discussions organized in Riga showed that practitioners comprehend the meaning of the near-future scenarios and they see the need to incorporate it in the existing regional planning and flood prevention plans. Economic losses were calculated on the basis of the size of the affected areas, land-use, water depth during floods and duration of flooding (Riga Against Flood!, 2012). During the BaltCICA seminars, the interdisciplinary group of experts proposed that the ranking of the flood-prone areas using MCDA approach should be done in the first place, as the primary task is to determine which of the areas will be worst affected by flooding. Cost-benefit analysis of the implementation of flood prevention measures is a next step, requiring further research.

6.5 Conclusions

The results of modelling revealed that Riga City has a rather high number of flood-prone areas that in most cases are not connected in terms of water management options. Accordingly, for developing flood prevention and climate change adaptation measures, there is a need to define comprehensive and acceptable criteria for planning purposes, including prioritization of the measures in terms of timing and investments needed. In close cooperation between the experts of the Riga municipality and the University of Latvia, possible adaptation options for urban spatial development are being prepared for integration in the comprehensive spatial plan, building and other regulations of the municipality. The Riga City case plays an important role for knowledge and policy transfer in the field of flood prevention and climate change adaptation for other Latvian municipalities and for preparation and improvement of respective national policies.

MCDA has been valued by city planners as a clearly comprehensible support tool for conducting discussions with stakeholders (inhabitants, investors and environmental protection organizations). MCDA can also play a role in knowledge transfer and social learning. Possibility of setting the criteria weights individually allows discussion participants to acquire a better understanding of the causal links between the environmental, economic and social processes. That sort of use, evaluation and sharing of different stakeholders' knowledge can improve common knowledge about flood causes and processes and prevention measures needed. Thus common goals as regards flood prevention in such a complex area as Riga City can be better achieved. Stakeholder involvement and their knowledge management are relevant as a contemporary flood risk management approach involves various dimensions and activities.

New concepts are required to support active participation of stakeholders in developing flood risk management plans in the sense of the EU Flood Directive (2007). Bottom-up approaches are likely to fulfil this requirement. The development of mutual trust and an open atmosphere turned out to be a crucial factor for proceeding with planning. Analysis of the city's sensitivity to floods and economic analysis is a necessary extension of the obligatory flood maps in order to understand the system's limits, enabling stakeholders to assess the risks more realistically. Harmonization with other directives and planning procedures (e.g., the EU Water Framework Directive 2007/60/EC) has to be performed at an early stage, introducing already planned synergetic measures as an element of flood risk management planning. As this is a process that requires extensive knowledge and resources, the expertise of different stakeholder groups and international experiences have to be considered. The Riga City case proves that, in spite of public finance cuts at the national level, the expert community at the local level is capable of attracting EU funding (BaltCICA and Life+ project Riga against flood!) and is continuing to work towards a safer urban environment. Thus, climate change adaptation measures are included in its everyday tasks and strategic aims.

Acknowledgements

The article is based on studies under the Latvian national research programme 'Climate Change and Waters' (KALME, 2006–2009), EU projects in the Baltic Sea region 'Towards climate change adaptation in the Baltic Sea region' (ASTRA, 2005–2007) and 'Climate Change: Impacts, Costs and Adaptation in the Baltic Sea Region' (BaltCICA, 2009–2012) as well as flood management solutions for Riga developed within the Life+ project 'Integrated Strategy for Riga City to Adapt to the Hydrological Processes Intensi-fied by Climate Change Phenomena 2010–2012 (Riga against flood!)'.

References

Albrechts, L., 2001. How to proceed from image and discourse to action: As applied to the Flemish Diamond. *Urban Studies*, 38 (4), pp.733–745.

Aligica, P.D., 2005. Institutional and stakeholder mapping: Frameworks for policy analysis and institutional change. *Public Organization Review*, 6, pp.79–90.

Avotniece, Z., Rodinov, V., Lizuma, L., Briede, A. and Kļaviņš, M., 2010. Trends in the frequency of extreme climate events in Latvia. *Baltica*, 23 (2), pp.135–148.

Biedriņš, A. and Ļakmunds L., 1990. *No Doles līdz jūrai*. Rīga: Zinātne.

Brázdil, R., Demarée, G.R., Deutsch, M., Garnier, E., Kiss, A., Luterbacher, J., Macdonald, N., Rohr, C., Dobrovolný, P., Kolář, P. and Chromá, K., 2010. European floods during the winter 1783/1784: Scenarios of an extreme event during the 'Little Ice Age'. *Theoretical Applied Climatology*, 100, pp.163–189.

Clark, W.C. and Majone, G., 1985. The critical appraisal of scientific inquiries with policy implications. *Science, Technology and Human Values*, 10 (3), pp.6–19.

Commission Directive 2007/60/EC of 26 November 2007 on the Assessment and Management of Flood Risks.

CSB (Central Statistical Bureau), 2011. *Central Statistical Bureau of Latvia Database*. [online] Available at: <www.csb.gov.lv> [Accessed 16 December 2011].

EC (European Commission), 2003. *Best Practices on Flood Prevention, Protection and Mitigation, European Communities*. [pdf] Available at: <http://www.floods.org/PDF/Intl_BestPractices_EU_2004.pdf> [Accessed 16 December 2011]

EC, 2012. *The EU Water Framework Directive – integrated river basin management for Europe*. [online] Available at: <http://ec.europa.eu/environment/water/water-framework/index_en.html> [Accessed 05 June 2012]

EEA (European Environment Agency), 2006. *The Changing Faces of Europe's Coastal Areas: EEA Report No 6/2006*. Luxembourg: Publications Office of the European Union.

EEA, 2010. *Adapting to Climate Change: The European Environment State and Outlook 2010*. Luxembourg: Publications Office of the European Union.

Eisenhardt, K. and Santos, F., 2012. Knowledge-based view: a new theory of strategy? In: Pettigrew, A.M., Thomas, H. and Whittington, R., eds. 2012. *Handbook of Strategy and Management*. London: SAGE Publications, pp. 139–164.

Ensor, J. and Berger, R., 2009. Community-based adaptation and culture in theory and practice. In: Adger, W.N., Lorenzoni, I. and O'Brien, K.L., eds. 2009. *Adapting to Climate Change: Thresholds, Values, Governance*. Cambridge: Cambridge University Press, pp. 227–239.

Freeman, R.E., 1984. *Strategic Management: A Stakeholder Approach*. Boston: Pitman Publishing.

German, L., 2010. Local knowledge and scientific perceptions: questions of validity in environmental knowledge. In: German, L.A., Ramisch, J.J. and Verma, R., eds. 2010. *Beyond the Biophysical: Knowledge, Culture, and Politics in Agriculture and Natural Resource Management*. Dordrecht, Heidelberg, London, New York: Springer Science + Business Media B.V., pp.99–125.

Glavovic, B.C., 2008. Sustainable coastal communities in the age of coastal storms: Reconceptualising coastal planning and ICM as 'new' naval architecture. *The Journal of Coastal Conservation*, 12 (3), pp.125–134.

Grimble, R. and Wellard, K., 1997. Stakeholder methodologies in natural resource management: A review of concepts, contexts, experiences and opportunities. *Agricultural Systems*, 55 (1), pp.173–193.

Haas, P.M., 1992. Introduction: Epistemic communities and international policy coordination. *International Organization*, 46 (1), pp.1–35.

Haas, P.M., 2004. When does power listen to truth? A constructivist approach to the policy process. *Journal of European Public Policy*, 11 (4), pp.569–592.

Harries, T. and Penning–Rowsell, E., 2011. Victim pressure, institutional inertia and climate change adaptation: The case of flood risk. *Global Environmental Change*, 21, pp.188–197.

Harrison, S.R. and Qureshi, M.E., 2000. Choice of stakeholder groups and members in multicriteria decision models. *Natural Resources Forum*, 24 (1), pp.11–19.

Heitza, C., Spaeter, S., Auzet, A.-V. and Glatrona, S., 2009. Local stakeholders' perception of muddy flood risk and implications for management approaches: A case study in Alsace (France). *Land Use Policy*, 26, pp.443–451.

Hillman, M., 2009. Integrating knowledge: The key challenge for a new paradigm in river management. *Geography Compass*, 3 (6), pp.1988–2010.

Holste, J.S. and Fields, D., 2010. Trust and tacit knowledge sharing and use. *Journal of Knowledge Management*, 14 (1), pp.128–140.

Hunt, J. and Shackley, S., 1999. Preconceiving science and policy: Academic, fiducial and bureaucratic knowledge. *Minerva*, 37, pp.141–164.

Hutter, G. and Schanze, J., 2008. Learning how to deal with uncertainty of flood risk in longterm planning. *International Journal of River Basin Management*, 6 (2), pp.175–184.

Johnson, C., Penning–Rowsell, E. and Parker, D., 2007. Natural and imposed injustices: the challenges in implementing 'fair' flood risk management policy in England. *The Geographical Journal*, 173 (4), pp.374–390.

Korczynski, M., 2000. The political economy of trust. *Journal of Management Studies*, 37 (1), pp.1–21.

Ludvigs, P., 1967. *Mūsu Latvijas ūdeņi*. Rīga: Grāmatu draugs.

Lugeri, N., Kundzewicz, Z.W., Genovese, E., Hochrainer, S. and Radziejewski, M., 2010. River flood risk and adaptation in Europe – Assessment of the present status. *Mitigation and Adaptation Strategies for Global Change*, 15, pp.621–639.

McFadden, L., Penning-Rowsell, E. and Tapsell, S., 2009. Strategic coastal flood-risk management in practice: Actors' perspectives on the integration of flood risk management in London and the Thames Estuary. *Ocean and Coastal Management*, 32, pp.636–645.

Meijerink, S., 2005. Understanding policy stability and change. The interplay of advocacy coalitions and epistemic communities, windows of opportunity, and Dutch coastal flooding policy 1945–2003. *Journal of European Public Policy*, 12 (6), pp.1060–1077.

Merz, B. and Thieken, A.H., 2005. Separating natural and epistemic uncertainty in flood frequency analysis. *Journal of Hydrology*, 309, pp.114–132.

Moskovkina, E., 1960. *Floodings on the Daugava River*. [Московкина, Е.Г. (1960) Паводки на реке Даугава. Изд. АН Латвийской ССР]

Mostert, E., Pahl-Wostl, C., Rees, Y., Searle, B., Tabara, D. and Tippett, J., 2007. Social learning in European river basin management: Barriers and fostering mechanisms from 10 river basins. *Ecology and Society*, 12 (1), Art.19.

Neuvel, J.M.M. and van den Brink, A., 2009. Flood risk management in Dutch local spatial planning practices. *Journal of Environmental Planning and Management*, 52 (7), pp.865–880.

Parker, D.J., 2007. *Systematisation of Existing Flood Risk Management Concepts* (FLOOD-ERA working paper). Enfield: Flood Hazard Research Centre.

Raadgever, G.T., Dieperink, C., Driessen, P.P.J., Smit, A.A.H. and Rijswick, H.F.M.W. van, 2011. Uncertainty management strategies: Lessons from the regional implementation of the water framework directive in the Netherlands. *Environmental Science & Policy*, 14, pp.64–75.

Reed, M.S., 2008. Stakeholder participation for environmental management: A literature review. *Biological Conservation*, 141 (10), pp.2417–2431.

Ridder, D., Mostert, E. and Wolters, H.A. eds., 2005. *Learning Together Manage Together*. [pdf] Osnabruck: University of Osnabruck. Available at: <http://alpsknowhow

.cipra.org/main_topics/decision_making/pdfs/harmonicop _learning_managing.pdf> [Accessed 05 June 2012].

Riga against flood!, 2012. *Integrated Strategy for Riga City to Adapt to the Hydrological Processes Intensified by Climate Change Phenomena*. [online] Available at: <http://www.rigapretpludiem.lv> [Accessed 05 June 2012].

Riga City Council, 2012. *Apkaimes: Statistika* [Neighbourhoods: Statistics]. [online] Riga: Riga City Council. Available at: <http://www.apkaimes.lv> [Accessed 05 June 2012].

Samuels, P.G., Morris, M.W., Sayers, P., Creutin, J.-D., Kortenhaus, A., Klijn, F., Mosselmann, E., von Os, A. and Schanze, J., 2009. Advances in flood risk management from the FLOODsite project. In: Samuels, P.G., Huntington, S., Allsop, W. and Harrop, J., eds. 2009. *Flood Risk Management: Research and Practice*. London: Taylor and Francis, pp. 433–443.

Shackley, S. and Wynne, B., 1995. Integrating knowledges for climate change: Pyramids, nets and uncertainties. *Global Environmental Change*, 5 (2), pp.113–126.

Smith, E.A., 2001. The role of tacit and explicit knowledge in the workplace. *Journal of Knowledge Management*, 5 (4), pp.311–321.

Stakle, P., 1941. *Hydrometric surveys of Latvian inland waters from November 1 to October 31, 1940*. Rīga: Jūrniecības pārvalde.

Stehr, N. and Ufer, U., 2009. On the global distribution and dissemination of knowledges. *International Social Science Journal*, 60 (195), pp.7–24.

Swain, C. and Tait, M., 2007. The crisis of trust and planning. *Planning Theory & Practice*, 8 (2), pp.229–247.

Tippett, J. and Griffiths, E.J., 2007. New approaches to flood risk management – implications for capacity building. In: Ashley, R., Garvin, S., Pasche, E. Vassilopoulos, A. and Zevenbergen, C., eds. 2007. *Advances in Urban Flood Management*. London: Taylor and Francis, pp.383–413.

Tran, P., Marincioni, F., Shaw, R., Sarti, M. and An, L.V., 2008. Flood risk management in central Viet Nam: Challenges and potentials. *Natural Hazards*, 46, pp.119–138.

Weber, N. and Christopherson, T., 2002. The influence of nongovernmental organisations on the creation of Natura 2000 during the European policy process. *Forest Policy and Economics*, 4 (1), pp.1–12.

Werff, P.E. van der, 2004. Stakeholder responses to future flood management ideas in the Rhine river basin: Nature or neighbour in hell's angle. *Regional Environmental Change*, 4, pp.145–158.

Werritty, A., 2006. Sustainable flood management: Oxymoron or new paradigm? *Area*, 38 (1), pp.16–23.

WMO (World Meteorological Organization), 2006. *Legal and Institutional Aspects of Integrated Flood Management*, WMO-No. 997. Geneva: World Meteorological Organization.

WMO, 2009. *Integrated Flood Management Concept Paper*. [pdf] Geneva: World Meteorological Organization. Available at: <http://www.apfm.info/pdf/concept_paper_ e.pdf> [Accessed 05 June 2012].

7 Climate Adaptation in Metropolis Hamburg: Paradigm Shift in Urban Planning and Water Management towards 'Living with Water'?

Joerg Knieling & Mareike Fellmer

Hafen City University Hamburg, Urban Planning and Regional Development, Germany

7.1 Introduction

The cities of the northern German waterfronts are densely populated areas and hence highly vulnerable to the impacts of climate change. With hindsight, it becomes obvious that adaptation to the natural dynamics of the sea level already has a long tradition on the coastline of the North Sea (Fischer, 2011). Dwelling mounds, such as those on North Frisian Holms, or water-orientated living, like houseboats in The Netherlands, show experiences in flood-adapted construction. But climate change is expected to create a new dimension of threat with regard to the coastline.

This contribution deals with the adaptation approach as it concerns the risks of climate change and how it affects urban waterfronts. It discusses how far the risk dimension of climate change leads to a shift of paradigm in the way flooding is tackled. For decades technical solutions, like dykes or removable walls, have dominated the discussion promising 100% security to people and the built environment. But the consequences of climate change are expected to be too far-ranging to be solved exclusively on a technical basis. Thus, a more holistic approach is discussed, emphasizing more thoroughly the relationship between society and the ecological system and searching for new ways of living with water and changing environmental conditions in an integrated way. This debate will be exemplified for Metropolis Hamburg and in particular for the case of Hamburg's Elbe Island, including an urban quarter with about 50 000 inhabitants.

7.2 Urban development and climate change in Hamburg

At the core of the Metropolitan Region of Hamburg is the Free and Hanseatic City of Hamburg which is situated 120 km south-east of the North Sea. The Metropolitan Region of Hamburg is also closely connected with the Baltic Sea. Affected by tides of the North Sea on a semi-daily basis, the area extends to a flood barrage in the city of Geesthacht, 30 km east of Hamburg. This shows that Hamburg does not only profit from the advantages of being a port city, but is also obliged to deal with the difficulties of its location. The most significant of these challenges is the threat of storm surges, a phenomenon which is likely to increase due to climate change (Woth, Weise & von Storch, 2005; Storch & Claussen, 2011).

The risk of storm surges has a direct influence not only in respect of dyke construction, but also for living on the floodplains. Against this background urban and regional planning in Hamburg has followed a strict

Climate Change Adaptation in Practice: From Strategy Development to Implementation, First Edition.
Edited by Philipp Schmidt-Thomé and Johannes Klein.
© 2013 John Wiley & Sons, Ltd. Published 2013 by John Wiley & Sons, Ltd.

pattern throughout the centuries: living has to be situated in the *Geest* (dry area) and working in the *Marsch* (wet area). The prominent 'Federplan' (Kallmorgen, 1968), designed in 1920 by Fritz Schumacher, head of the city's planning department, not only shows Hamburg's urban development along traffic axes – thus reminiscent of the lines of a feather-shape – but also identifies the Hamburg Elbe Island as being located in the *Marsch*.

In the context of its vision for city development from 2002, which identifies Hamburg as a growing city (FHH, 2002), and its urban development concept 'Leap across River Elbe' (FHH, 2005) the Senate has departed from this pattern some years ago. On Hamburg's Elbe Island in particular, new living and working areas have been planned in Wilhelmsburg (FHH, 2005; FHH 2007). This new orientation of urban development has not only been inspired by politics and administration, but also by interest groups, such as the Chamber of Commerce (HK Hamburg, 2004). Depending on the source of each study, concept or plan that comes into consideration, it is estimated that between 15 000 and 50 000 new inhabitants could live on Elbe Island in the near future (FHH, 2002; FHH, 2003; FHH, 2007). This would mean up to double the current population and would also correspond with an expansion of the necessary commercial and industrial areas. If new building projects were to be conducted on this scale, it would vastly increase the danger of damage by storm surges. However, none of the studies, concepts and plans referred to above makes any statements about how to deal with these risks.

In order to illustrate the extent of the threat, it is worth looking back at events that happened earlier. In 1962 Hamburg was hit by a strong storm surge which took the lives of more than 300 people. Concerning this, it seems particularly surprising that there are hardly any restrictions on building behind the dykes of Elbe Island. Living in front of the dykes is prohibited, and those who live behind are to a large extent reliant on the existing flood protection systems.

However, in the interests of the population the question arises, how can the growing risks resulting from climate change be dealt with? Which planning guidelines could help to minimize damage in case of an eventual dyke failure? Before debating a changing adaptation paradigm towards 'living with water', the next section describes key concepts and variables of adaptation.

7.3 Key concepts and variables of adaptation

Adaptation in the context of climate change is described as a '[...] process, action or outcome in a system [...] in order for the system to better cope with, manage or adjust to some changing condition, stress, hazard risk or opportunity' (Smit & Wandel, 2006, p.282). Hence, adaptation goes along with questioning existing paradigms and routines concerning planning. Vulnerability, resilience and adaptation capacity can be seen as basic variables of modern adaptation policies (Smit, Burton, Klein & Wandel, 2000; Adger et al., 2003; IPCC, 2007). Within this contribution, these concepts will be presented and characteristics of resilience will be deduced. Then, these characteristics will be demonstrated with regard to flood adaptation strategies, exemplified by Hamburg's Elbe Island.

As mentioned above, in 'risk society' (Beck, 1986) there is a search for solutions to cope with risk in a more holistic manner. The danger from natural hazards concerning people, nature or goods increases in the situation of environmental changes and a growing urban population (Bohle, 2008). In this context the concept of vulnerability is characterized by its distinct spatial and temporal scales, by scientific uncertainties regarding scenarios and models, and by the policy context (Füssel & Klein, 2006; Aerts et al., 2012). Therefore, the sensibility of cities and regions towards unpredictable changes depends on social, ecological, economic and political conditions. Regarding increasing storm surge risks, the concept of resilience is related to vulnerability. The notion of *resilience* is discussed in different ways and a transfer of the concept in practice is difficult to consider. One definition describes resilience as 'degree to which a system rebounds, recoups or recovers from stimulus' (Smit, Burton, Klein & Wandel, 2000, p.238). Other definitions characterize resilience as 'a return time to a steady-state following a perturbation' (Gunderson, 2003, p.34). According to these, and with respect to climate change, resilience can be seen as the capability of systems to undergo change. Concerning cities, resilience has been used in the context of their sensitivity to disasters and their ability to deal with natural disasters and uncertainties (Overbeck, Hartz & Fleischhauer, 2008; Newman, Beatley & Boyer, 2009). Resilient cities aim at built-in systems that help to adapt to change (Newman, Beatley & Boyer, 2009). Hence, resilience includes three integrated systems to enable people and nature to face disturbance: first, 'the amount of change a system can undergo [. . .] and still remain the same controls on function and structure', secondly, 'the degree to which the system is capable of self-organization', and finally, 'the degree to which the system expresses capacity for learning and adaptation' (Resilience Alliance, 2009). Therefore, the concept of resilience goes along with the adaptive capacity dealing with change and continued development. Against this background of diversity, coping ability, learning capabilities and responsiveness, the characteristics of flexibility and robustness are considered as the most relevant to enhance resilience (Folke et al., 2002; Gunderson, 2000; Zevenbergen, 2007).

• Flexibility is explained in terms of ecological adaptiveness and as living with environmental variability. Corresponding ways of life which cope with increased variability and the unpredictability of environmental conditions, like changes in seasonal cycles or water levels, are able to increase the adaptive capacity (Berkes & Jolly, 2001). Furthermore, the buffering capacity in flood-prone areas is seen as criteria of flexibility, for example, the capacity for water storage (Adger, Kelly & Ninh, 2001; Fankhauser, Smith & Tol, 1999). Thus, reserve areas become relevant to protect against extreme floods because they offer retention space for minimizing high water levels.

• Robustness means strengthening coastal protection to withstand severe storms and floods (de Bruin et al., 2009; Fankhauser, Smith & Tol, 1999). The built environment is described as robust as it can be adjusted to withstand a wider range of future weather conditions (Fankhauser, Smith & Tol, 1999, p.76). Hence, strong, less sensible building structures have become relevant criteria for the adaptation capacity.

To sum up, living with environmental dynamics as well as providing buffer capacities to withstand extreme events are relevant criteria for the development of resilient settlement structures. Urban systems are obliged to be pliable and compliant to achieve a sufficient adaptation capacity (Smit, Burton, Klein &

Wandel, 2000). In particular, these requirements are to be taken into account when waterfronts are newly developed or re-developed. After having introduced some key concepts of adaptation to climate change and having pointed out characteristics of resilience, the following section will discuss the case of Hamburg's Elbe Island and introduce the adaptation strategy of 'cascading flood compartments'. In the following section against the background of the Hamburg case the handling of storm surges in urban development and a change of paradigm from flood protection systems to risk management in coastal zones will be discussed.

7.4 Changing adaptation paradigms: From technical solutions to 'Living with Water'?

Today's flood management in Hamburg mainly relies on technical engineering solutions. The water management administration has developed highly qualified flood protection systems with excellent safety standards. Although technical installations for flood control provide protection, they create an illusory sense of security for living behind the dykes so that the potential danger is suppressed or nearly forgotten (Kron, 2003). Thus, flood protection traditionally has been focused on technical solutions like dykes, flood barriers or walls which contain the natural river bed. The recent debate about the consequences of climate change has taken into account the limits of existing danger prevention and especially a single protection line (e.g. Greiving, 2003; Aerts et al., 2008; Jolly, 2008). The faith in technical measures and their capability of containing nature has increasingly been questioned and there have been many approaches discussing the reduction of vulnerability in the event of flooding or in regard to the consequences of climate change (e.g. Few, 2003; Füssel, 2007; Beatley, 2009; Bouwer, Bubeck & Aerts, 2010; Dawson et al., 2011). Beatley (2009, p.13) highlights different key drivers for unsustainable coastal development: 'These include coastal population growth, larger demographic trends, a desire to be near to and enjoy the amenities of coastal living, and public policy and financial system that has largely encouraged and underwritten coastal risks. A limited understanding of the long-term (or even short term) risks and dangers of living in coastal environments further contributes to these vulnerable patterns of development, [. . .].' This change can be described as a new requirement for flood management; technical security with regard to flood protection is no longer able to cope with the exigencies of modern flood hazards. It constitutes fixed waterfronts and does not allow flexibility in floodplains. Therefore a change of paradigm from prevention to management of danger has taken place (Greiving, 2003). While the paradigm shift happened in some areas (e.g. Rotterdam opens its Waterfronts to the North Sea) other areas (e.g. Hamburg seals their Waterfronts off the River Elbe) still stick to safety. Risk analysis and risk assessment describe new ways of planning in coastal zones. This change becomes apparent when taking a closer look at the international debate on the topic. For example, in The Netherlands, where for a long time flood protection has also primarily been based on technical solutions such as raising levees and building flood barrages and walls, an increasing number of supplementary solutions and approaches have been considered (de Bruin et al., 2009). Besides technical solutions, new approaches for handling uncertainties in water management – for example flood insurances, flood risk mapping systems, general risk management systems, and urban concepts like 'room for the river', which include measures to give the natural flooding area back to the river – have been applied (McGranahan, Balk & Anderson, 2007; Aerts et al., 2008). Furthermore, lately, the development of new approaches has switched the attention to alternative ways of 'living with water' (Immink, 2004). Studies and projects concentrating on flood adaptive building in urban development have been accomplished (e.g. Flesche & Burchard, 2005; Zevenbergen, 2007; Veerbeek et al., 2008). Basically, four different types of adaptive building can be identified (e.g. Flesche & Burchard, 2005; Leven met water, 2012; Simons, 2008):

1. Swimming or floating houses, which are dependent on permanent high water levels; they are options for canals or the river themselves.
2. Dyke houses, which have a double function: on the one hand they build a dyke line, and on the other hand a dyke house has a housing function.
3. Amphibian houses, which need temporary high water. They flow with the dynamic of changing water levels.

4. Houses on stilts, which have a long tradition in flood protection and may safeguard their inhabitants in temporary flooded areas.

These examples of flood adaptive urban planning reflect the current debate on reopening water-related areas for urban development (e.g. Aerts et al., 2008; Beatley, 2009; Hill, 2012). Besides, they have been shaped to contribute to adaptation to climate change. Within these adaptation strategies the areas behind the dykes and near the river are the main focus of coastal protection. Furthermore, restricted and adjusted behaviour in the coastal zones becomes relevant. The adjusted focus on dyke zones rather than dyke lines is also a new approach for urban planning. In this context the difference between riverine and coastal zones is important. While riverine waterfronts are characterized by a dense spatial structure and a minor availability of space, coastal zones have to deal with higher flood intensity, have more space for retention areas or second dyke lines. Thus a close cooperation between water management and spatial planning is required. This includes more institutional flexibility and learning processes within the different stakeholders being involved in waterfront planning (Aerts et al., 2008).

Another challenge for adaptation deriving specifically from climate change is how to deal with its uncertainty and the inherent long-term perspective. Despite a wide range of climate simulations and predictions (Woth, Weisse & von Storch, 2005), it is still not possible to give definite forecasts of how sea levels will rise in the future; the predictions of sea level rise vary significantly and are uncertain (Aerts et al., 2008). Thus, urban planning is confronted with the challenge of protecting existing and prospective settlements appropriately without sufficient knowledge of potential threats. Against this background, strategies tackling storm surge risk can be divided into four categories of prevention (Egli, 2005; BMBVS, 2006):

1. Area prevention: either maintaining open spaces or conscious positioning of buildings and constructions.
2. Constructional prevention: the adaptation of buildings regarding potential flood dangers.
3. Risk prevention: in particular financial prevention in the form of private savings or insurance policies as well as emergency measures like mobile walls and sandbags.

4. Behaviour prevention: individual preparation for possible flood events.

All the above fields are relevant for urban planning and water management, especially area prevention and constructional prevention. The German Federal Building Code and the Regional Planning Act provide a legal framework to flood risk orientated urban development. Measures like the designation of flood risk areas can be found on different levels of planning. By designation within the binding land-use plan building exigencies and utilization limitations can be established within a specific land-use area. The measures mentioned below are provided to be integrated into the settlement structures with the help of the existing planning instruments of land-use planning and urban design (BauGB, 2006: § 5 and 9):

Land-use planning:

• Usage limitations as a safety measure for areas that in terms of flood control and drainage have to be kept free, for example, retention and flood endangered areas.
• Designation of areas that need to take into account specific precautionary measures against the forces of nature when being developed, for example, areas that need to be raised or embanked before further development is allowed.
• Application of a notation and display of flood endangered areas in the land-use plan.

Urban design:

• The fixation of specific types of use, for example, residential use, green space, or the positioning of certain buildings.
• Restrictions with relations to building and usage, for example, inadmissibility of cellars or space for flood protection systems.
• Minimum height above sea level for certain usages, for example, a minimum of 7.5 metres above sea level for residential housing.

These planning instruments allow spatial reserves and buffer areas for extreme weather events to be secured over long time periods. Nevertheless, regarding planning instruments, the fixity of planning as well as of the built environment becomes obvious. Urban infrastructure is relatively static (McGranahan, Balk & Anderson, 2007). Plans in regional and land-use planning or urban design are developed for a certain

timeframe as well as marked off areas and they are oriented on administrative settings, hardly on natural systems. Furthermore, plans have long-term validities so that changes are difficult. This underlines that climate change raises basic and important questions concerning the current praxis of adaptation to flooding as well as of spatial planning.

The following section deals with the potential of adaptation strategies to tackle extreme storm surges from a theoretical and conceptual point of view. Therefore, the concept of resilience will be exemplified.

7.5 Reflecting the practice of adaptation strategies: The case of Hamburg's Elbe Island

Against the background of climate change there has been a lively debate about adaptation strategies. In particular – apart from continuously raising the levees – more space for rivers and the reduction of possible damage potentials behind the levees have been demanded (e.g. Immink, 2004; Jeschke, 2004; Klijn, van Buuren & van Rooij, 2004; Garrelts, Lange & Flitner, 2008). In the passages above the conceptual framework of resilience has been described and characteristics of resilient urban development have been defined. In the following section the terms of flexibility and robustness as main requirements of resilience in the case of Hamburg's Elbe Island will be analysed.

Hamburg's Elbe Island is an urban quarter with about 50 000 inhabitants (see Figure 7.1). It is situated directly on the River Elbe which has developed a broad delta of various streams in this part of the city. Its limits are the Northern Elbe and the Southern Elbe. For the City of Hamburg this broad extension of the river Elbe is an important challenge, posing the question of how to link the different parts of the city which are situated north and south of the river Elbe. Furthermore, Elbe Island is exposed to severe storm flood risks – formally still being part of the North Sea that reaches that far inland. Currently, the island is protected by a circular embankment. However, in respect to future climate change, these levees, for technical reasons and because of the urban structure, cannot be raised indefinitely.

Figure 7.1 Elbe Island of Hamburg–Wilhelmsburg.
Source: Modified from © ATKIS GDB 2010 - LGV Hamburg.

Because of its important linkage-character for the city's urban development, Hamburg's Elbe Island has become the focus of ambitious urban development processes (FHH, 2005). There is the mentioned strategy called 'Leap across the River Elbe' as well as the International Building Exhibition 'IBA 2013' which aims at developing the quarter.

Against the background of limited flood protection from the embankment with regard to the consequences of climate change, with the 'system of cascading flood compartments' an unconventional strategy has been developed for Elbe Island to reach flood resilient waterfronts (Pasche et al., 2008; Knieling, Schaerffer & Tressl, 2009) (see Figure 7.2). At present, this strategy has the status of an unrealized concept and one that is out of step with current flood protection strategies in Hamburg.

The 'system of cascading flood compartments' combines area prevention with constructional solutions. The idea is to (re)build a second dyke-line of polders behind the primary dyke, thereby creating a system of different compartments. Polders are low-lying areas enclosed by embankments or dykes. In the case of a dyke overflowing, the system diverts the water to the hinterland. For Hamburg's Elbe Island new internal dykes within the existing ring-dyke could protect the most economically valuable areas (Aerts et al., 2008). Also, certain housing areas with a dense population would gain further protection against extreme floods. Furthermore, the residences within these compartments would get additional protection against flood risks by constructional prevention (Pasche et al., 2008). When expanding to the waterfronts, these areas could also be used for flood-secure types of housing like floating or amphibian homes. Alternative housing types can offer an additional protection to the ring-dyke of Elbe Island.

The strategy goes along with the change of paradigm in flood protection management (Greiving, 2003). Instead of only relying on a single dyke line, the area

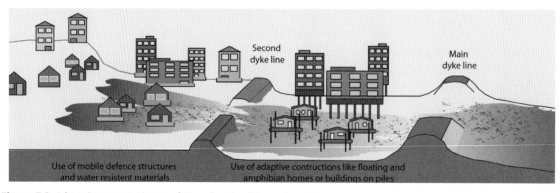

Figure 7.2 Adaptation strategy 'System of Cascading Flood Compartments'.
Source: Modified from Pasche et al., 2008.

behind the dyke becomes part of the adaptation strategy. Restrictions concerning building and usages and a new concept of 'living with water' complement the protection.

By conceptualizing adaptation strategies as a means of minimizing the vulnerability and supporting the resilience of Elbe Island, flexible infrastructures and space for water in case of a severe flood event are being discussed. Flexibility and robustness as attributes of resilience can be seen in different aspects of the 'system of cascading flood compartments'. With respect to flexibility, the flood protection system can be improved by a combination of primary dykes, secondary dyke lines, adapted building and preventive land use. According to these possibilities the 'system of cascading flood compartments' is more flexible than a single dyke line. When the sea level rise is more intensive than predicted a combination of main dyke line and flood compartments, as well as mobile protection elements in the dyke hinterland increases the flexibility and therefore the adaptive capacity of urban flood-prone areas. Buffer capacities, like water storage and polders, provide reserves for water in the case of an extreme flood (e.g. Fankhauser, Smith & Tol, 1999). In the above discussion about the robustness needed to withstand more severe storms and floods, the second dyke lines as well as constructional prevention have been emphasized. They enhance the robustness of the area against higher floods because adapted houses are more robust and less vulnerable against floods and the second dyke line protects more sensitive areas and provides more time for eventually required evacuations (Aerts et al., 2008). The type of building, the materials and the usage of the buildings are significant variables which can also have an influence on the adaptive capacity of the flood-prone area (Zevenbergen, 2007).

Hamburg's vision of a growing metropolis openly discusses the idea of living space on water. It points out that floating homes go along with the demand to live close to the water. Furthermore, the City of Hamburg aims at maintaining its maritime flair and image by implementing water-related types of living (FHH, 2006). Moreover, these water-related residential buildings could contribute to establishing a new identity of urban waterfronts on Elbe Island. Instead of separation from the water, the image could change to a new quality of 'living with water' by opening

and experiencing water areas in the city. According to Berkes and Jollys' (2001) notion of flexibility, living with a degree of environmental variability is of central importance for adaptive capacity. This criterion is integrated into the way people live with the different levels of water within the compartments and possible future risks. Therefore, the compartments presented offer multifunctional spaces; under normal conditions they can be used for living and during flood events convert into floodwater retention areas (Veerbeek et al., 2008). Thus, depending on water level, the compartments constitute adjustable elements for dealing with flood risks.

Furthermore, the cooperation of water authorities and house owners provides the potential to cope with flood risks and to react more flexibly to an uncertain future. Because of their significant potential for safeguarding buildings, private flood proofing measures like mobile walls and adapted building also contribute to resilient urban waterfronts (Veerbeek et al., 2008).

With regard to the ability to carry on in the face of disturbance, like extreme flood events, the presented strategy of a 'system of cascading flood compartments' is expected to be less vulnerable. Because the buildings within the flood compartments are to be adapted to high water levels, the damage will be lower than without the restrictions for buildings and usages behind the dykes. Nevertheless, the strategy is viewed critically by some local authorities because the population's trust concerning the main dyke line could be undermined. Furthermore, existing buildings in the area which have not been adapted sufficiently could lose value compared to flood adapted dwellings. This demonstrates that it is relevant to tackle seriously obstacles that can occur during the transition period of implementing the new strategy. Aspects like citizens' risk perception, acceptance of new types of housing, or willingness to accept the new orientation are relevant constraints on the implementation of such a 'system of cascading flood compartments' (Knieling, Schaerffer & Tressl, 2009).

7.6 Conclusion

Against the background of climate change waterfront cities are becoming more and more aware of new framework conditions for urban and regional planning and are exploring new approaches and solutions for adaptation. The impacts of climate change on waterfronts can be tackled in two different ways: first, regarding the growing threat of storm surges, it is possible to seal off areas from the sea by flood barrages and dykes; second, adapted settlement structures behind these dykes can provide additional flexibility and hence contribute to strengthening the resilience of cities. Both approaches have proven their value for storm surge endangered areas. This contribution's aim has been to describe potentials as well as restrictions of the second strategic approach.

The recent discussions about societies' handling of risks and about resilience of cities' respective regions refer to a more holistic management of climate risks reaching beyond mere technological approaches. Therefore, a change of paradigm is gaining attention as a main requirement for adaptation strategies: from mainly engineering-oriented flood protection to an integrated approach of risk management focusing on flood protection as well as on 'living with water' on the floodplain. Against this background, adaptation strategies for resilient waterfronts aim at flexible and robust systems, including an urban infrastructure which is constructed in a way that responds to the dynamics of nature. Water-related buildings, such as floating and amphibian houses, provide options that contribute to flood-secure settlement structures. Thus, the redevelopment of resilient waterfronts is confronted with the challenge of integrating the requirements of a flexible and robust urban structure to counter the uncertainties of future climate change.

In the case of Hamburg's Elbe Island, so far protection strategies have mainly been focused on a single dyke line. The newly developed 'system of cascading flood compartments' described above is expected to offer additional protection and an alternative paradigm of adaptation. It contains second and third dyke lines, a compartment structure providing flexible water storage capacity, and adaptive types of building. Furthermore, floating and amphibian homes as well as technical adaptation measures concerning single houses, such as removable walls, can be integrated into water-related settlement sites and provide future flood-secure living.

The existing instruments of spatial planning provide various possibilities to integrate elements of the described adaptation strategies, for example, to secure buffer areas for extreme weather events over long time periods or to rule specific quality standards concerning adaptive building. In any case, as far as planning instruments is concerned, the fixity of planning as well as of the built environment implies the disadvantage of being relatively static and, thus, of not meeting the requirement of flexibility. Further research should deal with this contradiction and consider how urban planning as well as building can integrate more elements of flexibility and dynamics. This includes the degree of responsiveness of settlement sites with regard to the expected changing requirements concerning future climate change.

In conclusion, on the one hand strategies like the 'system of cascading flood compartments' may contribute to a more appropriate understanding of future adaptation challenges and enhance the resilience of urban waterfronts. On the other hand, concerning the necessary shift of paradigms it seems to be of essential importance to analyse properly the dominant planning cultures and self-conceptions of flood adaptation in administration and politics. If the planning paradigm is requested to change from a safety-oriented to a risk management culture, the approach of 'living with water' can play a key role in the redevelopment of resilient water-related residential areas. Furthermore, in addition to flood protection barriers this approach offers the chance of floating homes and comparable water-related buildings which can initiate new urban identities with maritime flair and character. From this point of view, adaptation may not just be seen as a necessity but could become a potential for urban development regarding water-related sites. Further empirical research is required to analyse these potentials as the various restrictions deriving from regulative as well as emotional or cultural constraints.

To close, it has to be mentioned that the approaches described must also take into account the social environment of waterfront development. On the one hand, this final aspect underlines the complexity of urban development processes; on the other hand, it emphasizes the fact that the question of resilience is not only a technical or engineering issue, but needs a broad debate between both engineering and social sciences. In this field, there is quite a huge workload still to be handled as practice nowadays in many cases underlines.

References

Adger, W.N., Kelly, P.M. and Ninh, N.H. eds., 2001. *Living with Environmental Change. Social Vulnerability, Adaptation and Resilience in Vietnam*. London: Routledge.

Adger, W.N., Huq, S., Brown, K., Conway, D. and Hulme, M., 2003. Adaptation to climate change in the developing world. *Progress in Development Studies*, 3 (3), pp.179–195.

Aerts, J.C.J.H., Botzen, W., van der Veen, A., Krywkow, J. and Werners, S., 2008. Dealing with uncertainty in flood management through diversification. *Ecology and Society*, 13 (1), Art. 41.

Aerts, J.C.J.H., Botzen, W., Bowman, M.J., Ward, P.J. and Dircke, P., 2012. Introduction: Coastal cities and adaptation to climate change. In: Aerts, J., Botzen, W., Bowman, M.J., Ward, P.J. and Dircke, P., eds. 2012. *Climate Adaptation and Flood Risk in Coastal Cities*. New York: Earthscan, pp.1–8.

BauGB (Baugesetzbuch), 2006. Baugesetzbuch from 23.09.2004. BGBl. I: 2414. Changed by article 1 of the law from 21.12.2006. BGBl. I: 3316.

Beatley, T., 2009. *Planning for Coastal Resilience: Best Practices for Calamitous Times*. Washington, Covelo, London: Island Press.

Beck, U., 1986. *Risikogesellschaft – Auf dem Weg in eine andere Moderne*. Frankfurt am Main: Suhrkamp Verlag.

Berkes, F. and Jolly, D., 2001. Adapting to climate change: Social-ecological resilience in a Canadian Western Arctic community. *Conservation Ecological*, 5 (2), Art. 18.

BMBVS (Bundesministerium für Verkehr, Bau und Stadtentwicklung) ed., 2006. *Hochwasserschutzfibel – Bauliche Schutz- und Vorsorgemaßnahmen in hochwassergefährdeten Gebieten*. BMBVS: Bonn.

Bohle, H.-G., 2008. Leben mit Risiko – Resilience als neues Paradigma für die Risikowelten von morgen. In: C. Felgentreff and T. Glade, eds. 2008. *Naturrisiken und Soziale Katastrophen*. Berlin, Heidelberg: Spektrum, pp.435–441.

Bouwer, L.M., Bubeck, P. and Aerts, J.C.J.H., 2010. Changes in future flood risk due to climate and development in a Dutch polder area. *Global Environmental Change*, 20 (3), pp.463–472.

Dawson, R.J., Ball, T., Werritty, J., Werritty, A., Hall, J.W. and Roche, N., 2011. Assessing the effectiveness of non-structural flood management measures in the Thames Estuary under conditions of socio-economic and environmental change. *Global Environmental Change*, 21 (2), pp.628–646.

De Bruin, K., Dellink, R., Ruijs, A. et al., 2009. Adaptation to climate change in The Netherlands: an inventory of climate adaptation options and ranking of alternatives. *Climatic Change*, 95 (1–2), pp.23–45.

Egli, T., 2005. Vorsorge gegenüber Naturrisiken: Darlegung des Handlungsspielraumes am Beispiel der Hochwasservorsorge. In: H. Karl, J. Pohl and H. Zimmermann, eds. 2005. *Risiken in Umwelt und Technik. Vorsorge durch Raumplanung*. Hannover: Akademie für Raumforschung und Landesplanung, pp.64–66.

Fankhauser, S., Smith, J.B. and Tol, R.S.J., 1999. Weathering climate change: some simple rules to guide adaptation decisions. *Ecological Economics*, 30, pp.67–78.

Few, R., 2003. Flooding, vulnerability and coping strategies: local responses to a global threat. *Progress in Development Studies*, 3 (1), pp.43–58.

FHH (Freie und Hansestadt Hamburg), 2002. *Leitbild: Metropole Hamburg – Wachsende Stadt*. Hamburg: FHH.

FHH, 2003. *Fortschreibung des Leitbildes: Metropole Hamburg – Wachsende Stadt*. Hamburg: FHH.

FHH, 2005. *Sprung über die Elbe*. Hamburg: Freie und Hansestadt Hamburg, Behörde für Stadtentwicklung und Umwelt.

FHH, 2006. *Mitteilung des Senats an die Bürgerschaft: Schwimmende Häuser und Hausboote sowie Stellungnahme des Senats zu dem Ersuchen der Bürgerschaft vom 24. November 2004, 'Wohnen auf dem Wasser'*. [pdf] Available at: <http://boardhopper.de/bilder/sturm/Buergeschaftsbeschluss.pdf [Accessed 31 August 2012].

FHH, 2007. *Räumliches Leitbild – Entwurf*. Hamburg: Behörde für Stadtentwicklung und Umwelt.

Fischer, L., 2011. Küste – von der Realität eines mentalen Konzepts. In: L. Fischer and K. Reise, eds. 2011. *Küstenmentalität und Klimawandel – Küstenwandel als kulturelle und soziale Herausforderung*. Munich: oekom, pp.31–53.

Flesche, F. and Burchard, C., 2005. *Water House*. Munich: Prestel.

Folke, C., Carpenter, S., Elmqvist, T., Gunderson, L., Holling, C.S. and Walker, B., 2002. Resilience and sustainable development: Building adaptive capacity in a world of transformations. *AMBIO*, 31 (5), pp.437–440.

Füssel, H.-M. and Klein, R.J.T., 2006. Climate change vulnerability assessments: An evolution of conceptual thinking. *Climate Change*, 75 (3), pp.301–329.

Füssel, H.-M., 2007. Vulnerability: A generally applicable conceptual framework for climate change research. *Global Environmental Change*, 17, pp.155–167.

Garrelts, H., Lange, H. and Flitner, M., 2008. Anpassung an den Klimawandel: Siedlungsplanung in Flussgebieten. *RaumPlanung*, 137, pp.72–76.

Gill, S., Handley, J., Ennos, R. and Pauleit, S., 2007. Adapting cities for climate change: The role of the green infrastructure. *Built Environment*, 33 (1), pp.115–133.

Greiving, S., 2003. Im Hochwasserschutz ist ein Umdenken von der Gefahrenabwehr zum Risikomanagement erforderlich. In: I. Roch, ed. 2003. *Flusslandschaften an Elbe und Rhein. Aspekte der Landschaftsanalyse, des Hochwasserschutzes und der Landschaftsgestaltung*. Berlin: WFV, pp.129–143.

Gunderson, L.H., 2000. Ecological resilience – In theory and application. *Annual Reviews of Ecology and Systematics*, 31, pp.425–439.

Gunderson, L.H., 2003. Adaptive dancing: interaction between social resilience and ecological crises. In: F. Berkes, J. Colding and C. Folke, eds. 2003. *Navigating Social-ecological Systems. Building resilience for Complexity and Change*. Cambridge: Cambridge University Press, pp.33–52.

Handelskammer (HK) Hamburg, 2004. *Leben und Arbeiten im Herzen Hamburgs – Die Entwicklungsperspektive der Elbinsel*. Hamburg: HK Hamburg.

Hill, K., 2012. Climate-resilient urban waterfronts. In: Aerts, J., Botzen, W., Bowman, M.J., Ward, P.J. and Dircke, P., eds. 2012. *Climate Adaptation and Flood Risk in Coastal Cities*. New York: Earthscan, pp.123–144.

Immink, I., 2004. Established and recent policy arrangements for river management in The Netherlands: an analysis of discourses. In: B. Tress, G. Tress, G. Fry and P. Opdam, eds. 2004. *From Landscape Research to Landscape Planning – Aspects of Integration, Education and Application*. Dordrecht: Springer, pp.387–404.

IPCC (Intergovernmental Panel on Climate Change), 2007. *Climate Change 2007 – The Physical Science Basis. Working Group I Contribution to the Fourth Assessment Report of the IPCC*. Cambridge: Cambridge University Press.

Jeschke, A., 2004. Spatial planning as an integrative instrument in coastal protection management. In: G. Schernewski and T. Dolch, eds. 2004. *Geographie der Meere und Küsten*. Warnemünde: EUCC – The Coastal Union, pp.149–152.

Jolly, A., 2008. *Managing Climate Risk: A Practical Guide for Business*. London: Thorogood.

Kallmorgen, W., 1968. *Schumacher und Hamburg – Eine fachliche Dokumentation*. Hamburg: Christians.

Klijn, F., van Buuren, M. and van Rooij, S.A.M., 2004. Flood-risk management strategies for an uncertain future: Living with Rhine river floods in The Netherlands. *AMBIO*, 33 (3), pp.141–147.

Knieling, J., Schaerffer, M. and Tressl, S., 2009. *Klimawandel und Raumplanung – Flächen- und Risikomanagement überschwemmungsgefährdeter Gebiete. Beispiel der Hamburger Elbinsel*. Warnemünde: EUCC – The Coastal Union.

Kron, W., 2003. Hochwasserrisiko und Überschwemmungsvorsorge in Flussauen. In: H. Karl, J. Pohl and H. Zimmermann, eds. 2003. *Raumorientiertes Risikomanagement in Technik und Umwelt. Katastrophenvorsorge durch Raumplanung*. Hannover: Akademie für Raumforschung und Landesplanung, pp.79–101.

Leven met water, 2012. *Bouwen met Water*. [pdf] Available at: <http://www.levenmetwater.nl/projecten/bouwen-met-water/ [Accessed 31 August 2012].

LGV (Landesbetrieb Geoinformation und Vermessung), 2009. Stadtkarte. LGV G1-121, 1: 60000, Hamburg: FHH.

McGranahan, G., Balk, D. and Anderson, B., 2007. The rising tide: assessing the risks of climate change and human settlements in low elevation coastal zones. *Environment and Urbanization*, 19 (1), pp.16–37.

Newman, P., Beatley, T. and Boyer, H., 2009. *Resilient Cities – Responding to Peak Oil and Climate Change*. Washington, DC: Island Press.

Overbeck, G., Hartz, A. and Fleischhauer, M., 2008. Ein 10-Punkte-Plan, 'Klimaanpassung'. Raumentwicklungsstrategien zum Klimawandel im Überblick. *Informationen zur Raumentwicklung*, 6/7, pp.363–380.

Pasche, E., Ujeyl, G., Goltermann, D., Meng, J., Nehlsen, E. and Wilke, M., 2008. Cascading flood compartments with adaptive response. In: D. Proverbs, C. Brebbia and E. Penning-Roswell, eds. 2008. *Flood Recovery, Innovation and Response*. Southampton: WIT Press, pp.303–312.

Resilience Alliance, 2009. *Key Concepts: Resilience*. [online] Available at: <http://www.resalliance.org/576.php [Accessed 31 August 2012].

Schubert, D. and Harms, H., 1993. *Wohnen am Hafen. Leben und Arbeiten an der Wasserkante. Stadtgeschichte – Gegenwart – Zukunft. Das Beispiel Hamburg*. Hamburg: VSA-Verlag.

Simons, J., 2008. Wie will waar waterwonen? *Waterwonen*, 1, pp.13–16.

Smit, B. and Wandel, J., 2006. Adaptation, adaptive capacity and vulnerability. *Global Environmental Change*, 16, pp.282–292.

Smit, B., Burton, I., Klein, R.J.T. and Wandel, J., 2000. An anatomy of adaptation to climate change and variability. *Climatic Change*, 45, pp.223–251.

Storch, H. von and Claussen, M. eds., 2011. *Klimabericht für die Metropolregion Hamburg*. Berlin: Springer.

Veerbeek, W., Gersonius, B., Zevenbergen, C., Pyan, N., Billah, M.M.M. and Fransen, R., 2008. *Urban Flood*

Management Dordrecht. Proceedings of Workpackage 3: Resilient Building and Planning. Rotterdam: UfM Dordrecht.

Woth, K., Weisse, R. and von Storch, H., 2005. *Dynamical modelling of North Sea storm surge extremes under climate change conditions – an ensemble study.* [pdf] Geesthacht: GKSS. Available at: <http://www.hzg.de/imperia/md/content/gkss/zentrale_einrichtungen/bibliothek/berichte/ gkss_berichte_2005/gkss_2005_1.pdf [Accessed 28 August 2012].

Zevenbergen, C., 2007. *Adapting to Change: Towards Flood Resilient Cities.* Inaugural Address of Chris Zevenbergen, Professor of Flood Resilience of Urban Systems at UNESCO-IHE Institute for Water Education in Delft. The Netherlands.

8 Climate Change Adaptation Policy in Bergen: Ideals and Realities

O. Langeland, J.E. Klausen & M. Winsvold

Norwegian Institute for Urban and Regional Research (NIBR), Norway

8.1 Introduction

This chapter focuses on climate change adaptation in Bergen, a city with a rather ambitious and far-reaching climate change adaptation policy. It discusses the extent to which the Bergen adaptation policy can be said to be coherent and comprehensive and integrated into other municipal policy fields. In order to do so an examination of the comprehensiveness of the city's climate policy has been accomplished and potential blind spots have been identified. Moreover, it examines the climate projects in which Bergen participates to see if they are sufficiently coordinated to achieve synergy effects, that is to see if all relations between existing projects are taken into consideration in order to achieve learning and knowledge transfer and if supplementary coordination has been needed. Finally, it was assessed whether all relevant actors and stakeholders have been included in the process so as to ensure participation and legitimacy.

There are probably several reasons why Bergen has an active climate change adaptation policy. The exposed location of the city at the west coast of Norway, directly facing the North Sea, and several extreme weather events over the past few years have undoubtedly increased awareness related to the possible future impacts of climate change (Amundsen, Berglund & Westskog, 2010; Hjeltnes, 2011) and also speeded up the implementation of adaptation measures. The Bergen region is characterized by fjords, mountains and islands and its population is used to heavy rain, strong wind, river flooding, landslides and high waves. Bergen is the wettest city in Europe and annual precipitation is expected to increase over the next decades. According to climate change scenarios, challenges pertaining to landslides, river flooding and extreme weather events are predicted to intensify (Hanssen-Bauer et al., 2009). Bergen may also have to cope with sea level rise and storm surges during the century. If the worst scenarios prove right sea level rise and storm surges may have devastating impacts on the city's infrastructure, transport system and tunnels, buildings and sewage system and cultural heritage sites. In particular, the fact that the city's world heritage site, Bryggen, the old wharf of Bergen, which mirrors the cultural identity of the city, is threatened by sea level rise is important for increasing awareness and the will to act.

Other explaining factors regarding implementation of climate change adaptation measures may be that Bergen as a large city has sufficient resources (financial, human etc.), capacity to use external expertise, is highly involved in different networks related to climate change issues and has dedicated individuals in political and administrative positions to take care of climate issues (Hjeltnes, 2011; Dannevig, Rauken & Hoversrud, 2012). The city takes part in many climate-related projects on the local, regional, national and international levels. Bergen has also organized its climate policy in a way that is supposed to ensure internal and external coordination of the several projects

Climate Change Adaptation in Practice: From Strategy Development to Implementation, First Edition.
Edited by Philipp Schmidt-Thomé and Johannes Klein.
© 2013 John Wiley & Sons, Ltd. Published 2013 by John Wiley & Sons, Ltd.

and activities in which the city is involved. All these factors may contribute to the city's *urban capability*[1] so it can deal with climate change issues in a more efficient manner. In sum, exposed position, negative experiences and large capabilities together with sufficient political will, may explain why Bergen seems to be at the forefront on climate change adaptation policy in Norway.

The case study is based on several data sources. Primary data is collected from interviews and workshop discussions with representatives from the municipality, the business sector, trade organizations and environmental organizations. In addition, secondary data from local, regional and national plans and documents have been used. The regional climate change scenarios are based on scientific reports (Hanssen-Bauer et al., 2009) and official white papers (Det Kongelige Miljøverndepartement, 2012).

The chapter is organized as follows: The next section presents a brief history of the city and the socio-economic context. The third section gives an overview of the geography of the city and outlines the main challenges related to climate change. The fourth section presents the main purposes for the BaltCICA project in Bergen focusing on knowledge transfer, learning and coordination. The foresight approach is presented followed by an overview of climate strategies and adaptation measures including projects, plans and documents, key actors and agencies as well as stakeholder involvement. The last section concludes the chapter and summarizes key messages and issues from the Bergen case. It looks into what lessons have been learnt, the transferability of results and findings, strengths and weaknesses of adaptation strategies, projects, governance arrangements and forms of interactions.

8.2 History, context and conditions

Bergen was founded early in the year 1000 AD (Hansen, 2005). The city has always been an interna-

tional shipping and trade city and towards the end of the 13th century, Bergen became one of the Hanseatic League's most important bureau cities (Kloster, 1952). From around year 1000 until 1830 Bergen was the largest city in Norway (Hartvedt, 1999). Today Bergen is the second largest city in Norway with a population of 264 000 inhabitants and the city is the administrative centre of Hordaland County. Bergen Metropolitan Area has a population of approximately 383 000 (Statistics Norway, 2012). Together with the other large cities in Norway Bergen has experienced a relatively strong population growth during the past decade. The population in the city region grew from approximately 332 000 in 2000 to 383 000 in 2011 (Statistics Norway, 2012). The population density in the city region is 568/km^2 and in the city it is 2550/km^2.

Bergen specializes in oil and gas manufacturing, and knowledge-intensive services, particularly in financial and business services. Bergen international airport is the main heliport for the huge Norwegian North Sea oil and gas industry. The city also has a stronghold in maritime businesses and activities such as aquaculture and marine research, with the Institute of Marine Research (IMR) as the leading institution. Bergen's inter-municipal harbour is by far Norway's largest port and one of Europe's largest ports, according to the inter-municipal company Port of Bergen (BOH, 2012). The city promotes itself as the capital of Western Norway and the 'Gateway to the Fjords of Norway'. Therefore, tourism, and particularly cruise tourism is an important source of income for the city. Bergen is also an important cultural hub in its region. The city was one of nine European cities holding the title of European Capital of Culture in the Millennium year, and it hosts the headquarters of TV 2, Norway's largest commercial television channel.

Bergen has a parliamentary form of government. A seven-member City Government holds the executive power, but relies on support from a majority among the members of the elected City Council. Each of the seven City Government commissioners is in charge of a City Department covering one specific policy area. The City Department of Climate, Environment and Urban development is responsible for urban development, environmental affairs, climate, cultural heritage, roads and transport, water and sewerage and social housing. In 2008 a specific climate section

---

[1] Urban capability is used as a synonym to the concept 'dynamic capability' (Teece, 2000; 2009) and it refers to how cities utilize their resources, organize knowledge transfer and learning processes, practice their different governance modes and exercise knowledge management.

consisting of seven employees was established within this department. The climate section coordinates the work on climate, environment and energy in the municipality and plays the main role in Bergen's active climate policy (Bergen Kommune, 2012).

Situated in a narrow space between the mountains and the sea, central Bergen has limited space for expansion and development. The main parts of economic, social and cultural activities are concentrated in the city centre, and population growth and increasing tourism activities put heavy pressure on green areas in the centre, particularly in the hillsides and on the coast line. The concentrations of buildings and activities close to the sea and the mountains make the city very exposed to flooding from sea level rise and to landslides from the mountain side. The topography of the city and the existing road network also imply that significant traffic goes through the city centre which is characterized by narrow streets, medieval traits and historical buildings. This results in air pollution, CO_2 emissions and reduced quality of living for the city residents.

8.3 Geography and climate challenges

Bergen municipality occupies the majority of the Bergen peninsula and it covers an area of 465 km^2. The city is sheltered from the North Sea by several islands, see Figure 8.1. The city centre is situated among a group of mountains known as the seven mountains but only a few of them are located within the borders of the Bergen municipality. The landscape is very hilly with little continuous lowland. Both the city centre and the city districts are surrounded by mountains and sea. Half of the area is situated more than 160 meters above sea level. Around 40% is covered by forest, 23% is densely populated area, 4.4% is fresh water and 3.5% is arable land (Bergen Kommune, 2012).

Bergen has a temperate oceanic climate with relatively mild winters and cool summers. The annual mean temperature is 7.7 °C and the temperature changes relatively little throughout the year. This makes Bergen one of the warmest cities in Norway.

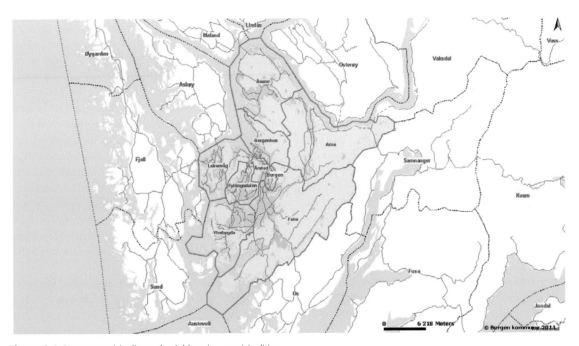

Figure 8.1 Bergen municipality and neighbouring municipalities.
Source: Bergen municipality

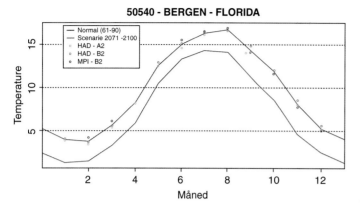

Figure 8.2 Annual mean temperature in Bergen 2071–2100.

Source: Norwegian Meteorological Institute, 2012.

The effect of the Gulf Stream used to be the usual explanation for this, but recent research indicates that warm wind from the west may be more important than warm water from the Gulf Stream (Seager, 2006; Kirkebøen, 2012).

Scenarios indicate that the annual mean temperature in western Norway from the late 1990s to 2100 will increase between 1.9 and 4.2 °C and generally this will occur in the winter time (Hanssen-Bauer et al., 2009). Figure 8.2 shows changes in annual mean temperature for Bergen in the late part of the century, the period 2071–2100 (red line) compared to the reference period 1961–1990 (green line). All projections in the scenario show that the weather will be warmer, and that in the winter time the temperature will increase from close to zero to up to 4–5 °C.

Precipitation is also expected to increase towards the end of the century with more than 40% in western Norway, particularly in autumn and winter. Bergen has already got an annual precipitation of 2250 mm, but precipitation varies a lot within relatively short distances in the city, from approximately 1500 mm to 3500 mm per year.

An increase in days with heavy rainfall is also expected in western Norway, where Bergen is situated. Table 8.1 shows the amount of precipitation within a medium, low and high scenario. Annual changes for this region in the medium scenario indicate that there will be approximately 80% increase in days with heavy rainfall, and that in the autumn there will be more than twice as many days with heavy rainfall as in the reference period. According to the medium scenario the amount of precipitation in days with heavy rainfall will increase approximately 15%

annually with an increase of almost 20% in autumn. Such an increase of days with heavy rainfall may increase the vulnerability related to flooding and landslide considerably.

Being located between mountains with rather steep hills, the city is highly exposed to landslides. These may occur when the slope of the hill is over 30 degrees, and in Bergen large parts of the hillsides are more than 40 degrees. Increased precipitation, and particularly heavy rainfall, accumulation of water and processes of freezing and defrosting and human activities related to digging and fill in on the mountain side are the most important releasing causes for landslides. A lot of settlements, houses and institutions are situated in the hillsides of Bergen. In the urban area Fjellsiden (the mountain side) approximately 100 houses are estimated to be exposed to landslide (Farstad, Olsen & Ingvaldsen, 2012). In addition, schools and different public institutions such as nursing homes, kindergartens and sport grounds are exposed in the same area (Bergen Kommune, 2012). Figure 8.3 shows the risk zones for landslide in central Bergen and in the larger city region. The black dots indicate rock fall from mountains and as can be seen both the central city area and many of the fjords in the Bergen region are very exposed to such slides. There have been two landslide accidents with fatalities in the city in past years.

According to Vasskog, Drange and Nesje (2009) the sea level is expected to rise by approximately 75 cm in Bergen during the 21st century. In addition to the estimated sea level rise, the storm surge in Bergen may increase up to 221–276 cm by the end of the century. If the estimated sea levels rise and the storm

Table 8.1 Relative changes (%) in number of days with heavy rainfall and relative change (%) in precipitation amount in days with heavy rainfall in western Norway, period 1961–1990 to 2071–2100. Medium, low and high scenario

Region	Season	1961–1990 to 2071–2100 Changes (%) in number of days with heavy rain			1961–1990 to 2071–2100 Changes (%) in amount of precipitation on days with heavy rain		
		M	L	H	M	L	H
Sunnhordland og Ryfylke	Annual	79.9	24.3	175.7	14.2	3.2	26.1
(Western Norway)	Winter (DJF)	89.8	24.9	204.9	11.9	−1.4	28.2
	Spring (MAM)	81.0	5.4	186.8	14.1	−0.4	31.4
	Summer (JJA)	47.3	−5.8	93.5	10.8	−2.4	22.3
	Autumn (SON)	117.2	36.3	231.9	18.5	5.3	34.3

Source: Hanssen-Bauer et al., 2009.

surge proves right, it will have significant impacts on the city's infrastructure, transport system, tunnels and sewage system and cultural heritage sites. The predicted sea level rise will also inundate buildings related to settlements and industries, historical sites, quays and port facilities and fish farms and wetlands. Most port facilities in the region will be flooded and useless. Also a sea level rise of 75 cm will make most of the quays unfit for mooring many vessels and thereby hit both the domestic and cruise traffic hard. Therefore, several adaptation measures have been discussed in order to prevent damaging impacts. These include

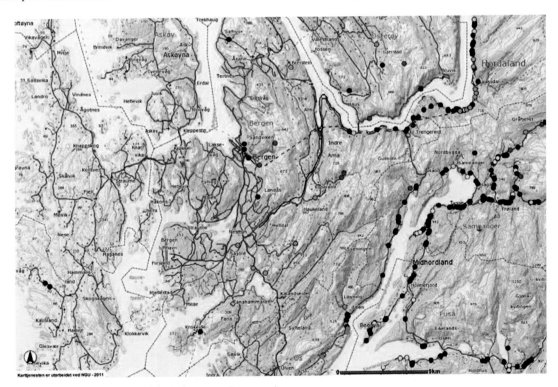

Figure 8.3 Areas exposed to landslides in the greater Bergen region.
Source: Geological Survey of Norway.

movable tide barriers, local barriers alongside the quay and up-lift of critically situated buildings (Grieg Foundation, Visjon Vest & G.C. Rieber Fondene, 2009).

A newly published report from Simpson et al. (2012) indicates less sea level rise than the report from Vasskog, Drange and Nesje (2009), see Figure 8.4. Due to increased land uplift the sea level rise in Bergen at the end of century is expected to be only about 20 cm according to calculations based on IPCC's climate models. If this scenario is correct, sea level rise is not a big future threat for Bergen. However, Simpson et al. (2012) also explore a high-end scenario in order to define an upper bound on 21st century regional sea level change for Norway. In this scenario sea level rise in Bergen may vary from 25 up to 114 cm and if the highest level occurs the problems deduced from the prediction in Vasskog, Drange and Nesje (2009) may again be real. Due to large uncertainties in the scenarios and in order to be prepared also for worst case scenarios, the expected climate changes also seem to call for an active climate policy for Bergen in the coming years.

These observations have been presented in order to direct the reader's attention to the key climate related issues that Bergen is expected to face during the course of the century. We now turn our attention to the BaltCICA project, which was initiated to support the adaptive capacities of the participating cities and regions.

8.4 Knowledge transfer, learning and coordination

The main aim in the BaltCICA project has been to work with local and regional partners in order to make regions and municipalities better able to cope with a changing climate, that is, to contribute to increased urban capability and adaptation capacity. Since municipalities are playing a main role in climate change adaptation in Norway (CICERO, 2011; Harvold et al., 2010), and particularly in areas such as land-use planning and water supply, it was important to focus on the political-administrative apparatus in Bergen and on how the city had organized their climate change policy, internally and externally. This implied, amongst other factors, focusing on organizational learning (Levitt & March, 1988; March, 1991) in order to understand if and how the municipality could learn from experiences and how adaptation measures were developed and implemented and by which governance modes.

Although the BaltCICA project had an aim of contributing to the development of adaptation strategies it soon became apparent that Bergen's existing climate policies were already quite ambitious. The city was engaged in so many projects and climate-related activities that it could probably not manage any more in the short term. As a consequence, the BaltCICA project

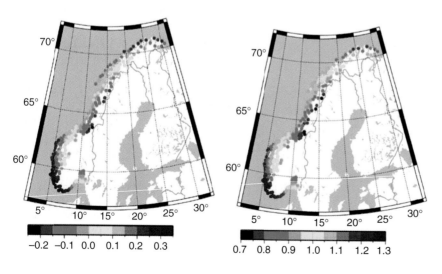

Figure 8.4 Estimated sea levels along the Norwegian coast in the period 2090–2099 in relation to 1980–1999. IPCC-based scenario (left) and high-end scenario (right). *Source*: Simpson et al., 2012.

refrained from suggesting new projects or implementing new adaptation measurements. The city's high level of activity was seen as quite demanding in terms of *coordination* of the various activities, and the challenge was how to achieve *knowledge transfer and learning* between actors, sectors and levels.

Coordination turned out to be the single most important issue when assessing Bergen's climate change adaptation strategies. Winsvold, Stokke, Klausen and Saglie (2009, p.476) distinguish between three forms of coordination, related to three different modes of governance: 'a hierarchical mode in which coordination of different actors is ensured through formal regulations, command and control; a market mode in which coordination of actors is ensured by the price mechanism of the market; and a network mode of governance in which coordination ideally happens through arguing and bargaining among the involved actors'. The main focus in the Bergen study was on the hierarchical mode and the network mode since these two types seem to be the most ones used most frequently.

8.4.1 Climate adaptation policy in Bergen – a foresight approach

A foresight approach[2] to climate change adaptation was used in the BaltCICA project. Foresight is used in futures studies and often describes activities such as critical thinking concerning long-term developments, debate and effort to create wider participatory democracy and shaping the future, especially by influencing public policy (Georghiou et al., 2008). Foresight is *future-oriented*, it is a *participatory* and dialogue-based approach and it should be *policy relevant* for planning purposes and priority setting. Scenario methods are widely used in foresight analyses in many countries for policy-making in different fields, for instance within an industry or technology, in a territory as in regional planning and decision-making ('regional foresight') or in the field of climate change adaptation policy. In Bergen the foresight process focused on a broad range of adaptation problems, but questions of economic and legal responsibility for climate change damages was particularly illuminated.

The use of foresight in climate change adaptation processes has a lot in common with the participatory vulnerability assessment approach (Smit & Wandel, 2006) that has been implemented in various local contexts, including the Canadian arctic (Ford & Smit, 2004), Samoa (Sutherland, Smit, Wulf & Nakalevu, 2005) and the US–Mexico border (Vásquez-León, West & Finan, 2003). According to Sutherland, Smit, Wulf & Nakalevu (2005), participatory vulnerability assessment requires 'intensive participation of community members to identify climatic conditions relevant to the community and to assess the effectiveness of adaptive strategies' (p.12). Whereas the approach as presented in these studies has been put to use not least in the context of rural or indigenous communities, our study brings fairly similar ideas to bear on a highly developed Western European urban setting. There seems to be an increasing recognition about the importance of cities for climate change adaptation (Betsill, 2001; Kousky & Schneider, 2003; Bulkeley, 2010). Yet the inherent complex and many faceted nature of climate change adaptation politics calls for a correspondingly nuanced approach to politics and governance, not least as regards the position of cities in networks and as multilevel governance systems (Eakin & Lemos, 2006; Monni & Raes, 2008; Seto, Sánchez-Rodríguez & Fragkias, 2010). The foresight approach is useful not just for identifying vulnerabilities but also for linking this analysis with political processes and actual governing.

The various partners in BaltCICA used this approach for various purposes, including mapping, data collection, information transmission and interaction – in some cases, with actual decision-making following suit, in other cases as parts of planning or strategy building exercises. As stated, our ambition in Bergen was not to initiate new plans and decisions, rather to consolidate and connect the several existing processes. Following this, the BaltCICA foresight activities had a focus on learning and coordination. This section will describe how these activities proceeded, conceptualized with a five-stage generic foresight model as illustrated in the model in Figure 8.5.

The first stage in the Bergen study was to define problems, see how the climate policies were organized and which actors were involved. The expected result from this process was to get an overview of available resources and adaptation strategies.

---

[2]A more detailed description of the foresight approach is given in Klausen, Winsvold and Langeland (in press).

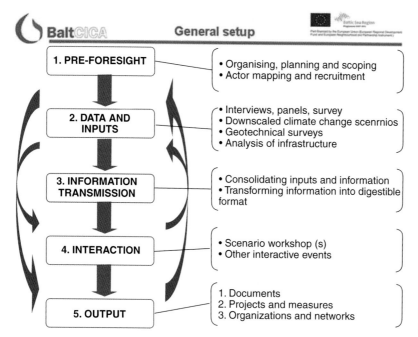

Figure 8.5 The BaltCICA general setup for adaptation foresight.

A survey carried out in 2009 revealed that adaptation primarily focused on climate problems already affecting the city, such as flooding by river, precipitation and landslide. Sea level rise and extreme weather events were important issues on the adaptation agenda in the longer term. Blind spots in Bergen's climate change adaptation strategy were also identified at this stage. It turned out that the distribution of economic and legal accountability for climate related damages was largely undetermined. As already presented earlier in the chapter several actors and stakeholders are involved in various projects and networks in the climate adaptation strategies in Bergen. They represent private sector interests in urban development such as land owners, developers, house owner organizations, residents' associations and insurance companies and, public actors from local, regional and national levels and, actors from the research community. Although Bergen city has a high activity level regarding climate change adaption it also became apparent that the awareness of the need for climate change adaptation was not very high among all stakeholders. Surveys and meetings indicated that it was a challenge to some stakeholders, such as developers, to make them think about what consequences climate change might have in general and for their own organiza-

tion. This finding is in line with Berkhout, Hertin and Gann (2006), who point out that business organizations may face several obstacles in learning how to adapt to climate change impacts. That could be due to weak and ambiguous signals of climate change and to uncertainty about the benefits of different adaptation measures. The result may be a wait-and-see strategy or that the organization tries to 'externalise risks associated with climate change through insurance and collaboration' (Berkhout, Hertin & Gann, 2006, p.151).

8.4.2 Climate strategies and adaptation measures

To get an overview of the climate strategies and adaptation measures in Bergen a review of research reports on Bergen City's climate adaptation projects, municipal plans and risk and vulnerability analysis (RVA) was carried out. Bergen was the first municipality in Norway to work out a climate plan in year 2000. The plan has been revised several times since then, the last time in 2009. The main aim in the year 2000 plan was to reduce greenhouse gas emissions including CO_2. This target is still valid. In 2007, environment and energy were implemented in the land use part of the municipality master plan. Reduction in greenhouse gas and use of energy is thereby included in the land use and transport policy of the municipality (Lyng, 2009). Adaptation has gradually become a more important part of the climate issues in Bergen during the last decade. Some specific events have contributed to this, particularly in autumn 2005 when four people died in two landslides and a heavy flooding in the Nestun River which caused severe damage. Repeatedly storm surges on Bryggen also added to an increased awareness of extreme weather events and, importantly, these events were related to changes in the climate, which underscored the need for climate change adaptation strategies and measures (Hjeltnes, 2011).

Several stakeholders have been involved in the preparation of the climate plan, such as municipal agencies, municipal companies, the County Governor, the County Government and civil organizations. The work has been organized in different working groups focusing on the topics of land use and transport, stationary energy, consumption pattern and waste. Together with adaptation to climate change these are now the main issues in the plan. Bergen has also been an active participant in the process of developing a new climate plan for Hordaland County in which the city is situated (Hordaland Fylkeskommune, 2012a). This plan which was decided upon in 2010 is a regional plan for 2010–20 and it is founded on the National Planning and Building Act. The plan focuses on emission of greenhouse gases, energy and adaptation to climate change and is accompanied by its own action programme.

Bergen city has implemented several adaptation and mitigation measures over the past few years. The climate change projects in which the city participate aim at building knowledge about local climate models, registration of climate vulnerability and drawing up forecasts and scenarios to help identify what adaptation measures will be necessary in the Bergen area. This work is linked to national and international research communities. The Bjerknes Centre for Climate Research which is located in Bergen is a particularly important scientific partner for the city in its climate policy work. In recent years incidences of extreme weather have resulted in loss of human life and material assets. Risk and vulnerability assessment, therefore, have been carried out in connection with the land-use part of the municipal master plan. The RVA analyses have been focusing on the risk of floods,

powerful winds, high tides, large waves, extreme precipitation and earth and rock slides in Bergen. This knowledge is employed to reduce the potential consequences of accidents and disasters, and it also plays an important role in urban planning and in the processing of building applications. Different types of risk maps have been drawn up, mapping, for example, local precipitation, floods, water levels and land-slide risk areas in the whole of the city. These maps are used in planning processes and developments to reduce the risk of unforeseen incidents.

The most important adaptation measures which are implemented and planned for are listed below:
- RVA;
- Information campaigns;
- Methods and guidelines for stakeholders;
- Measures for reducing damaging from sea level rise and flooding;
- Measures to protect fresh water;
- Measures to protect cultural heritage.

The most important plans for the next three years are:
- Risk management plans;
- Standards for building new infrastructure.

8.4.3 Climate strategies – networking and projects

Networking and involvement in projects are significant parts of Bergen's climate strategies. The city participates in several national and international networks on climate change adaptation. At the regional county level there is a Regional Climate Council and a Climate Network consisting of representatives from Hordaland County, Bergen municipality and the other municipalities in Hordaland County. The Climate Council is a meeting place for politicians from the County, a forum for discussing common climate measures and to exchange experiences and ideas so as to achieve synergy effects across municipalities and administrative levels (Hordaland Fylkeskommune, 2012b). The Climate Council is a continuation of the former Regional Climate Panel headed by the City of Bergen which had representatives from the regional council and Business Region Bergen. The Regional Climate Panel has been the driving force behind the region's climate adaptation efforts, and addressed issues such as regional climate work, transport analyses and climate challenges that represent a potential for new commercial development. The panel also makes up a Learning and Action Alliances (LAA) in the Interreg project MARE in which Bergen participates (MARE, 2012). The LAA is a social learning framework for integrating climate change adaptation better into urban planning in a context characterized by very long-term planning horizon and multi-actor settings and governance levels (Ashley et al., 2011).

Bergen also participates in a Climate Forum which is a meeting place for actors from the business community, authorities, organizations and research institutions. The aim of the forum is to increase dissemination of knowledge on climate change and adaptation policies in Bergen and Western Norway by building bridges between the research community, the business community and society at large. On the national level Bergen is part of the programme Cities of the Future which is a cooperation between the 13 largest Norwegian cities and the government (Regjeringen, 2012). In addition to climate change adaptation, the network focuses on greenhouse gas emission reduction in land use and transport, consumption and waste and energy and buildings. Bergen also participates in several international climate networks. The city is a member of the European Climate Forum (EFC). It has signed The Covenant of Mayors, an initiative of the European Commission for cities to reduce their carbon dioxide emissions, it is a member of the organization United Cities and Local Governments (UCLG) aimed at reducing greenhouse gas emissions, and part of the International Council of Local Environmental Initiatives (ICLEI), which works for sustainable development.

In order to acquire knowledge on climate change and possible adaptation measures, Bergen city participates in numerous national and international research projects and municipality-led projects on climate change adaptation. At the national level the city has participated among other things in NORADAPT (CICERO, 2011). This project maps the vulnerability of climate change in eight Norwegian municipalities, and it seeks to work out individual adaptation strategies for each municipality. Bergen has also collaborated with Stavanger and Kristiansand on the energy solutions of the future through the EnergiMiljø i Sørvest project (Berrefjord & Thomassen AS, 2010). These cities wish to take the lead in developing sustainable and environmental efficient transport solu-

tions and forward-looking use of energy. In addition to the research projects, Bergen carries out several local projects focusing on infrastructure, water supply, wastewater management and natural watercourses. The city also has a particular school projects related to waterway management, and a project focusing on climate change and human rights.

At the international level Bergen city is a partner in the Interreg project MARE (MARE, 2012) which focuses on implementation of local adaptation measures to reduce and adapt to flood risk in the North Sea region. Bergen was also a case study in the large EU-project Techneau (TECHNEAU, 2011) focusing on technology solutions for drinking water supply. The city is an associated partner in the Espon Climate project which focuses on climate change and territorial effects on regions and local economies (ESPON, 2012) and, in the Interreg project BaltCICA which focuses on climate change adaptation in the Baltic Sea Region.

8.4.4 Stakeholder consultations on blind spots

The BaltCICA project gathered information through a series of meetings with municipal staff and through interviews with stakeholders representing different sectors and interests in urban development. An important aim for this data collection was to use the information as input to sort out if and how the city would handle blind-spot issues on legal and economic responsibility which was detected as a blind spot in the early stage of the foresight process. The municipality and various stakeholders discussed climate change impacts and tried to assess the consequences of climate change with regard to the following aspects: legal, economic, knowledge, competence, capacity, resources, cooperation, coordination and responsibility, and to come up with ideas for measures and solutions to the perceived problems. The interaction between the municipality and stakeholders on economic and legal responsibility aimed at making stakeholders in urban development affected by climate change *aware* of questions relating to economic responsibility, risk and cost; and at making stakeholders from different sectors *discuss* among themselves how they could better cooperate in order to meet the hazards posed by climate change to the built environment.

8.4.5 Legal rules and behaviour

The discussions on legal and economic aspects concentrated on the need for legal rules and how the insurance industry could have an influence on climate change adaptation. Legal rules, insurance premiums and investments can motivate human behaviour to act in accordance with a climate-friendly development, both at an individual and at a collective level. The main functions of legal rules are to contribute to preferred behaviour, to solve conflicts of interests and to ensure that certain values are taken care of (Hall, 2002). Laws should make a predictable framework by telling who is responsible for what, and what the responsibility consists of. On the climate field legal rules are supposed to take care of general values such as resilience in order to ensure that climate damages are reduced as much as possible and, to contribute to economically efficient solutions and a just cost sharing if and when damages occur.

Legal rules may imply a duty to prevent or hinder climate damages (ex ante) or regulate responsibility and compensation when damages happen (ex post). Ex ante rules may include claims on RVA, building codes and so forth. RVA are compulsory for all municipalities, and areas identified as vulnerable cannot be used for real estate development, according to land-use plans. Ex post rules focus on who is responsible for what and what can be compensated and how. Ex post and ex ante rules are interlinked in the way that ex ante duties normally will set off duties ex post.

Legal rules may also comprise claims on insurance, particularly for high-risk events and large uncertainty as with climate change damages. The insurance industry may be significant for climate change adaptation because insurance premiums induce investors to take climate exposure into account (Mills, 2003). The insurance industry also invests several hundreds of billions kroner each year in different projects and is thereby in a position to indirectly affect the climate. Green investment products can reduce damages and create incentives for climate-friendly actions. Preventing damages, particularly fire damages, will reduce CO_2 emissions significantly. Insurance premiums may hamper or stimulate specific investments in housing and industrial building to the benefit of the climate and also invest in firms that develop new climate technology (Mills, 2009).

To sum up: Sorting out who is responsible for planning, financing and implementing adaptation measures in order to prevent climate change damages and, who is responsible for compensating for such damages if and when they occur, is of vital importance for carrying out a comprehensive and efficient climate policy. As long as such questions are not adequately clarified, adaptation measures may prove difficult to implement. Uncertainties regarding who should pay for which adaptation measures (ex ante), what should be compensated, and by whom, when climate related damages occur (ex post), is an obstacle for carrying out climate change adaptation strategies. Legal rules that contribute to *resilience* and a *fair sharing* of costs related to both preventing and compensating for negative impacts of climate change may help cities overcome such obstacles.

8.5 Climate adaptation policy in Bergen – coherent and comprehensive?

Bergen has an ambitious climate policy but is the city able to fulfil its visions and is the policy coherent and comprehensive enough? It would exceed the range of this chapter to assess this policy in any wide-ranging manner, but we can raise some questions which may help us to understand better how it works. As mentioned, Bergen is involved in several projects and activities, some large international projects and several national and local projects. The city also makes use of different coordinating mechanisms or modes of governance in its climate policy. Climate policy is a very complex issue and the policy is organized along both vertical and horizontal lines. Traditional hierarchical regulation is used for internal coordination alongside networking activities for external coordination and knowledge transfer. The question is how this administrative architecture functions – is the internal coordination between different departments satisfactory, do different projects interact with each other so as to increase learning, do external networks promote learning and awareness rising of common interests and transfer of knowledge between relevant stakeholders?

Internally the climate policy should be coordinated by the climate department, but there may be several rocks in this ocean. Hierarchical organization may be suitable for internal coordination within departments, but it can be an obstacle for horizontal coordination between departments (Winsvold, Hjeltnes, Klausen & Langeland, in press). Bureaucratic barriers may sometimes be a larger hindrance than lack of resources and capacity. The departments in Bergen municipality are aware of this danger and they emphasize the need for inter-departmental coordination and close communication between, for instance, the planning department and the water department in the climate change policy (Hjeltnes, 2011). However, these departments are cornerstones in the municipal administration with their own distinct culture and, it is probably not that easy to bring together such strong actors and ensure that they adapt to a common policy. Hierarchical organization is not always the best way to utilize local or tacit knowledge either (Lam, 2000), and such knowledge is often a prerequisite for climate change adaptation (Berkes & Jolly, 2001). Solving complex issues requires inputs of various forms of knowledge as well as an environment which stimulates collective learning. These features are not among the normal characteristics of hierarchies (Winsvold, Stokke, Klausen & Saglie, 2009). The municipality's extensive networking through panels and forums provides it with various knowledge and ground-up expertise from actors in the business community (developers) and to some extent also from civil society. However, according to one of the informants from an environmental NGO the municipality

has often been reluctant to take advice from external actors.

With regard to the issue of network and stakeholders in Bergen, there has been a special focus from the municipality to expand the scope of existing networks. It was regarded as important to include the business community more efficiently, to encourage preventive actions with the developers and the municipality, to motivate developers and the business community to initiate actions and cooperate more closely with the municipality. The institutional context for learning and coordination through established networks, such as the regional climate panel and the climate forum and, the development of new forms of stakeholder involvement and so on, is depicted in Figure 8.6. The purpose of this relational context is to raise awareness about specific issues and expand the scope of existing networks in climate change adaptation, to coordinate and establish routines for knowledge sharing and learning across municipal sectors and between the municipality and other stakeholders and, to use search strategies through network and stakeholder involvement to identify areas that are not yet covered by Bergen City's adaptation strategy.

Networking and stakeholder involvement may bring about vital knowledge but since networks are based on voluntary agreements and coordinating takes place through consensus, there is still a way to go in order to bring about legitimate and authoritative

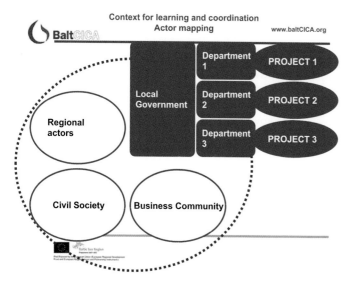

Figure 8.6 Context for learning and coordination.

decisions. A general lesson to be learned is that climate change adaptation policy is complex and involves actors from different spheres of society – representatives from political-administrative institutions, actors from the private business sector and from civil society. All these actors contribute with knowledge and resources necessary for an efficient and legitimate climate change adaptation policy, but the complex interplay between them may also lead to an ambiguous distribution of responsibility. The complexity of climate change adaptation requires cooperation, integrated approaches, multi-actor and multilevel governance, and multidisciplinary scientific knowledge as well as local knowledge. Scientific knowledge from various natural and social sciences is needed for understanding and dealing with causes and impacts of climate change. Local knowledge is often needed for working out feasible adaptation measures, involvement of relevant stakeholders, and therefore necessary both in order to ensure that all relevant knowledge is brought to the fore, and to ensure legitimate decisions.

8.6 Preliminary conclusions

Since climate change may affect geographical areas across political and administrative borders, both horizontal and vertical coordination is demanded. All these issues are at stake in the climate change adaptation policy in Bergen. In this chapter we have raised the question if Bergen has been taking on too much with its ambitious climate policy and if the city has organized and carried out its strategies in a way that ensures knowledge transfer, learning and stakeholder involvement. To put it briefly: Has Bergen an efficient and coherent climate adaptation policy? Although a comprehensive answer to this question needs to be found elsewhere, some lessons can be learned from the Bergen case. One is that adaptation to climate change is not primarily a technical issue, but a question of organizing complex social processes. Sufficient natural science knowledge of climate change is a necessary first step but only a first step, it is essential also to pass several other barriers before adaptation measures can be implemented. A second step is to create awareness and get an understanding of possible climate change impacts among relevant stakeholders in the society. This will normally imply social

science knowledge but also local knowledge and/or layman knowledge. Different stakeholders can contribute to such bottom-up knowledge by identifying gaps in the adaptation policies and thereby pave the way for a more stakeholder-driven adaptation strategy which may be more legitimate and efficient. However, efficiency also requires internal coordination between departments and sectors in the city administration.

A second lesson is that networking may improve external cooperation between stakeholders from different sectors and stimulate knowledge transfer and learning across sectors. However, in order to be efficient climate change adaptation policies also must be institutionalized and integrated into the planning system and, most importantly, the different departments in the city administration must cooperate closely. An optimal mix of hierarchy and network may contribute to maintaining interests for climate change adaptation over time and may also best secure legitimacy, visibility and mobilization. The final message, then, is that when dealing with complex issues it is necessary to use different governance modes, that is, it is necessary to work with climate change adaptation also through market mechanisms and networks, not just hierarchies. When implementing adaptation measures one must draw on experiences and knowledge from economic actors like developers and insurance companies and from different civil society actors. Actors in the market and in networks may have concurrent interests with political authorities, the question is how to best utilize the experience and knowledge from markets and networks in the climate change adaptation work.

References

Amundsen, H., Berglund, F. and Westskog, H., 2010. Overcoming barriers to climate change adaptation – a question of multilevel governance? *Environment and Planning C: Government and Policy 2010*, 28 (2), pp.276–289.

Ashley, R.M., Blanskby, J., Newman, R. et al., 2011. Learning and action alliances to build capacity for flood resilience. *Journal of Flood Risk Management*, 5 (1), pp.14–22.

Bergen Kommune, 2012. *The Climate Section*. [online] Available at: <https://www.bergen.kommune.no/omkommunen/avdelinger/klimaseksjonen> [Accessed 5 September 2012].

Berkes, F. and Jolly, D., 2001. Adapting to climate change: social-ecological resilience in a Canadian western Arctic community. *Conservation Ecology*, 5 (2), Art.18.

Berkhout, F., Hertin, J. and Gann, D.M., 2006. Learning to adapt: Organisational adaptation to climate change impacts. *Climatic Change*, 78, pp.135–156.

Berrefjord & Thomassen AS, 2010. *Storbyprosjektet energi og miljø*. [online] Available at: <http://www.bandt.no/storby/StorbyprosjektetEnergiMiljo.swf> [Accessed 3 September 2012].

Betsill, M.M., 2001. Mitigating climate change in US cities: opportunities and obstacles. *Local Environment*, 6, pp.393–406.

BOH (Bergen og Omland havnevesen), 2012. *Gods*. [online] Available at: <http://www.bergenhavn.no/index.cfm?id=256048> [Accessed 5 September 2012].

Bulkeley, H., 2010. Cities and the governing of climate change. *Annual Review of Environment and Resources*, 35 (1), pp.229–253.

CICERO (Center for International Climate and Environmental Research – Oslo), 2011. *NORADAPT – Community Adaptation and Vulnerability in Norway*. [online] Available at: <http://www.cicero.uio.no/projects/detail.aspx?id=30182&lang=en> [Accessed 5 September 2012].

Dannevig, H., Rauken, T. and Hoversrud, G. 2012. Implementing adaptation to climate change at the local level. *Local Environment*, 17 (6–7), pp.597–611.

Det Kongelige Miljøverndepartement, 2012. *Meld. St. 21 (2011–2012): Melding til Stortinget – Norsk klimapolitikk*. [pdf] Oslo: Det Kongelige Miljøverndepartement. Available at: <http://www.regjeringen.no/pages/37858627/PDFS/STM201120120021000DDDPDFS.pdf> [Accessed 4 September 2012].

Eakin, H. and Lemos, M.C., 2006. Adaptation and the state: Latin America and the challenge of capacity-building under globalization. *Global Environmental Change*, 16, pp.7–18.

ESPON, 2012. *ESPON CLIMATE – Climate Change and Territorial Effects on Regions and Local Economies in Europe*. [online] Available at: <http://www.espon.eu/main/Menu_Projects/Menu_AppliedResearch/climate.html> [Accessed 5 September 2012].

Farstad, E., Olsen, L.H. and Ingvaldsen, K., 2012. *Presenterer ROS II: Anbefaler sikring i Salhus og Fjellsiden*. [online] Available at: <https://www.bergen.kommune.no/aktuelt/tema/risikokartlegging/article-62507> [Accessed 17 September 2012].

Ford, J. and Smit, B., 2004. A framework for assessing the vulnerability of communities in the Canadian Arctic to risks associated with climate change. *Arctic*, 57, pp.389–400.

Georghiou, L., Cassingena Harper, J., Keenan, M., Miles, I. and Popper, R. eds., 2008. *The Handbook of Technology Foresight*. Cheltenham: Edward Elgar.

Grieg Foundation, Visjon Vest and G.C. Rieber Fondene, 2009. *Regional Havstigning – Prosjektrapport*. Bergen: Grieg Foundation, Visjon Vest and G.C. Rieber Fondene.

Hansen, G., 2005. *Bergen c 800 – c 1170: The Emergence of a Town*. Fagbokforlaget, Bergen.

Hanssen-Bauer, I., Drange, H., Førland, E.J. et al., 2009. *Klima i Norge 2100. Bakgrunnsmateriale til NOU Klimatilpasning*. Oslo: Norsk klimasenter.

Hartvedt, H.G., 1999. *Bergen Byleksikon*. Oslo: Kunnskapsforlaget.

Harvold, K., Innbjør, L., Kasa, S., Nenseth, V., Saglie, I.-L., Tønnesen, A. And Vogelsang, C., 2010. *Responsibility and Political Measures When Adapting to Climate Change*. Oslo: CIENS.

Hjeltnes, K.J., 2011. *Fra erkjennelse til handling Undertittel: Klimatilpasning i Bergen kommune*. Master thesis, University of Oslo.

Hordaland Fylkeskommune, 2012a. *Klimaplan for Hordaland 2010–2020*. [online] Available at: <http://www.hordaland.no/Hordaland-fylkeskommune/Regionalutvikling/Miljo/filer/Klimaplan-for-Hordaland> [Accessed 3 September 2012].

Hordaland Fylkeskommune, 2012b. Klimarås for Hordaland. [online] Available at: <http://www.hordaland.no/Hordaland-fylkeskommune/Regional-utvikling/Miljo/klima/Klimarad> [Accessed 5 September 2012].

Kirkebøen, S.E., 2012. Forskere mener at golfstrømmen er overvurdert. *Aftenposten*, [online] 29 Aug. Available at: <http://www.aftenposten.no/viten/Mener-at-Golfstrommen-er-overvurdert-6977249.html> [Accessed 5 September 2012].

Klausen, J.E., Winsvold, M. and Langeland, O. (in press). The role of knowledge and knowledge transfer for climate adaptation.

Kloster, R., 1952. *Castle and City: Through Historical Bergen*. Bergen: Published for Turisttrafikkomiteen i Bergen, Hordaland Sogn og Fjordane and F. Beyer in Commission F. Beyer.

Kousky, C. and Schneider, S., 2003. Global climate policy: Will cities lead the way? *Climate Policy*, 3, pp.359–372.

Lam, A., 2000. Tacit knowledge, organizational learning and societal institutions: An integrated framework. *Organization Studies*, 21 (3), pp.487–513.

Levitt, B. and March, J.G., 1988. Organizational learning. *Annual Review of Sociology*, 14, pp.319–340.

Lyng, A.M., 2009. *Klima- og energihandlingsplanen for Bergen*. [online] Available at: <https://www.bergen.kommune.no/aktuelt/tema/klimaogmiljo/article-40144> [Accessed 3 September 2012].

March, J., 1991. Exploration and exploitation in organizational learning. *Organization Science*, 2, pp.71–87.

MARE (Managing Adaptive Responses to changing flood risk), 2012. *About.* [online] Available at: <http://www.mare-project.eu/about> [Accessed 5 September 2012].

Mills, E., 2003. Climate change, insurance and the buildings sector: technological synergisms between adaptation and mitigation. *Building Research & Information*, 31 (3–4), pp. 257–277.

Mills, E., 2009. *From Risk to Opportunity. Insurer Responses to Climate Change.* CeresReport.

Monni, S. and Raes, F., 2008. Multilevel climate policy: the case of the European Union, Finland and Helsinki. *Environmental Science & Policy*, 11 (8), pp.743–755.

Norwegian Meteorological Institute, 2012. *Fremtidsklima.* [online] Available at: <http://www.met.no/Bergen.9UFRLYWI.ips> [Accessed 3 September 2012].

NOU (Noregs offentlege utgreiingar), 2010. *Tilpassing til eit klima i endring – Samfunnet si sårbarheit og behov for tilpassing til konsekvensar av klimaendringane.* Oslo: Servicesenteret for departementa Informasjonsforvaltning.

Regjeringen, 2012. *Framtidens byer.* [online] Available at: <http://www.regjeringen.no/nb/sub/framtidensbyer/om-framtidens-byer.html?id=548028> [Accessed 3 September 2012].

Seager, R., 2006. The source of Europe's mild climate. *American Science*, 94 (4), p.334.

Seto, K., Sánchez-Rodríguez, R. and Fragkias, M., 2010. The new geography of contemporary urbanization and the environment. *Annual Review of Environment and Resources*, 35, pp.167–194.

Shavell, S., 2002. Law versus morality as regulators of conduct. *American Law and Economics Review*, 4 (2), pp.227–257.

Simpson, M., Breili, K., Kierulf, H.P., Lysaker, D., Ouassou, M. and Haug, E., 2012. *Estimates of Future Sea-Level Changes for Norway. Technical Report of the Norwegian Mapping Authority.* Kristiansand: Norwegian Mapping Authority.

Smit, B. and Wandel, J., 2006. Adaptation, adaptive capacity and vulnerability. *Global Environmental Change*, 16 (3), pp.282–292.

Statistics Norway, 2012. *StatBank Norway – 02 Population.* [online] Available at: <http://statbank.ssb.no//statistikkbanken/default_fr.asp?PLanguage=1> [Accessed 17 September 2012].

Sutherland, K., Smit, B., Wulf, V. and Nakalevu, T., 2005. Vulnerability in Samoa. *Tiempo*, 54, pp.11–15.

TECHNEAU (Technology Enabled Universal Access to Safe Water), 2011. *Home.* [online] Available at: <http://www.techneau.org/> [Accessed 3 September 2012].

Teece, D.J., 2000. *Managing Intellectual Capital.* New York: Oxford University Press.

Teece, D.J., 2009. *Dynamic Capabilities & Strategic Management.* New York: Oxford University Press.

Vásquez-León, M., West, C.T. and Finan, T.J., 2003. A comparative assessment of climate vulnerability: agriculture and ranching on both sides of the US–Mexico border. *Global Environmental Change*, 13, pp.159–173.

Vasskog, K., Drange, H. and Nesje, A., 2009. *Havnivåstigning. Estimater av framtidig havnivåstigning i norske kystkommuner.* Revised version. Tønsberg: The Directorate for Civil Protection and Emergency Planning (DSB).

Winsvold, M., Stokke, K.B., Klausen, J.E. and Saglie, I.L., 2009. Organizational learning and governance in adaptation in urban development. In: W.N. Adger, I. Lorenzoni and K. O'Brien, eds. 2009. *Adapting to Climate Change: Thresholds, Values, Governance.* Cambridge: Cambridge University Press.

Winsvold, M., Hjeltnes, K.J., Klausen, J.E. and Langeland, O., (in press) Climate change adaptation through hierarchies and networks in the city of Bergen.

9 Adaptation to Climate Change in the Smeltalė River Basin, Lithuania

Egidijus Rimkus, Justas Kažys, Edvinas Stonevičius & Gintaras Valiuškevičius

Department of Hydrology and Climatology, Vilnius University, Lithuania

9.1 Introduction

The regular inundation of the southern part of Klaipėda city due to flash floods in the Smeltalė River and Baltic Sea level fluctuations are the main problem of the Klaipėda city case study area. Floods are natural phenomena. However, some human activities and climate change contribute to an increase in the likelihood and adverse impacts of flood events (Directive 2007/60/EC, 2007). The main goal of the project is to assess climate change impacts on the Smeltalė River hydrologic regime and to initiate local community adaptation actions in the Smeltalė River basin.

At the beginning of the BaltCICA project the specific problems related to climate change in the case study area were presented to the local community. Later in the project assessment of climate change impacts and introduction of particular climate change scenarios were performed. After that, various adaptation measures and feasibility studies were presented to the local community. At the end of the project local interest groups and stakeholders chose the best climate change adaptation measures to implement in the case study area.

The project has joined the scientists, the stakeholders and the local interest groups together to tackle the problems related to climate change in Klaipėda city. Fruitful collaboration actions should be continued in the future. The gained experience could be implemented in other regions of Lithuania and the Baltic Sea countries which face climate change impacts on a local level.

9.2 Case study area

Smeltalė is a small river in the Lithuanian Coastal lowland region. The river falls into the Curonian Lagoon (Klaipėda Strait) within the Klaipėda city area (Figure 9.1). The Smeltalė River 20.9 km long of which 15 km are regulated, and it has a basin area of 120.1 km^2 (Gailiušis, Jablonskis & Kovalenkovienė, 2001). The average annual runoff in the Smeltalė River Basin reaches 12 l/(s × km^2). Maximum discharge formation in the coastal lowland region rivers is primarily related to heavy rain. It is estimated that during the flash floods more than 170 l/(s × km^2) can reach the Smeltalė River.

However, stationary hydrological measurements in the Smeltalė River have never been carried out and all hydrological characteristics of river stream and the surrounding territory are presented using results of indirect calculations (Gailiušis, Jablonskis & Kovalenkovienė, 2001).

Climate Change Adaptation in Practice: From Strategy Development to Implementation, First Edition.
Edited by Philipp Schmidt-Thomé and Johannes Klein.
© 2013 John Wiley & Sons, Ltd. Published 2013 by John Wiley & Sons, Ltd.

Figure 9.1 Case study area – the lower reaches of the Smeltalė River.
Source: Georeference Background by National Land Service under the Ministry of Agriculture of the Republic of Lithuania.

Inundation in the Klaipėda city area generally occurs due to two main reasons:
• flash flooding caused by heavy precipitation or sudden snowmelt;
• flooding caused by high water level in the Baltic Sea and the Klaipėda Strait.

Usually extreme flooding forms under mixed weather conditions. Mostly it happens when the water level in the Klaipėda Strait is high while heavy precipitation falls on the basin area. Heavy precipitation is often accompanied by strong winds, which leads to wind surge in the Baltic Sea and the Klaipėda Strait.

One of the key features of the Smeltalė River Basin is the two very distinctive parts of the basin. The lower reaches of the basin are heavily urbanized while the remaining area of the basin is rural. The urbanized catchment area is 18 km^2 (~15% of the total basin area). The rural catchment area covers 102 km^2 (~85% of the total basin area). Approximately 70% of the urban catchment area is occupied by residential and industrial areas, while 70% of the rural basin area is covered by agricultural areas.

The lower reaches of the Smeltalė River area are very flat. Consequently, this area is extremely vulnerable to inundation during heavy precipitation events.

Water flowing to the river from the storm drainage network in the lower reaches of Smeltalė is an additional key factor which strongly contributes to the mentioned meteorological conditions. Klaipėda city covers more than one sixth of the basin area. Consequently, the storm drainage water quickly reaches the river channel and sharply increases the peakflood discharge. In the city territory, the Smeltalė River is crossed several times by sewerage pipes and bridges and in the case of large-scale flooding they serve as barriers for flowing water what leads to an increase in the inundated area.

High-water level in the Smeltalė lower reaches leads to several problems:
• large inundated areas;
• interruption in exploitation of small ship harbour in the river's lower reaches;
• significant amount of pollutants and sediments from rain drainage network.

9.3 Climate variability and changes in Klaipėda city

9.3.1 Changes in the past

Permanent daily meteorological measurements in Klaipėda city started in 1881. The mean annual air temperature in Klaipėda increased by 1.1 °C during 1881–2010 (Figure 9.2a). An especially warm period was observed in 1997–2009 when the mean annual temperature each year exceeded 7 °C. On the other

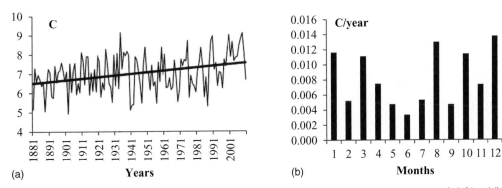

Figure 9.2 Mean annual air temperature changes in Klaipėda (1881–2010) (a) and monthly air temperature trends (°C/year) (b).

hand, in 2010, due to low winter temperature, the mean annual temperature was slightly lower than the long-term average for the 1971–2000 climate base line.

The characteristic of annual mean air temperatures mostly depends on the cold season (November–March) because its temperature can fluctuate more than in the warm season (April–October). Figure 9.2b clearly shows that the most significant increase in temperatures was in the October–March period. Warm season months trend rates are also positive, but smaller. Several extraordinarily warm August months at the end of the analysed period led to a high positive temperature trend.

There was a negligible decrease of annual precipitation amount (average 0.15 mm per year) in Klaipėda during the period of measurement (1925–2010) (Figure 9.3a). The number of days with precipitation also decreased. On the other hand, seasonal differences are obvious. The precipitation amount in

the cold season (especially in November–January) increased (trend value reaches 0.3 mm/year) while in the warm period it slightly decreased (Figure 9.3b). Positive tendencies of the recurrence of daily and three-day heavy precipitation events were also determined (Figure 9.4).

The probability of heavy precipitation events increased in summer while the total precipitation amount slightly decreased. It means that the percentage of heavy precipitation in the total annual precipitation amount also increased.

During the observation period the maximum daily precipitation amount (73.9 mm) was recorded on 28 July 1988. The three-day maximum rainfall value (128.2 mm) was reached between 9 and 11 August 2005.

Daily and three-day annual maxima probabilities were calculated using the Generalized Extreme Value (GEV) distribution (Table 9.1). 10-, 30- and 100-year return periods were analysed. This continuous

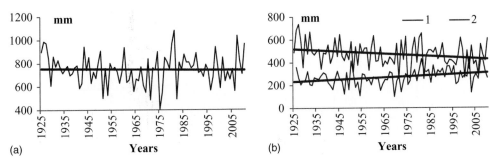

Figure 9.3 Changes of annual (a) and seasonal (1 – cold period (November–March), 2 – warm period (April–October) (b) amount of precipitation in Klaipėda in 1925–2010.

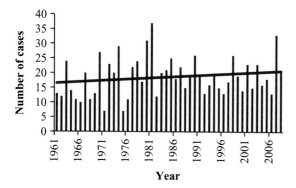

Figure 9.4 Annual number of cases in Klaipéda in 1961–2008 when daily precipitation amount exceeded 10 mm.

Table 9.1 10, 30 and 100-year return level of daily and three-day precipitation amount (mm) calculated using the Generalized Extreme Value (GEV) distribution (reference period 1961–2008)

Heavy precipitation duration	10-year return level	30-year return level	100-year return level
Daily precipitation	49	64	83
Three-day precipitation	72	93	122

probability distribution combines the Gumbel, Frechet and Weibull distributions used to model extreme events (Kotz & Nadarajah, 2000). The GEV distribution is widely used for the approximation of short-term (up to several days) extreme precipitation amounts.

During the period 1950–2010 distinct tendencies of decrease of snow cover parameters were observed

(Figure 9.5). Changes in the number of days with snow cover are very prominent while a decrease in snow thickness is not statistically significant. Thick but unstable snow cover in Klaipéda can form during the short-term one to three-day synoptic process.

9.3.2 Future projections

Output data of the regional climate model CCLM (COSMO Climate Limited-area Model) were used in this study. The modelling is based on the IPCC A1B and B1 emissions scenarios (Nakicenovic et al., 2000) where B1 is a low-emissions scenario (considered to be the 'best case') and A1B is a relatively high-emissions scenario.

Model outputs indicate that air temperature are likely to increase in Klaipéda in the 21st century by 2.7 °C (B1 emissions scenario) to 4.2 °C (A1B emissions scenario) (Figure 9.6a). Air temperature will rise during the whole year. However, the most intensive shift is expected in winter (up to 6 °C). The largest air temperature changes are predicted for January.

According to the CCLM model outputs, annual precipitation will increase in the Klaipéda by 17–22% in the 21st century (Figure 9.6b). The A1B scenario projects the largest changes in winter and spring. The precipitation amount in winter would increase by 38% (mostly in January). It is likely that precipitation during the cold period of the year will rise more rapidly due to more frequent advection of warm and moist air masses. Under the B1 emissions scenario the highest growth of precipitation amount is expected in autumn (about 26%), meanwhile the positive changes in winter and spring will be slightly lower (up to 20%). Summer precipitation changes in Klaipéda will be insignificant. Negative changes are very likely in the

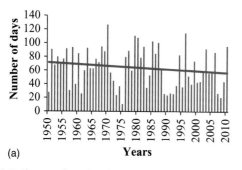

(a)

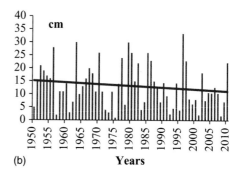

(b)

Figure 9.5 Changes of number of days with snow cover (a) and maximal snow depth (b) in Klaipéda in 1950–2010.

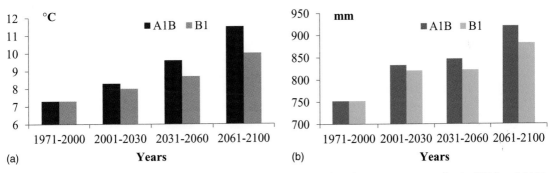

Figure 9.6 Projected air temperature (a) and precipitation amount (b) in Klaipėda in the 21st century according to CCLM model A1B and B2 emissions scenarios.

second part of summer. A decrease of the precipitation amount and a rise in air temperatures will lead to a possible strengthening of droughts in this part of the year.

CCLM model projections show an increase of daily heavy precipitation events (> 10 mm) in the 21st century (about 30%). The largest changes are expected in autumn. According to the CCLM model outputs, the recurrence of three-day heavy precipitation events (> 20 mm) will also increase significantly (up to 35%) and their frequency will reach 13 cases per year (in 1971–2000 the mean value was 9.6 cases per year). Both B1 and A1B scenarios project large positive and statistically significant changes of heavy precipitation event frequency. In autumn, the positive changes will be most distinct; meanwhile reoccurrence of such heavy precipitation events in summer will probably remain the same during the 21st century.

The projections of five-year recurrence of one-day rainfall (mm) according to A1B and B1 scenarios were used to estimate the changes of heavy rain events in the 21st century. The calculations of flash flood peak discharges were based on these projections of five-year return period of daily rainfall (Figure 9.7).

According to the A1B emissions scenario a five-year return level of daily precipitation amount in the seacoast region will decrease in the middle of the 21st century and will exceed current levels at the end of the century (Figure 9.7). Meanwhile the B1 scenario projects a slight decrease of these values in the second part of the 21st century.

Changes of three-day annual maximum precipitation amount will be very similar. According to the A1B

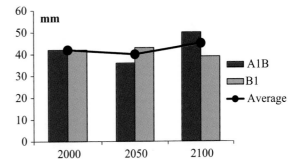

Figure 9.7 Five-year return level of daily precipitation amount (mm) according to A1B and B1 emissions scenarios calculated using the GEV distribution.

scenario 10-year's return level will reach 89 mm at the end of the 21st century and according to the B1 scenario 68 mm (average value for reference period 1971–2000 is 71 mm). 100-year return level will reach 138 and 98 mm respectively (at the end of the 20th century it was 108 mm).

Snow cover duration and maximum thickness is likely to decline in future due to an increase in winter temperatures. At the same time, an increase in winter weather variability is also very likely. Therefore, probability of short-term but quite thick snow cover will remain.

9.4 Flash floods in the Smeltalė River

During intensive rainfall the surroundings of the Smeltalė River are usually inundated. The area of inundated territory can be relatively large, because the lower reaches of Smeltalė River are almost flat.

One of the most important factors in determining the hydrological regime of the Smeltalė River is the rain drainage network. The rain drainage network reduces the time runoff concentration and as a result the peak discharges of floods are higher. The rain drainage network drains only one fifth of the total Smeltalė River catchment area, but it is very dense and leads to a significant hydrological response. The rain drainage network is one of the main sources of sediments and pollutants in the lower reaches of the Smeltalė River. Studies have found that after heavy rain the water quality downstream from the drainage network significantly decreases (Nacionalinių Projektų Rengimas and Menhyras Ltd., 2007). The large increase of sediment in the river bed was also observed. The extent of inundation in the lower reaches of the Smeltalė River is also affected by the sewage collector, which intersects the river. When the water level reaches the bottom of collector it starts to act as a dam.

The prolonged periods of high water in the Smeltalė River are harmful to a fishing boat harbour, which is established in the Smeltalė River entry. The Nemunas Street Bridge is on the way from the port to the sea. The fishermen complain that the space under the bridge becomes too narrow for their boats during high flood events. The large amount of sediment in the port section of the river is also harmful. The port operating organization has to remove part of the sediment from the river. The removed sediment must be stored in special dump, because large parts of the sediment come from the rain drainage network and are polluted

(Nacionalinių Projektų Rengimas and Menhyras Ltd., 2007).

Projections of five-year recurrence precipitation according to the A1B and B1 emissions scenarios were used to estimate the changes in five-year recurrence flash flood peak discharges. The runoff formation conditions in urban and rural areas of the Smeltalė River differ from each other. Therefore, different methods were used to estimate the peak discharge of urban and rural parts of the catchment. The peak discharge of five-year return level flood and its changes in the 21st century in the rural part of the Smeltalė River were estimated with the TR-20 model (SCS, 1965).

The TR-20 model can be calibrated by two parameters: SCS (Soil Conservation Service) curve number and time runoff concentration. Time of concentration can be calculated by the relief and geometry of river catchment and by properties of the river network. SCS curve number can be estimated by the soil type and land use. The SCS curve number of the river catchment is usually very uncertain. The uncertainty analysis revealed that inaccurate estimation of the Smeltalė River catchment SCS curve number may produce an uncertainty of peak discharge from −25% to +40% (Figure 9.8a).

The model results are very sensitive to land use changes in the catchment. In recent years a lot of people have moved from Klaipėda city to new suburban areas in the Smeltalė River catchment. It is possible that the peak discharges of the Smeltalė River upper reaches will increase due to land use changes if the same tendency of people migration remains. The sensitivity analysis of the TR-20 model suggests that 20 km² of new private houses built in the rural area

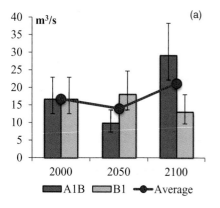

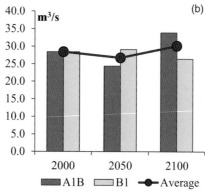

Figure 9.8 Projections of five-year flood peak discharge in the Smeltalė River (vertical bars represent the uncertainty) (a) and in the Klaipėda city rain drainage network (b) according to A1B and B1 emissions scenarios.

of the Smeltalė catchment could increase the peak discharge of flash floods up to 40%.

More than 80% of the urban part of the Smeltalė River catchment is covered by the rain drainage network. The peak discharge of flash floods in the drainage network was estimated according to STR 2.07.01-2003 procedure, which is described in Lithuanian building regulations. The calculation showed that the peak discharge from drainage network will likely decrease in the middle of the 21st century and slightly increase at the end of the 21st century (Figure 9.8).

In both rural and urban parts of the Smeltalė River catchment the projections of peak discharge follow the pattern of intensive precipitation projections. Consequently, the problems associated with flash floods are likely to increase at the end of the 21st century.

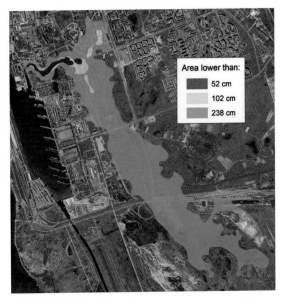

Figure 9.9 The area lower than 52, 102 and 238 cm in the southern part of Klaipėda city.
Source: Georeference Background by National Land Service under the Ministry of Agriculture of the Republic of Lithuania.

9.5 The effect of high sea level on the lower reaches of the Smeltalė River

Klaipėda city is a coastal town. Two rivers, the Danė and the Smeltalė, have their entries in the territory of Klaipėda city. The lowest parts of Klaipėda city are in the valleys of the Danė and the Smeltalė Rivers. High water in the Baltic Sea and the Klaipėda Strait causes high-water level in the rivers. Some low-lying parts of the city in the Danė and the Smeltalė valleys have been inundated in the past. Such situations have occurred during strong storm surges.

Climate change projections according to CLIMBER model (Petoukhov et al., 2000; Ganopolski et al., 2001) show continued increase in sea level near Klaipėda city. Projections of annual sea level rise near Klaipeda vary between 20 cm (B1m scenario) and 90 cm (A1FIh scenario) until the end of the 21st century. The six scenarios average is 52 cm which is the most likely estimate.

The annual maximum water level in the Baltic Sea near Klaipėda city every year exceeds the annual water level by 50 cm or more. In the case study area the territory lower than 50 cm is relatively small. Only the Smeltalė River valley and small fishing boat harbour can be inundated if the water level is near 50 cm (Figure 9.9).

At the end of the 21st century a storm surge of 50 cm can cause many more problems. Together with a higher sea level of 52 cm a storm surge of 50 cm would result in a water level of 102 cm. The area below 102 cm extends to residential territory and covers a large part of the Smeltalė River lower reaches (Figure 9.9). Such water level can cause a variety of problems. The inundation of territory is just the most obvious one.

High water level in the sea and the river can be the cause of drainage network overflow, fishing boat harbour operation disruptions, intrusions of salt water to the Smeltalė River and ground water. The sea level rise projections and the historical storm surge records suggest that at the end of the 21st century water level could exceed 102 cm at least once in a year.

The highest storm surge observed in the Baltic Sea near Klaipėda city was in 1967. The water level during this storm surge was 186 cm. The same storm surge at the end of the 21st century could result in a water level near 238 cm. During such an extreme event the sea could inundate a large part of the southern part of Klaipėda city (Figure 9.9).

9.6 Possible adaptation measures to high water levels in the Smeltalė River

Study results show that the flooding situation in the Smeltalė River basin would not improve in the 21st century. It is likely that the Smeltalė River lower reaches will be inundated more frequently because of climate change. Both heavy precipitation and high sea level are likely to lead to high water levels in the Smeltalė River. Consequently, different measures should be implemented to deal with the different causes of high-water level.

Possible adaptation measures to manage heavy precipitation events can be divided into:

1. Direct (technical) measures for water balance control – to divert precipitation water somewhere else, enlarge capacity of river bed, etc.

2. Indirect measures for water balance – to control urbanization processes in the rural part of the basin, regulate relative size of urbanized part of the basin, etc.

One of the most efficient direct measures to water balance control – to keep maximum water runoff in the river bed, by:

1. Deepening the Smeltalė River and affluent river channels.

2. Implementing high embankments and dykes in riversides.

3. Elevating bridges in the flood-prone territories.

Implementation of these measures guarantees long-term effects in the flood-prone urban part of the Smeltalė River Basin under intense climate change processes. Moreover, the measures secure exploitation of small ship harbour (less silt accumulation during summer heavy precipitation events; elevated bridges guarantee boats traffic during high water levels). Runoff caused by heavy rain is similar in urban and rural Smeltalė River Basin parts despite that the urban area is a few times smaller than the rural area. It means that regulation is more efficient in the urban part.

Migration trends from Klaipėda city to the rural areas of the Smeltalė River Basin are obvious. Developing settlements will change land use and it will increase maximum runoff formed by heavy precipitation events. Indirect measures are long term, but implementation should be started right now (foreseen in the Klaipėda city development plan).

A few adaptation measures to deal with flooding due to high Baltic Sea water levels are possible. The most efficient measure in the case of a storm surge induced water level of 50 cm could be low-level embankments on both riversides of the flood-prone area. If the water level rose to 102 cm, the flood would reach the northern part of the river-side which is a residential district (Figure 9.9). Higher embankments are needed to protect the flood-prone territory. Also, the bridges should be elevated (up to two meters from the recent base). If the water level reached 238 cm, the inundated territory would be very large. Such extreme events can be expected only once per few decades. Dykes on both riversides should be around three meters high in the Smeltalė river. Also the Smeltalė River reaches should be deepened and embanked.

9.7 Assessment of the efficiency of possible adaptation measures in the Smeltalė River

The projections of sea level rise and flash flood peak flow changes were presented to the Klaipėda city municipality administration in a first scenario workshop. The stakeholders proposed four adaptation measures to high water level and surplus sediments in the Smeltalė River:

1. Enlargement of the river channel capacity beside the Klaipėda city water collector (sewage) place.

2. Installation of silt settlers (ponds) near sewage collection system.

3. Elevation (two meters) of the bridge in Nemunas Street.

4. Sea gate (with pump system) construction protecting the Curonian lagoon from flooding.

The result of the workshop was the initiation of a feasibility study to assess the costs and the efficiency of adaptation measures. The feasibility study was prepared by 'Vilniaus hidroprojektas' company. The feasibility study results were presented to stakeholders at a second scenario workshop. The feasibility of four measures proposed by the stakeholders was estimated

(Vilniaus Hidroprojektas Ltd., 2010). Also two alternative measures to control high water levels in the Smeltalė River were proposed by the authors of the feasibility study:

1. Grounded embankments on both watersides of the river.
2. Complex embankment and dyke systems on both watersides of the river.

The stakeholders have decided that the most suitable measure for the Smeltalė River lower reaches will be complex embankment and dyke systems on both watersides of the river. This measure is the most expensive, but also the most suitable adaptation measure concerning efficiency, implementation costs and long-term durability (due to climate change impacts). Moreover, embankments should be incorporated in the recreation area which will be established in the area.

9.8 Quantitative assessment of adaptation measures efficiency

It becomes common practice that the assessment of climate change projects invokes specific measure selection methods based on quantitative (mostly economic) approaches (Hallegatte, Hourcade & Dumas, 2007) beside scientific research and recommendations.

Cost-Benefit Analysis (CBA) allows choosing possible measures assessing only economical parameters (European Communities, 2009). Meanwhile, Multi-Criteria Decision Analysis (MCDA) also involves social, ecological, aesthetical and other measuring criteria (Geldermann & Rentz, 2007). The CBA could be treated as part of the MCDA.

One of the project aims was to implement the CBA and the MCDA in the Klaipėda city case study area. The project partners and stakeholders were asked to fill in a questionnaire from which they were able to assess adaptation measures using quantitative methods and to identify the main indicators of the feasibility of adaptation measures in the Smeltalė River lower reaches.

The majority of respondents assumed that climate change adaptation process costs (43%) and benefits (42%) should be detailed in monetary terms. A number of the respondents suggested assessing only the main costs (mainly direct flooding damage) and adaptation measures benefits could be calculated approximately. Almost equal numbers of respondents thought that the benefits could be estimated by the theoretical approach (36%) or suggested that only primary benefits should be estimated (33%). A damage cost assessment scheme for the Smeltalė River Basin flood events was made according to respondents' answers (Figure 9.10). General criteria for flood-prone areas were developed during the BaltCICA project meetings. The most feasible and most important criteria are related to infrastructure and building value assessment. Meanwhile, ecological and social criteria are quite difficult to assess.

The respondents gave very different answers concerning the importance of the six adaptation measures in flood-prone areas of the Smeltalė River Basin. However, the most favourable answer was the first alternative – grounded embankments on both watersides of the river.

The last question was attributed to adaptation measures weights, that is, to rank them according to the importance (from 1 – not important, to 5 – very important). Nine adaptation measure criteria were presented in the questionnaire: efficiency (functioning properties), aesthetic view, exploitation consumption, friendly to environment, implementation costs, juridical opportunities, technical opportunities, time consumption and wide practical properties. Respondents' mean weights of importance for all criteria were quite similar – from 3.1 (for *time consumption*) to 4.3 (for *efficiency* and *environmental friendliness*).

Nine different MCDA methods (Fülöp, 2005) were used to assess the adaptation measures. According to the MCDA the best adaptation measure appears to be the second alternative – complex embankments and dykes system on both watersides of the river (Table 9.2). Meanwhile, the most favourable measure (first alternative) gleaned from the questionnaire is the best choice using only *Maximin* and *disjunctive* methods. The MCDA suggests that other adaptation measures are not very suitable in flood-prone areas of the Smeltalė River Basin.

Questionnaire results show that the local community is capable finding the best adaptation measure by themselves. However, the CBA and the

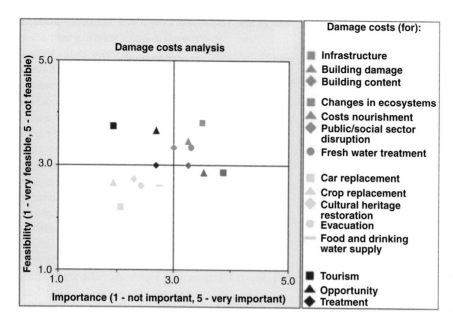

Figure 9.10 Damage cost assessment scheme for the Smeltalė River Basin flooding events; 3.0 level corresponds to average importance (x axis) and feasibility (y axis) conditions.

Table 9.2 Best adaptation measures for flood events, chosen by different MCDA methods. Six measures were determined by a feasibility study

Possible adaptation measures to reduce floodings in the southern part of Klaipėda city	Pros & Cons	Maximin	Maximax	Conjunctive	Disjunctive	Lexicographic	SMART (WSM)	SMART (WPM)	PROMETHEE II
							MCDA methods		
Water collector (sewage) capacity enlargement		•							
Installation of silt settlers (near sewage collectors)				•					
Elevation (2 metres) of the bridge under the River in Nemunas street									
Sea gate (with pump system) construction protecting from flooding					•				
Grounded embankments on both watersides of the river		•			•				
Complex embankments and dykes system on both watersides of the river	•	•	•		•	•	•	•	•

MCDA methods should be used as additional instruments for assessing quantitative results, avoiding inequitability and helping to convince local stakeholders.

9.9 Implementation of possible adaptation measures in the Smeltalė River

According to the results of the feasibility study and the stakeholders' decision made in the second scenario workshop, the Klaipėda city municipality initiated the preparation of the implementation of the adaptation measures in the Smeltalė River lower reaches. The project was prepared by 'Uostamiesčio projektas' company. On 30 November 2011, 'The Smeltalė River park' project was discussed in the Klaipėda city municipality meeting. The recreation area of the Smeltalė River park will be developed on the river embankments. The dykes proposed in the Smeltalė River park plan should protect flood-prone areas. The project will be implemented in several stages. First of all, the dykes and walking paths will be constructed. After that the rest of the infrastructure will be implemented due to Klaipėda city detailed plan requirements.

9.10 Conclusions

Close cooperation between scientists, local stakeholders and other community members is a key factor leading to implementations of climate change adaptation measures. Scenario workshops and other public events are very important in order to ensure such cooperation. Only the climate adaptation measures may be of interest to local authorities, which helps prevent already existing threats. In such cases projected changes of magnitude and reoccurrence of climate change induced threats should be taken into account. The easiest way to adapt to climate change is implementation of adaptation measures into existing development plans of local municipalities. CBA and the MCDA methods are very useful instruments in the decision-making process. These methods help to make decisions as well as making a qualitative assessment of decisions which were already made.

References

Council Directive 2007/60/EC of 23 October 2007 on the assessment and management of flood risks.

European Communities, 2009. *The economics of climate change adaptation in EU coastal areas.* [pdf] Luxembourg: Office for Official Publications of the European Communities. Available at: <http://ec.europa.eu/maritimeaffairs/documentation/studies/documents/report_en.pdf> [Accessed 5 May 2012].

Fülöp, J., 2005. *Introduction to Decision Making Methods.* [pdf] Available at: <http://academic.evergreen.edu/projects/bdei/documents/decisionmakingmethods.pdf> [Accessed 5 May 2012].

Gailiušis, B., Jablonskis, J. and Kovalenkovienė, M., 2001. *Lietuvos upės. Hidrografija ir nuotėkis.* Kaunas: Lietuvos energetikos institutas.

Ganopolski, A., Petoukhov, V., Rahmstorf, S., Brovkin, V., Claussen, M., Eliseev, A. and Kubatzki, C., 2001. CLIMBER-2: a climate system model of intermediate complexity. Part II: model sensitivity. *Climate Dynamics*, 17, pp.735–751.

Geldermann, J. and Rentz, O., 2007. Multi-criteria decision support for integrated technique assessment. In: J.P. Kropp and J. Scheffran, eds. 2007. *Advanced Methods for Decision Making and Risk Management in Sustainability Science.* New York: Nova Science, pp.257–273.

Hallegatte, S., Hourcade, J.C. and Dumas, P., 2007. Why economic dynamics matter in assessing climate change damages: Illustration on extreme events. *Ecological Economics*, 62, pp.330–340.

Kotz, S. and Nadarajah, S., 2000. *Extreme Value Distributions: Theory and Applications.* London: Imperial College Press.

Nacionalinių Projektų Rengimas and Menhyras Ltd., 2007. *Klaipėdos miesto paviršinių nuotekų šalinimo ir valymo sistemų plėtros koncepcija.* Vilnius: Public Enterprise, 'Nacionalinių projektų rengimas', UAB 'Menhyras' (Ltd.).

Nakicenovic, N., Alcamo, J., Davis, G., de Vries, B., Fenhann, J., Gaffin, S., Gregory, K., Grübler, A., Jung, T.Y., Kram, T., La Rovere, E.L., Michaelis, L., Mori, S., Morita, T., Pepper, W., Pitcher, H., Price, L., Riahi, K., Roehrl, A., Rogner, H.H., Sankovski, A., Schlesinger, M., Shukla, P., Smith, S., Swart, R., van Rooijen, S., Victor, N. and Dadi, Z., 2000. *IPCC Special Report on Emissions Scenarios*. Cambridge: Cambridge University Press.

Petoukhov, V., Ganopolski, A., Brovkin, V., Claussen, M., Eliseev, A., Kubatzki, C. and Rahmstorf, S., 2000. CLIMBER-2: a climate system model of intermediate complexity. Part I: model description and performance for present climate. *Climate Dynamics*, 16, pp.1–17.

SCS (Soil Conservation Service), 1965. *Computer Program for Project Formulation: Hydrology*. United States: SCS, Engineering Division.

Vilniaus Hidroprojektas Ltd., 2010. *Potvynių poveikio Smeltalės upėje švelninimo priemonių efektyvumo bei jų į gyvendinimo kaštų vertinimo galimybių studija*. Vilnius: UAB "Vilniaus hidroprojektas' (Ltd.).

10

The Geological Structure of Pyynikinharju Esker and the Local Effects of Climate Change

Jussi Ahonen[1], Tuire Valjus[1] & Ulla Tiilikainen[2]

[1] Geological Survey of Finland (GTK), Espoo, Finland
[2] City of Tampere, Urban Development, Tampere, Finland

10.1 Introduction

10.1.1 General information

In cooperation with the City of Tampere, the Southern Finland Office of the Geological Survey of Finland (GTK) has assessed the effects of climate change in the Pyynikinharju area. The research was carried out as part of the BaltCICA project studying the effects of climate change, climate change adaptation practices and their costs in the Baltic Sea Region. The Pyynikinharju area was selected as an object of study in the BaltCICA project due to its special features. Pyynikinharju is an esker formation, which is located near the centre of Tampere and is situated on a neck of land and bordered by Lake Näsijärvi and Lake Pyhäjärvi. The water level difference between the lakes is exceptional. This is particularly true when considering that water infiltrates to the esker formations fairly easily. The detailed structure of Pyynikinharju was not previously known. This means that there was no clear understanding of the area's sensitivity to changes in land use or the changing conditions caused by climate change.

This chapter presents the results of geological studies carried out in connection with the project as well as a structural 3D model based on the results. In addition to geological studies, scenarios on the effects of climate change in the area were created and adaptation and preparation practices for land use were developed, for example, in a workshop organized for experts in various fields.

10.2 Description of the study area

10.2.1 General description of the area

The study area covers the central and western part of the neck of land between Lake Näsijärvi and Lake Pyhäjärvi. The area is located 2.5 km west of the centre of Tampere, mainly in a built environment. The settlement is concentrated in the Pispala district on top of the esker and in the gentle south-western slope, while the largest traffic routes such as the railways follow the shores of Lake Näsijärvi.

Thanks to its central location, beautiful landscapes and its unique built environment, Pispala is a renowned residential area. Its buildings mainly consist of unique old houses. Pispala has been defined as a *built cultural* environment of national importance in the national inventory of the National Board of Antiquities (RKY, 2009). However, over the past few decades, so many new buildings have been constructed in Pispala that a decision was made to modernize the Pispala detailed plan in 2007. The objective of planning is to protect the cultural environment of Pispala, but also to enable the development or the construction of new and additional buildings in the district. The old Santaharju industrial area on the

Climate Change Adaptation in Practice: From Strategy Development to Implementation, First Edition.
Edited by Philipp Schmidt-Thomé and Johannes Klein.
© 2013 John Wiley & Sons, Ltd. Published 2013 by John Wiley & Sons, Ltd.

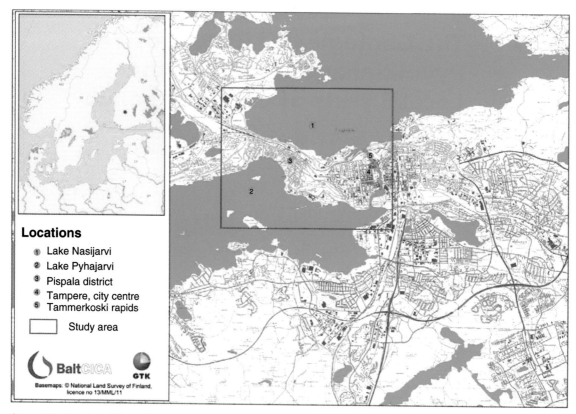

Locations

① Lake Nasijarvi
② Lake Pyhajarvi
③ Pispala district
④ Tampere, city centre
⑤ Tammerkoski rapids

▢ Study area

BaltCICA   GTK

Basemaps: © National Land Survey of Finland,
licence no 13/MML/11

Figure 10.1 Location of the study area.
Source: GTK, Jussi Ahonen.

northern slope of the Pispalanharju Esker will be zoned for residential use. In the west, the study area will be bound to the groundwater area of the Hyhky waterworks. The location of the study area has been illustrated in Figure 10.1 and there is a more detailed one in Figure 10.4.

10.2.2 General geological features of the study area and the history of the lakes

The long winding Pyynikinharju Esker was formed at the junction of the glaciers as the glacial waters melted away more than 11 000 years ago (Virkkala, 1962). It mainly consists of rocks, gravel and sand. In the western part of the study area, the formation meets the Nokia Esker, located to the west and southwest. The lowest point of the top of the esker has been located approx. 110 m a.s.l. at Hyhky (Virkkala, 1962). The highest point is located in Pispala, 160 m a.s.l. Two

small timber drive tunnels have been made across the esker at its lowest point, as well as the cut-and-cover tunnel of the Vaitinaro junction on the Nokia motorway.

Today, the water level of northern Lake Näsijärvi is approximately 95 m above sea level and the water level of southern Lake Pyhäjärvi is approximately 77 m above sea level. A difference of 18 m in water surface level is large for Finland, considering the distance of the lakes. After the Ice Age, once Lake Näsijärvi shrank, its waters flowed from the lake's northern point towards the north into the Gulf of Bothnia via the Lapua River. Since the outlet was located at the tip of Lake Näsijärvi where post-glacial land uplift was more extensive, the water level rose slowly at the tip near Tampere, at least 5.5 m. Approximately 7500 years ago, a new streambed was formed approximately three kilometres east of the study area. The Tammerkoski rapids formed as a result of this

Figure 10.2 The Tammerkoski rapids, formed approximately 7500 years ago, flow through the centre of Tampere.
Source: Photo: Lentokuva Vallas Oy, City of Tampere.

process (Figure 10.2) now unite Lake Näsijärvi and Lake Pyhäjärvi. The banks bordering what was then Lake Näsijärvi can be seen a few metres above the current water surface level of the current Lake Näsijärvi (Virkkala, 1962).

The discharge of the waters of Lake Näsijärvi caused a water surface level rise by a few metres in Lake Pyhäjärvi (Virkkala, 1962). The post-glacial land uplift originally causing the natural disturbance was at its maximum immediately after the ice sheet melted away (Kukkonen, 2003). The uplift is an ongoing process and the current annual rebound is 5 mm in the Tampere region (Kukkonen, 2003).

10.2.3 Changes caused by nature and human activities

At the centre of the study area, 500 m northwest of the Tahmelanlähde spring, there has been at least one serious natural disaster on a gentle slope declining towards Lake Pyhäjärvi (Alhonen, 1988). In 1598,

a great landslide caused widespread destruction in fields (Alhonen, 1988). Further east, the discharge of groundwater has occurred more freely on the slope. The pond-like Tahmelanlähde spring was previously a relatively copious spring of clear water.

It can be clearly seen how humans change the natural balance. The changes to the Tahmelanlähde spring are probably caused by the fact that in built environments, smaller amounts of rainwater infiltrate the soil compared to the natural state of esker formations. On the one hand, the roads and railways following the shores of Lake Näsijärvi (Figure 10.3) strengthen the slope on the neck of land between the lakes, protecting it from phenomena such as lakeside erosion. Structures built across the esker may weaken it and require control to some extent.

10.2.4 Land-use planning

There are several ongoing land-use planning projects in the study area and in its vicinity (Figure 10.4). The

Figure 10.3 Pispala and Santalahti and the traffic routes across the esker.
Source: Photo: Lentokuva Vallas Oy, City of Tampere.

Ongoing detailed planning processes:

1) Niemenranta, a new housing area

2) Simola Fields, a new housing area

3) Pispala, a cultural heritage housing area (protection plan)

4) Santalahti, a new housing area

5) Rantaväylä, a traffic tunnel

6) Ranta-Tampella, a new housing area

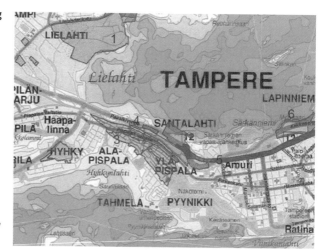

Figure 10.4 Ongoing detailed plans in the study area and in its vicinity in 2011.
Source: City of Tampere, Jussi Ahonen.

Pispala and Tahmela detailed plans are being reformed on the Pispala Esker. On the northern slope of the esker, the Santalahti small industry and workplace area is being zoned for housing. The BaltCICA study area also covers projects such as the development plan project for the Simola fields at Hyhky and the zoning project of the new Niemenranta housing area at Lielahti.

The Pispala detailed plan will be reformed gradually stage by stage. The plans for stages I and II are pending, and stage III will be launched in 2013.

10.2.5 Previous studies of the study area

Plenty of previous research material was available for the study. The geological maps of Tampere (Virkkala, 1962; Kukkonen, 2003) gave a good overview of the geological features of the area. Several studies have been published on the development of Lake Näsijärvi and Lake Pyhäjärvi (Alhonen, 1988; Tikkanen & Seppä, 2001).

More than 3000 drilling points were found in the City of Tampere database. In addition, the Finnish Transport Agency's drilling data (road and railway studies) were available in paper form for the study area. Most of the drillings carried out in the area were light drillings, mainly weight sounding. Only a few of the earlier drillings reached the bedrock level.

In addition, several technical study reports on the area were available. A 2010 report by Ramboll Finland Ltd estimated the lake water infiltration in Lake Näsijärvi and its flow to the Hyhky water point (Taipale & Pulkkinen, 2010). The study analysed features such as the oxygen isotope composition of water samples collected from different locations. They could be used to define the percentage of surface water in each water sample (Taipale & Pulkkinen, 2010).

10.2.6 Predicted climate change in the study area

The Tampere Central Region Climate Strategy 2030 was approved by the Tampere City Council in May 2010. The climate strategy defined a common vision of the central region climate policy, reduction commitments for various branches of activity as well as an operational programme used by the municipalities to reduce greenhouse gas emissions and to prepare for the consequences of climate change (Joint Authority of Tampere City Region, 2010).

The definition of the adaptation measures for the climate strategy was based on general climate change scenarios made for the Tampere Central Region until the year 2100. The scenarios were based on studies by VTT Technical Research Centre of Finland and provided the following information:

- the average annual temperature will increase by $4°C$;
- the maximum temperature will increase by $5°C$;
- the minimum temperature will increase by $10°C$;
- the average annual wind speed will remain unchanged;
- the maximum wind speed will increase by 5%;
- the annual precipitation will increase by 20%;
- the six-hour maximum precipitation will increase by 25%;
- the five-day maximum precipitation will increase by 30%;
- the six-hour maximum snowfall will increase by 15%;
- the maximum water-equivalent of the snow cover will decrease by 40%;
- the duration of snow cover will decrease by 60 days.

10.3 Research and modelling

10.3.1 Geological studies

Ground investigations were done in connection with the project to specify the soil/geological features of the study area.

10.3.2 Drillings

Soil drillings and observation well installations were performed in the study area between 2009 and 2011. The drillings were performed by the City of Tampere Measurement and Geotechnology Services using a GM200 drill rig.

The total amount of soil drillings was at 15 locations and 510 m. In addition, a total of seven groundwater observation wells were installed (Figure 10.5). In connection with the drillings, the stratigraphy of the soil was observed, and a total of 53 soil samples were taken. After the installation of the groundwater observation wells, the groundwater surface levels were measured in both the new and existing groundwater wells.

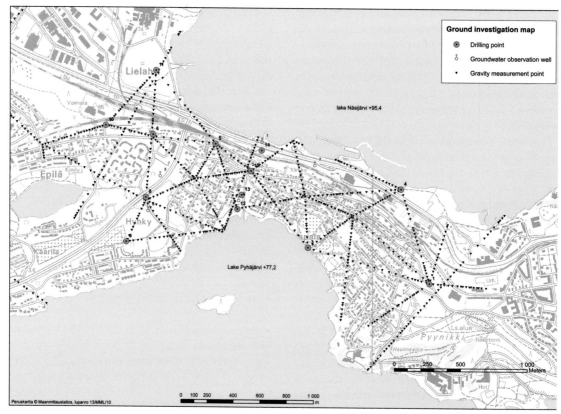

Figure 10.5 Ground investigations on the study area.
Source: GTK, Jussi Ahonen.

10.3.3 Gravity measurements

Gravity measurements can be used to examine the thickness and volume of geological units whose density differs from the units surrounding them (Valli & Mattsson, 1998).

The gravity measurements were performed in the study area in 2009. There were a total of 37 lines, and their total length was 19.7 km. In addition, three (3.6 km) lines measured in 2008 in the same area were re-interpreted in this context.

The location of measured gravity profiles is illustrated on Figure 10.5 and an example of the interpretation on Figure 10.6. The measurements were taken with a Worden gravimeter, using a 20 m point interval. The ends of gravity lines were bound to bedrock outcrops and drilling points, and VRS-GPS equipment was used to determine the elevation of the line end points. On the aligning bases, the distance between the reference points is less than one kilometre, and some

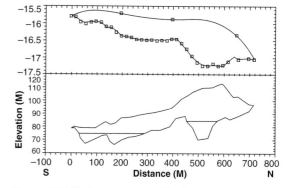

Figure 10.6 The interpretation profile of gravity measurement line 25 at the narrowest point of Pyynikinharju Esker. At the bottom of the figure, the upper curve depicts the elevations of the ground surface and the lower curve shows the elevation of the bedrock surface subject to the interpretation. The horizontal groundwater surface is only partly visible. Lake Näsijärvi is located north (N) of the line and Lake Pyhäjärvi south (S) of the line.
Source: GTK, T. Valjus, Jussi Ahonen.

bases overlap crosswise. This is how a sufficient level of accuracy can be ensured when interpreting gravity measurements.

10.3.4 Modelling and visualization

The collected data were initially processed with ArcMap software (version 9.3) and surface models of the results were created. The ground surface elevation model was based on the National Land Survey of Finland's laser scanning material, the bedrock surface elevation model on the rock outcrops, drilling data and gravity measurements, and the groundwater surface model on groundwater wells and natural water surface measurement data.

The surface models created were used as the basis for the structural 3D model created using the GEOCOM's Surpac software (version 6.1). Geological units based on drillings and other research material were determined for the model. According to the conclusion, the soil of the study area was divided into four main geological units for the model. The colour referring to the unit concerned in the modelling and cross sections is mentioned in parenthesis.

* bedrock (red);
* ground moraine – the bottom moraine layer detected in connection with the drillings (violet);
* sand and gravel formation + made-up ground – core of the esker (green);
* silt and clay layers – the layers of silt, clay, peat and slurry covering and bordering the esker (blue).

The definition of the units was based on observation data obtained from over 3000 geotechnical investigation points including new drillings made in this project.

The model consists of 5 × 5 × 1 m blocks. Each block is part of one of the previously mentioned geological units. The City of Tampere coordinates and elevation system was used in the model. The internal structure of the model can be examined with the aid of cross sections. The model itself is in Surpac format.

10.3.5 Geological structure of the Pyynikinharju Esker and a 3D model

Due to the bedrock shear zones across the study area, the bedrock surface is relatively low at the modelled area. The shapes of the bedrock are covered by the compact ground moraine which particularly occurs at the bedrock depressions. In places, the ground moraine layer rises above the level of the groundwater surface and thus has a local effect on the flow of groundwater. The extent of the ground moraine remained unclear, since most of the previous drillings had not reached this layer.

The Pyynikinharju Esker mainly consists of layers of well-sorted sand and gravel. On the sides of the esker clay and silt can be found. On the side of Lake Pyhäjärvi, the layers of clay and silt are located at a level that allows them to act as groundwater flow barriers. As a result of this, in the clay-covered areas artesian groundwater conditions occur. The study area contains several springs, with the Tahmelanlähde spring being the most significant. Areas with artesian groundwater are particularly sensitive to changes in the environmental conditions. In connection with land development, it is particularly important that the groundwater pressure is not released in an uncontrolled way.

Details of the structure of the Pyynikinharju Esker are presented in the cross sections below (Figures 10.8 to 10.13). The location of the cross sections is presented in Figure 10.7. The elevation ratio of the cross sections has not been emphasized; both the horizontal and the vertical axis have the same scale. The geological units in the model are presented with different colours, and the groundwater surface is shown as a thin line. The drilling points in the cross sections are presented with round symbols and vertical profiles. In examining the model and the cross sections, it should be considered that the soil structure and the stratigraphy of soil is known only at the drilling points and that for the remaining parts, modelling is based on geological interpretation.

10.3.6 Reports related to land-use planning

Spring mapping and report on the suitability for construction of the allotment garden area

An allotment garden area called the Pispala herb garden is located in the stage III planning area, at Tahmela on the shores of Lake Pyhäjärvi (Figure 10.14). In the 1998 Tampere inner city master plan, the area was allotted for single family dwellings.

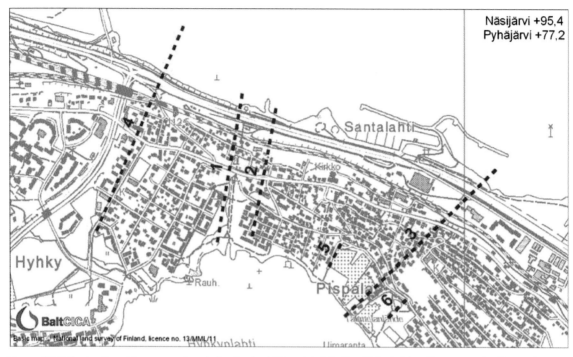

Figure 10.7 The location of the cross sections in the structural 3D model of the Pyynikinharju Esker. The cross section figures above are in the north-south direction, beginning on the shores of Lake Näsijärvi and ending on the shores of Lake Pyhäjärvi.

The first results based on the GTK soil modelling were published in autumn 2010. Some of the study area's groundwater wells had not yet been installed at the time, but the initial results indicated that groundwater was very close to the surface in some parts of the Lake Pyhäjärvi shoreline. Based on the observations, an investigation on the allotment garden area's suitability for construction and a mapping of the springs

in the entire area were ordered before launching the detailed planning (Jokinen, 2011).

Both investigation and mapping were done in spring 2011. Based on the report on the allotment garden area's suitability for construction, for which ground penetrating radar was used, the area is divided into a poorly constructible peat area, an average silt area and a readily constructible sand area on the shore

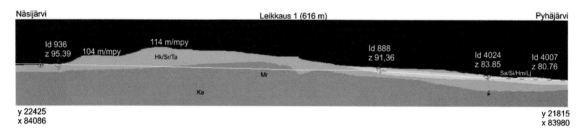

Figure 10.8 Cross section 1 at the narrowest part of the Pyynikinharju Esker. The southern slope gently slopes downward towards Lake Pyhäjärvi. The groundwater surface gradually slopes downward towards Lake Pyhäjärvi though the ground moraine

rises above the groundwater surface (white line) in some places. At this point, the bedrock surface does not rise above the groundwater surface. For a definition of soil units, see Section 10.3.4.

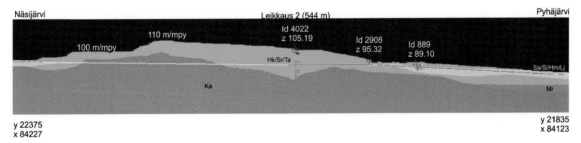

Figure 10.9 Cross section 2 is located east of cross section 1. In places, the bed rock surface rises above the groundwater surface, acting as a structure directing the flow of groundwater. Groundwater can still flow relatively freely through the esker at this point as the flow bypasses the uneven rock elevations.

(Jokinen, 2011). Based on the results, the groundwater surface level of the area is close to the ground surface and, in places, higher than the ground surface, which means that the area contains pressurized groundwater. Based on spring mapping, there are several active springs in the area and in its vicinity.

According to the report on the area's suitability for construction, artesian groundwater causes construction risks: an impact on the flow of groundwater may result in changes in the groundwater surface level in built areas and thus cause a risk of subsidence

and possibly even an increase in the risk of collapse along steep slopes (Jokinen, 2011). Since changes in the flow of groundwater are expected in connection with climate change, environmental authorities should examine the risks related to groundwater in more detail in the allotment garden area.

Workshop on adaptation practices

Adaptation measures have been planned during the project in conjunction with the City of Tampere Land Use Planning Department, experts in geotechnology,

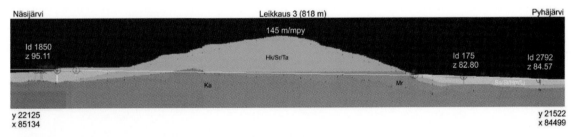

Figure 10.10 Cross section 3 at the highest elevation of the Pyynikinharju Esker. At this point, the filling bank of the shores of Lake Näsijärvi is at its widest. At the southern end of the cross section, the groundwater surface rises very close to the ground surface.

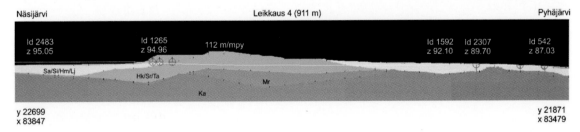

Figure 10.11 Cross section 4 is located at the western tip of the study area. At this point, thick layers of clay and silt are observed at the southern tip of the cross section.

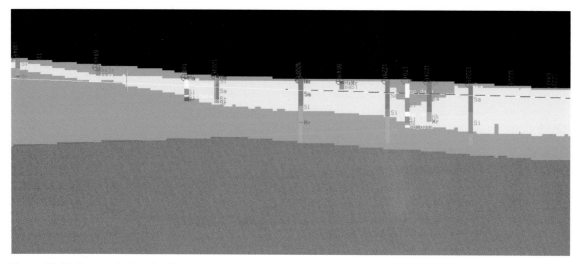

Figure 10.12 Cross section 5 at the allotment garden area. At this point, groundwater is artesian and its pressure level increases close to the ground surface, but it does not reach it (cf. Figure 10.9).

water supply and the environment as well as the Geological Survey of Finland.

The City of Tampere and GTK organized a Balt-CICA workshop entitled 'Climate Change Adaptation Practices in Urban Environment Planning – Case: the Pyynikki-Epilä Esker' ('Ilmastonmuutokseen sopeutumiskeinot kaupunkiympäristön suunnittelussa – case Pyynikin-Epilän harju') in Tampere on 18 October 2011. The workshop was intended for experts and authorities dealing with soil, groundwater and land-

use planning. Based on the material produced in the framework of the BaltCICA project, the workshop discussed the potential risks caused by climate change to soil, groundwater and built environment of the Pyynikki-Epilä esker area and defined risk adaptation practices (Figure 10.15).

Twenty-four experts in various fields participated in the workshop. The workshop defined increasing precipitation, potential increase in frequency of extreme natural phenomena (rain storms) as well as a decrease

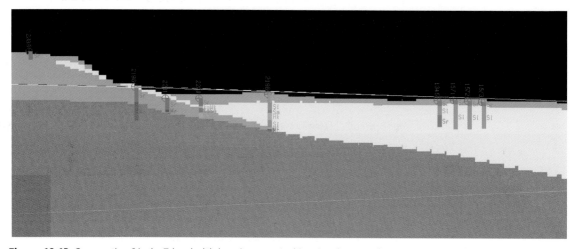

Figure 10.13 Cross section 6 in the Tahmelanlähde spring area. At this point, the groundwater is artesian and pressure level increases up to the ground surface level or above it.

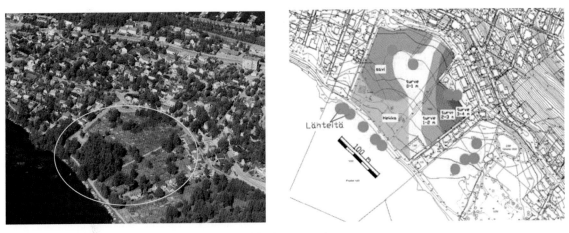

Figure 10.14 The Pispala allotment garden area and suitability for construction.
Source: Photo: Lentokuva Vallas Oy, City of Tempere. Map: City of Tampere, Petri Jokinen.

in frost and water content of the snow cover as phenomena causing the risks related to climate change in the esker area. Changes in land use and other human activities (construction, excavation, etc.) were also regarded as potential causes for risks.

Potential changes in the quality and level of the groundwater surface, problems caused by increasing surface runoff and changes in the area with artesian groundwater were considered the most significant risks. Changes in the quality of the water in Lake Näsijärvi may also cause risks. Rise of water level will cause risks in the downstream of Lake Pyhäjärvi's water system.

Only a part of the study area (Hyhky) is located in the groundwater extraction area. This means that instead of the quality of groundwater, the potential changes in the level of the groundwater surface and in its flow direction became the most important

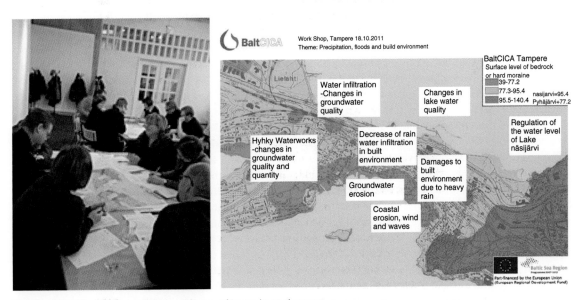

Figure 10.15 The Tampere workshop and some of its results on the map.
Sources: Photo: GTK, J. Klein. Map: City of Tampere, Jussi Ahonen.

questions at the workshop. The increase in precipitation in connection with climate change may have a positive effect on the esker's groundwater balance if the rain water is infiltrated to the esker. The risk is that increasing precipitation leads to more heavy rainfall events. In that case, the esker will not be able to absorb the water completely. Supplementary construction in the esker area and the related decrease in absorbent surface area could also affect the recharge of groundwater.

10.4 Results

10.4.1 Risk factors due to the geological structure of the Pyynikinharju Esker

Based on the 3D model, large-scale natural disasters are unlikely to occur today. For instance, the uncontrolled flow of water through the esker's old timber drive tunnels or along the transport corridor is not likely in the future based on the current settlement and control, even during exceptionally high floods combined with hard northern or north-eastern winds. These excavations have, of course, opened and weakened the structure of the Pyynikinharju Esker compared to its state before the excavations.

In the long term, the flooding of Lake Näsijärvi will not cause more danger than today, since the lake's discharge streambed is located at its least elevated tip. Potential changes can be dealt with where necessary by regulating the draining of water. Thousands of years ago, nature tested the resistance of the long-winding Epilä-Pyynikki Esker as the water levels slowly rose for several metres (the transgression of the southern tip of Lake Näsijärvi) and rapidly descended for a few metres (the formation of a new discharge streambed towards Lake Pyhäjärvi).

However, careless human activities can lead to minor risk factors, for instance on the gentle lower slope of the Pyynikki area that slopes downward towards Lake Pyhäjärvi and where the artesian groundwater sealed by the clay layer is attempting to discharge to the surface. If such an area becomes marshy, it indicates a prolonged phreatic groundwater discharge. Unguarded drilling, pilework and excavations may result in leaking holes, slippage and in the worst case, minor landslides.

10.4.2 Adaptation and preparation practices in land-use planning

Based on the workshop results, a list of the central adaptation and preparation practices was made:
- *Continuous monitoring of the balance and quality of groundwater* (e.g. inspections at regular intervals) – reactions to changes.
- *Precipitation monitoring in the esker area* – installation of rain gauges.
- *Local surface runoff guidelines* for Pispala, to be issued in connection with the development plan or as a separate project.
- A *further report on the artesian groundwater* in the Pispala allotment garden area: Mapping the risks caused by the construction and excavation measures and guidelines for the use of the area based on the mapping.
- *A more detailed mapping of the other areas that could contain artesian groundwater* and guidelines for the use of the areas mentioned above and for land-use planning.
- *Guidelines for the inhabitants*: A printed instruction leaflet and an online version as well as information through other channels on subjects such as methods for surface runoff management and groundwater. The inhabitants will be informed at workshops where the guidelines will also be drafted.
- *Examining the extent of shore infiltration at Lake Näsijärvi* and estimating whether the infiltration should be limited due to the quality of groundwater.
- *Reviewing the instructions for controlling Lake Näsijärvi* – have potential flood risks along Lake Pyhäjärvi water system and the lower course of the Kokemäenjoki River been taken into consideration?
- *Informing the authorities and planners* of the risks and adaptation practices (training events, updating planning instructions, an instruction leaflet on the adaptation practices).

Since the implementation of these measures was beyond the timeframe of the BaltCICA project, it was agreed that the City of Tampere Urban Development and Sustainable Community Unit would be in charge of programming further measures.

10.4.3 Future measures

Future measures are planned with respect to land-use planning, surface runoff management and climate change adaptation at the regional level.

The update of the Pispala detailed plan is for the Pispala area in the drafting stage. The planning areas cover the central parts of the Pispalanharju Esker and include the Tahmela allotment garden area.

The authority work meeting of the City of Tampere and the Pirkanmaa Centre for Economic Development, Transport and the Environment held in January 2012 dealt with the results on the BaltCICA study area. At the work meeting, participants agreed to prepare a report on the groundwater and surface runoff in the Pyynikinharju Esker area intended for zoning. A plan for surface runoff for the entire area would be made based on this report, and guidelines and regulations on groundwater and surface runoff required for the plan would be determined. Recommended measures and guidelines can also be added to the Pispala building system guidelines, which will be drafted in connection with the plan.

The groundwater and surface runoff report will also examine potential areas with artesian groundwater on the shores of Lake Pyhäjärvi. Based on the report, the potential limitations and recommendations for land use in the Pispala allotment garden area can be determined before the detailed planning is launched for the area.

In connection with the Santalahti detailed plan, a plan for surface runoff will also be made, and regulations on surface runoff will be indicated in the plan.

In connection with the Pispala detailed plan, the initial results of the BaltCICA project have been presented to the local inhabitants in the detailed plan participant group and in the Pispala district magazine. The central questions of modernizing the Pispala detailed plan are related to how to protect the old buildings in the area while assessing the right to erect new buildings. The results of the reports on groundwater and surface runoff will be communicated to the citizens. Climate change adaptation measures will be developed with the involvement of the inhabitants of the area. The management of surface runoff is an important subject for various reasons. The condensing urban structure changes urban hydrology and causes a risk of floods due to surface runoff. An amendment of the Water Services Act is pending, and the reform is likely to transfer the liability for the management of surface runoff to the municipalities. Comprehensive plans for surface runoff can prevent floods, protect groundwa-

ter and bodies of water and help prepare for climate change.

The Tampere inner city surface runoff programme will be made and the catchment area report will be updated with the help of an external consultant. The material produced in the framework of the BaltCICA project will be used for creating the surface runoff programme.

In addition the City of Tampere Urban Development and Sustainable Community Unit has started to coopere with the Tampere University of Technology with the aim being to prepare for floods caused by surface runoff and to identify the information needs of citizens.

The effects of previous extreme weather conditions (such as rain storms and winter floods) on surface runoff management will be examined in more detail (overflow of wastewater treatment plants or pumping stations, blocked rescue routes and erosion).

The cooperation will help recognize central operator's practices and development requirements. In addition, the need for preparation planning or for increasing cooperation between various sectors will be outlined.

The research work will also help estimate whether the authorities are adequately prepared for floods caused by surface runoff and how the inhabitants of the municipality should be informed of surface runoff. The Tampere Central Region Climate Strategy 2030 will focus on methods for controlling climate change. The City of Tampere unit for sustainable community is launching regional adaptation programme work as an extension of the climate strategy. Its main focus will be on preparing for the risks caused by climate change and adapting to the extreme weather conditions caused by climate change.

10.5 Conclusions

The BaltCICA project provided new information on the geological structure of the Pyynikinharju Esker as well as the risk caused by the climate change in the esker area. The project was particularly important to the esker area land-use planning projects and more generally to Tampere urban planning as it pointed out the importance of adapting to climate change. In Finland, Tampere is a forerunner in the field of climate change control, for example, through the ECO2 project (Tampereen kaupunki, 2012), but in the past there has been less emphasis on questions related to adaptation both in the regional climate strategy and in the City of Tampere land-use and public area maintenance planning.

The Tampere adaptation workshop brought up the need for dialogue with the inhabitants of the esker area on the potential risks caused by climate change and human activities and on the preparation and adaptation practices related to these risks. It was also considered necessary to issue a citizen's leaflet to be distributed to the inhabitants of Tampere on a larger scale. In connection with the Pispala detailed plan, dialogue with the inhabitants is already possible today. A more extensive adaptation workshop and a citizen's leaflet are likely to be possible as part of the regional adaptation programme work.

Familiarizing the authorities and operators concerned with the climate change adaptation practices is a challenge that the City of Tampere will have to consider in the future. The workshop brought up measures such as internal training events and the update of planning guidelines.

Acknowledgements

The authors of the report wish to thank the following people involved in the project. Geologists Hilkka Kallio, Sakari Kielosto and Juha Reinikainen of GTK participated in the geological studies and background reports. Geophysicist Tuire Valjus was in charge of the gravity measurements and their interpretation. The drillings were performed by the City of Tampere measurement services, with Sakari Oittinen and Harri Ruhala as well as Petri Jokinen from the City Planning Services in charge of the drilling procedures.

Maria Åkerman, Kari Hietala, Veikko Vänskä and Jouko Seppänen from the City of Tampere and Johannes Klein and Hilkka Kallio from GTK were involved in planning the scenarios and the adaptation and preparation practices.

References

Alhonen, P., 1988. Tampereen luonnonympäristön kehitysvaiheet. In: P. Alhonen, U. Salo, S. Suvanto and V. Rasila, 1988. *Tampereen historia I*. Tampere: Tampereen kaupunki, pp.4–50.

Joint Authority of Tampere City Region, 2010. *Tampereen kaupunkiseudun ilmastostrategia 2030*. Tampere: Tampereen Kaupunkiseutu, SITO, VTT Technical Research Center of Finland.

Jokinen, P., 2011. *Rakennettavuusselvitys, Pispalan ryytimaat*. Tampere: Tampereen Infra & Geowork Oy.

Kukkonen, M., 2003. Maaperäkartta. *Selitys 2123 09 Tampere*, 1:20000, Espoo: Geologian tutkimuskeskus.

RKY, 2009. *Sites of National Importance for Built Heritage, Finland National Board of Antiquities*. [online] Available at: <http://www.rky.fi> [Accessed 3 September 2012].

Taipale, T. and Pulkkinen, T., 2010. *Vaitinaron vesistötäyttö, Pohjaveden isotooppitutkimus*. Espoo: Ramboll Finland Oy.

Tampereen kaupunki, 2012. ECO2 – Eco-efficient Tampere. [online] Available at: <http://www.eco2.fi> [Accessed 3 September 2012].

Tikkanen, M. and Seppä, H., 2001. Post-glacial history of Lake Näsijärvi, Finland, and the origin of the Tammerkoski Rapids. *Fennia*, 179 (1), pp.129–141.

Valli, T. and Mattsson, A., 1998. Gravity method – an effective way to prospect groundwater areas in Finland. In: A. Casas, *Proceedings of the IV Meeting of the Environmental and Engineering Geophysical Society (European section)*. Barcelona, Spain 14–17 September 1998. Madrid: Instituto Geográfico Nacional, pp.185–188.

Virkkala, K., 1962. Suomen geologinen kartta. *Maaperäkartan selitys 2123*, 1:100000, Helsinki: Geologinen tutkimuslaitos.

11

Climate Change and Groundwater: Impacts and Adaptation in Shallow Coastal Aquifer in Hanko, South Finland

Samrit Luoma[1], Johannes Klein[1,2] & Birgitta Backman[1]

[1]Geological Survey of Finland (GTK), Espoo, Finland
[2]Aalto University, Espoo, Finland

11.1 Introduction

The shallow groundwater aquifer in Hanko area, in southern Finland, mainly consists of glaciofluvial Quaternary sand and gravel sediments, deposited during the deglaciation of the last Weichselian-glacial stage and Holocene. Aquifers are related to eskers or ice-marginal end moraine complexes (Fyfe, 1991; Saarnisto & Saarinen, 2001). The aquifer is an important freshwater source for Hanko town and local industries. The aquifer is sensitive to climate change because the groundwater table is close to the ground surface, the aquifer extends to the sea shore, and water pumping levels are at the moment below sea water level. The potential effects of climate change can cause risks for the groundwater resources of the Hanko coastal aquifer. Not only the change of recharge patterns (influenced by precipitation and temperature), but also sea level rise, make the coastal aquifer more vulnerable than other aquifers on the mainland. Moreover some groundwater risk areas already exist in the area, which make the aquifer more vulnerable to groundwater contaminations. In order to safeguard future water supply and groundwater resources management under climate change conditions, a comprehensive understanding of the hydrogeology and groundwater flow system of the aquifer is needed.

This study aimed to investigate the effects of climate change on the availability of the groundwater resources by evaluating changes in groundwater recharge and water tables under different future climate change scenarios using groundwater flow models. Three-dimensional groundwater flow models both as steady-state and transient flow models were developed to assess the impacts of climate change on the shallow groundwater aquifer in Hanko. The results from the groundwater flow model will provide information for local authorities for groundwater management in Hanko in the future.

11.2 The study area

The study area is located on the Hanko peninsula, southernmost of Finland, approximately 120 km southwest of Helsinki (Figure 11.1). The main groundwater area consists of Quaternary sand and gravel deposits. The total area of the Hanko peninsula covers about 116 km^2 with approximately 10 000 inhabitants. The economy consists of 60% services and 38% industry. The town has a busy port and chemical industry. Additionally, the town is also a popular summer resort with a coastline of approximately 130 km, of which 30 km are sandy beaches.

Climate Change Adaptation in Practice: From Strategy Development to Implementation, First Edition.
Edited by Philipp Schmidt-Thomé and Johannes Klein.
© 2013 John Wiley & Sons, Ltd. Published 2013 by John Wiley & Sons, Ltd.

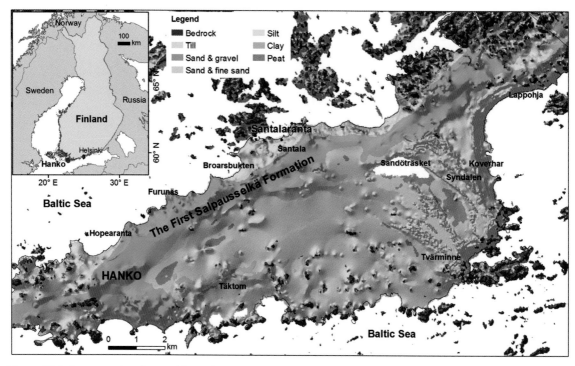

Figure 11.1 Quaternary deposit map of the study area.

11.2.1 Geological and hydrogeological background of Hanko aquifer

The shallow groundwater aquifer of Hanko belongs to the Quaternary First Salpausselkä end moraine complex. The primary sand and gravel formation deposited in front of the ice sheet into deep water in the end of the Weichselian glacial stage over a time span of 200 years, 11 250–9 900 years ago (Fyfe, 1991; Saarnisto & Saarinen, 2001). The form of this primary sand and gravel formation was quite narrow and low (Fyfe, 1991). When the ice sheet withdrew from the area, the deep water sitting formation was covered by glaciolacustrine sediments, fine sand and clay layers, of the Ancylus Lake and Littorina Sea. The sea level has been regressive since glacial time caused by the isostatic land uplift, the Salpausselkä formation was exposed to sea waves and wind (Kielosto, Kukkonen, Sten & Backman, 1996). Therefore, littoral sand deposits and fine sand wind deposits covered the clay and fine sand sediments in places. The bedrock in the Hanko area consists mainly of Precambrian crystalline igneous and metamorphic rock strongly fractured along preferential directions. The bedrock is mainly covered by Quaternary deposits, with some outcrops in the area.

The complicated geological structure of the First Salpausselkä formation results in complicated groundwater conditions. The permeability of formation material varies due to the changes in the grain size and thickness of the different layers. The structure of the aquifer in the northwest part is quite different from the southeast part. Sand and gravel layers alternate with clay and silt layers. In many places the geological stratigraphy from the soil surface downwards consists of the following: first sand and gravel layers with high hydraulic conductivity, then interbedded silt and clay layers with low conductivity, followed by sand and gravel layers with high hydraulic conductivity (Kielosto, Kukkonen, Sten & Backman, 1996; Breilin, Paalijärvi & Valjus, 2004).

11.3 Data

This study was to assess the impact of climate change on groundwater resources in the Hanko aquifer by utilizing three-dimensional (3D) groundwater flow models implemented in the MODFLOW code to evaluate the groundwater flow direction and water budget under current conditions and climate change scenarios based on the IPCC emission scenarios A1B and B1 (IPCC, 2000). The results from this study provide the 'Hanko Water and Wastewater Works' a knowledge base for groundwater management. Data originating partly from previous studies were collected by the BaltCICA project.

Available data was also collected from the Geological Survey of Finland (GTK), other research institutes, organizations and local industries. Compiled data consists of (1) geological structure, stratigraphy, and layer thickness data, (2) hydrogeological data, (3) geophysical data, (4) various kinds of monitoring data including precipitation, snow thickness, snow-water equivalent, air temperature, groundwater quantity and quality, groundwater pumping rate from the water intake plants on Hanko peninsula, and sea level data.

Additionally the following data were produced:

- Additional gravimetric data for the clarification of the formation structure of the aquifer.
- Groundwater monitoring data concerning groundwater level and groundwater quality in 12 groundwater observation wells including groundwater level, temperature and electric conductivity (EC) hourly during one year monitoring period.
- Chemical analysis of water samples from six wells in Santalanranta area and one sea water sample from the sea nearby the Santalanranta area (Figure 11.1).
- Slug-test in five observation wells for the estimation of the aquifer's hydraulic conductivity in the well screen interval.

Details of data and methods utilized in different data categories are explained in the following sections.

11.3.1 Current climate data

Time series of daily total precipitation and temperature data during the period of 1963–2010 (48 years) measured at the Tvärminne weather station are available from the Finnish Meteorological Institute (FMI),

and daily sea level data during the period of 1971–2010 (40 years) is available from the Marine Research Programme, Finnish Meteorological Institute. The sea level data consists of the daily rate for three measurements including the average, minimum and maximum sea water levels. A ten-year set (1991–2010) of daily snow-water equivalent data was received from the Finnish Environment Institute (SYKE). This data consists of measurements and simulations of the snow-water equivalent data from the station in the Santala area.

During 1971–2000, the average daily minimum, mean and maximum temperatures of the year are 2.9, 5.7 and 8.4 °C, respectively (Drebs et al., 2002). The lowest temperature was recorded at −33.9 °C on 10 January 1987 and the highest temperature was + 31.1 °C on 11 July 1983 (Figure 11.2). The lowest daily mean temperatures occur in January and February, and the highest daily mean temperatures occur during July and August. Total annual average precipitation was 620 mm. High precipitation mostly occurs in January and March as snow and in October and November as rainfall. The lowest average precipitation was recorded in May. Trend analysis was carried out for the temperature and precipitation data during the period of 1968–2010 using the Mann-Kendall, and Sen's non-parametric tests. Trend analyses show increasing trends with the rates of 0.0419 °C/yr for mean temperature and 0.0054 mm/yr for mean precipitation.

In Hanko area, groundwater recharge is generated by rain and snowmelt water. Figure 11.3 presents a flow chart of the recharge (r(t)) estimation process in this study. In winter and spring, the 'potential recharge' – pr (t), as input data for the recharge in the groundwater flow model consists of an estimate of snow melt from snow on the ground (Δ SWE (t)) and the precipitation for days with surface temperatures above −2 °C, assuming that precipitation at these days occurs as rain and can potentially infiltrate into the aquifer (Hänninen & Äikää, 2006; Sutinen, Hänninen & Venäläinen, 2008). Figure 11.2 shows that the amount of monthly mean potential recharge in spring and early winter, where the mean surface temperature is above −2 °C, is higher than the amount of monthly mean precipitation. Recharge − r (t) rate distribution contains temporal (t) and spatial (x) variations. How much of the available water (potential recharge)

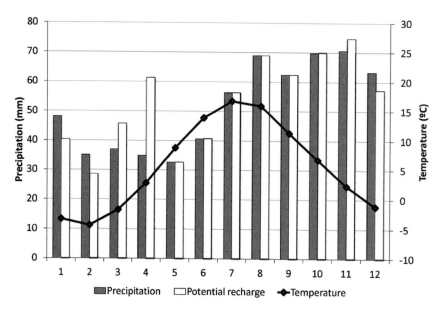

Figure 11.2 Monthly mean precipitation (grey bar, in mm, as rain and snow), monthly mean potential recharge (white bar, in mm, as rainfall and snowmelt water from snow water equivalent as described in Figure 3), and monthly mean temperatures (blue line, in °C) in the Tvärminne weather station in Hanko during the period 1971–2000.

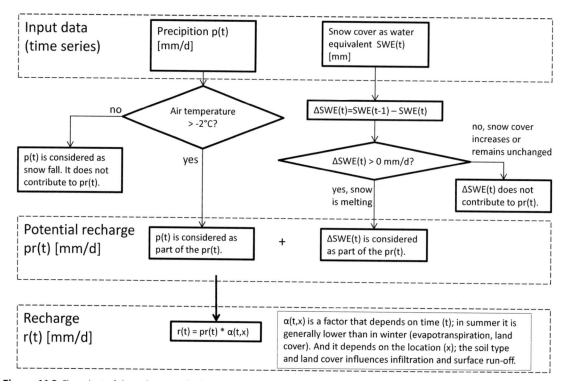

Figure 11.3 Flow chart of the recharge evaluation process used in this study.

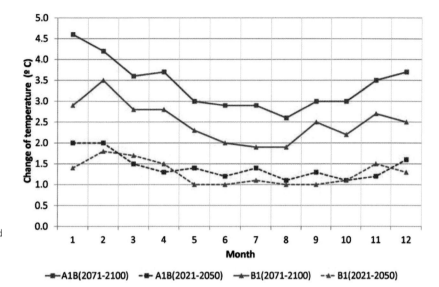

Figure 11.4 Changes of mean temperature for scenarios A1B and B1 for the periods 2021–50 and 2071–2100 compared to 1971–2000.

actually infiltrates, depends on soil type and condition, topography, land cover and evapotranspiration. In Figure 11.3 this is indicated as factor α (t, x). The recharge process is described more in detail in Section 11.4.2.

11.3.2 Future Climate Scenarios

The CLM (Climate Limited-area Modelling-Community) climate model data for the entire Baltic Sea Region was received from the World Data Center for Climate, Hamburg (2011). The data has a grid resolution of 0.2 × 0.2 Decimal Degrees. Two greenhouse gas emissions scenarios, A1B and B1 were used. These regional climate simulations contain 100 years data from the period 2001–2100 with a control run, C20, from period 1960–2000 available for comparison with the current climatic data for Hanko from FMI.

The A1B emissions scenario expects a possible future world of very rapid economic growth, global population peaking in mid-century and rapid introduction of new and more efficient technologies with a balance across all energy sources. The B1 scenario imagines a convergent world with rapid change in economic structures towards a service and information economy and the introduction of clean and resource-efficient technologies.

The Delta approach was used for the transfer process between the current climate and the scenario data

(e.g. Jylhä, Tuomenvirta & Ruosteenoja, 2004; Veijalainen et al., 2010; Okkonen, 2011). Two climatic parameters (temperature and precipitation) of those three data sets (C20, A1B and B1) were grouped into three time spans: 1971–2000 for C20, 2021–50 and 2071–2100 for A1B and B1 scenarios. Plots of changes of mean monthly temperature and mean monthly sum precipitation for scenarios A1B and B1 compared to temperature and precipitation for the period 1971–2000 are presented in Figures 11.4 and 11.5.

Temperature data

In Hanko, mean temperatures increase for both scenarios and both periods compared to current temperatures (Figure 11.4). Overall, the winter mean temperature increases more than in the summer. The highest mean temperature in A1B (2071–2100), would have a mean temperature increase 2.6 to 4.6 °C, with an annual mean of 3.4 °C increase compared to the period 1971–2000. The maximum increase of mean temperature takes place in winter with an increase of 3.7 to 4.6 °C and the minimum increase takes place in summer with an increase of 2.6 to 2.9 °C. In the period 2021–50, the mean temperatures in both scenarios A1B and B1 are similar with the mean temperature increase 1.0 to 1.4 °C in summer, and 1.3 to 2.0 °C in winter with an annual mean of 1.3 to 1.4 °C increase from the current.

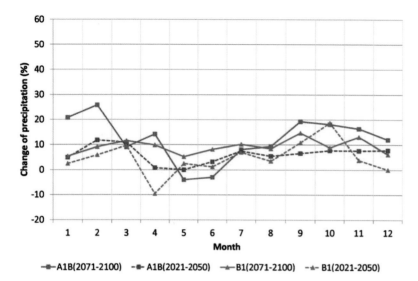

Figure 11.5 Changes of mean monthly sum precipitation for scenarios A1B and B1 for the period 2021–50 and 2071–2100 compared to 1971–2000.

Precipitation data

Unlike the temperature data, precipitation varies a lot between the scenarios (Figure 11.5). Overall, the annual sum precipitation increases 5 to 12% compared to 1971–2000, from the lowest in B1 (2021–50) to the highest in A1B (2071–2100). Seasonally, generally all scenario data, except data in April to June, show an increase of about 2 to 26% with the highest increase of precipitation in autumn and winter (Figure 11.5). In April to June the precipitation decreases by 0 to 10%, except for B1 (2071–2100) where the annual sum increases approximately 9% compared to 1971–2000.

Potential recharge

Potential recharge represents the amount of water potentially available for infiltration into the aquifer as recharge. Change in potential recharge directly affects the actual recharge. Overall, the average annual potential recharge in all scenarios increases by 8 to 15% compared to 1971–2000. Seasonally in all months, except in April to June, potential recharge increases (Figure 11.6). The highest increase occurs in winter (January and February) with dramatic increases of 18 to 55%. In spring, the snowmelt starts earlier and causes a potential recharge increase of 9 to 14%. Potential recharge has the lowest value in

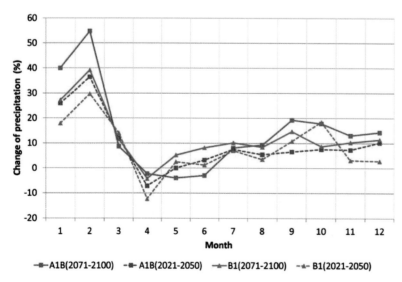

Figure 11.6 Changes of the scenarios mean of monthly sum potential recharge data A1B and B1 during the periods 2021–50 and 2071–2100 from the reference period.

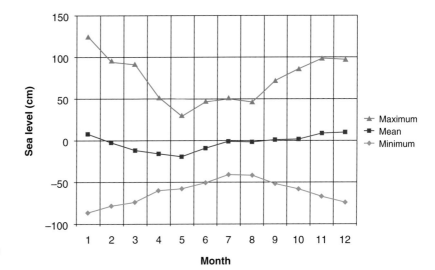

Figure 11.7 Monthly minimum, mean, and maximum sea level data during 1971–2010 in Hanko.

summer during May and June, and in the A1B scenario for 2071–2100 the potential recharge decreases about 3 to 4% from the current climate.

Based on the scenarios data, recharge is expected to increase because an increase of potential recharge during most of the year. In the winter months, when the highest increase is expected, temperatures are still low and evapotranspiration is hence small, that is, a large part of the potential recharge infiltrates into the aquifer and can contribute to the actual recharge.

11.3.3 Sea levels 1971–2010

The summary of monthly mean sea level data during 1971–2010 in Figure 11.7 shows the seasonal variation of sea level in the Hanko area. Overall the mean sea level varies between −0.62 and +0.74 m, with an average mean sea level of -0.03 m. Low sea levels take place in spring and summer. In autumn the mean sea level stays well around zero metres, and rises up above zero during the winter. In winter and autumn the sea level shows a high variation of up to 2.11 m. While, in late spring (April and May) and summer, the sea level is more stable and the differences of the maximum and minimum sea level range between 0.88 and 0.98 m. In addition, storm surges can push the sea level high up from normal sea level, for example, the Hanko area experienced the highest sea water level since 1887 at +1.24 m N60 during the storm surge on 9 January 2005. Trend analysis was carried out for the sea level data during the period of 1971–2010

using the Mann-Kendall and Sen's non-parametric tests. Assuming a linear trend, the sea level in the Hanko area was decreasing 1.33 mm/yr during the last 40 years (1971–2010). This value is considerably smaller than the value of 3 mm/yr relative land uplift (or sea level decrease) mentioned by Ekman (1996) or Kakkuri (1997). It has to be considered, however, that the time series taken into account in the articles starts only in 1971 and is rather short. Based on sea level observation starting in 1887 the decreasing trend in Hanko (and in the whole of Finland) is considerably slower after 1970 than before. In some cases (e.g. Hamina, in the eastern part of Gulf of Finland.) the observed sea level indicates a shift from decreasing to rising sea level in relation to the bedrock-bound reference (Johansson, Kahma, Boman & Launiainen, 2004). While the long-term sea level in the Baltic Sea can be well explained by such factors as land uplift and development of global mean sea level, the recent deviation from a steady linear trend can be explained by the water balance of the Baltic Sea and a strong correlation with the NAO (North Atlantic Oscillation) index (Johansson, Kahma, Boman & Launiainen, 2004; Johansson, Kahma & Boman, 2003).

11.3.4 Sea level rise scenarios

Sea level rise (SLR) data in the Baltic Sea area for the period 2000–2100 with the baseline 1995 (A1FI, A1B, B1, high and medium regionalized) stem from the CLIMBER model with compensation of the vertical

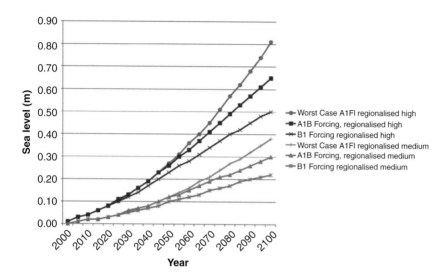

Figure 11.8 Sea level change scenarios during the period 2000–2100 in Hanko.

land movement component to create relative SLR scenarios for the provinces in the Baltic Sea Region. By the end of the 21st century, sea level in Hanko area may rise from 22 cm to 81 cm from the baseline in 1995 depending on the scenario (Figure 11.8). The CLIMBER model is described in detail in Petoukhov et al. (2000) and Ganopolski et al. (2001).

11.3.5 Observed groundwater level

Groundwater monitoring was carried out for 12 wells along the west coast of the main groundwater area during March 2009 and March 2010. The measurement was done by installing Schlumberger Mini-Diver data logger and pressure transducer in 12 wells which were divided into two groups based on the measurement types. The first group monitored groundwater level, temperature, and electrical conductivity (EC) in four observation wells, which are located close to the coastline, and the second group monitored groundwater level, and temperature in eight wells further in the mainland. All data (groundwater level, temperature and EC) were measured hourly. Groundwater levels from three monitoring wells during 1 April 2009 and 31 March 2010 plotted with the daily precipitation in the Hanko area are presented in Figure 11.9.

Groundwater level distributions vary spatially and temporally. They generally follow the ground topography from high groundwater level in the central-east mainland area gradually decreasing close to the sea level in the south-west. A steep hydraulic gradi-

ent is shown along the coastline in the north-west area. In the Santala area in the central-east of the study area, the fluctuations of groundwater levels – as in the entire area – show high water levels twice a year, during spring and late autumn to early winter and low water levels during summer and late winter (Figure 11.9). Sea water level has a direct impact on groundwater level variation along the coastline (Backman, Luoma, Schmidt-Thomé & Laitinen, 2007). In Hopearanta area, groundwater levels are generally low, with the average groundwater levels three to four meters a.s.l. and ground surfaces between 10 to 12 m a.s.l. The correlation between pumping rate of the Santalanranta water intake plant and groundwater level of observation wells around the pumping station and close to the sea shore, and sea level data was investigated for the mentioned period using the SPSS statistical software. Groundwater level has a negative correlation with the pumping rate, the correlation coefficients range between -0.61 and -0.82 ($p < 0.001$), and has a positive correlation with the sea level data, the correlation coefficients range between 0.38 and 0.53 ($p < 0.001$).

11.3.6 Water Intake Plant and Pumping Rate

The water intake plant in the Hanko area consists of five pumping stations of the Hanko Water and Wastewater Works and other pumping wells of local industries and private owned wells. The amount of

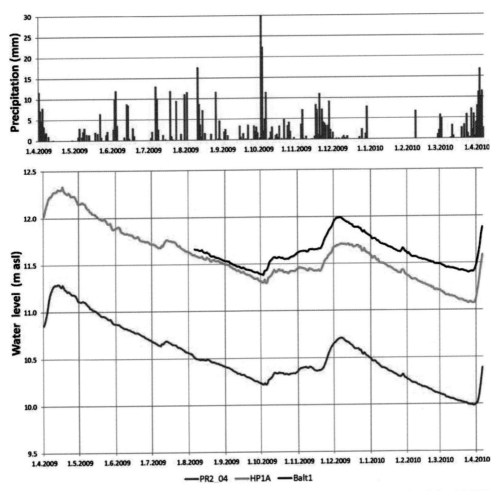

Figure 11.9 Hydrograph of groundwater level in the observation wells in Santala area and daily precipitation during 1.4.2009 – 6.4.2010.

groundwater pumping rates during 2009 provided by the Hanko Water and Wastewater Works was on average 3900 m³/d. The average pumping rate received from the Santalanranta water intake plant was 1355 m³/d. The daily pumping rates of the investigation period were available as input parameters for the groundwater flow model.

11.4 3D geological and groundwater flow models

The three-dimensional geological model and groundwater flow models, both as steady-state and transient models for the aquifer in Hanko, were constructed from the previously mentioned data, utilizing the integration of ArcGIS/ArcInfo and Groundwater Modelling Software (GMS™).

11.4.1 The 3D geological model

The shallow groundwater aquifer in Hanko consists of the glaciofluvial Quaternary sediments, deposited during the deglaciation. It naturally contains high heterogeneities and complexities. To provide a geological framework for the groundwater flow model and a quantitative understanding of groundwater flow characteristics in the Quaternary aquifer, which is essential for the long-term groundwater resource planning and management, a 3D geological model of the Quaternary deposit in the Hanko aquifer was constructed utilizing a great variety of geological and geophysical information, including Quaternary map, topography Digital Elevation Model (DEM), stratigraphy, groundwater table data, gravimetric survey and ground-penetrating radar (GPR) data. The software used was a combination of ArcGIS/ArcInfo and 3D geologic module from GMS, which was used also for the 3D visualization of the Quaternary geological deposit. Bedrock surface was identified and surface data of each horizon was interpolated using ordinary kriging and Inverse Distance Weighting algorithms in ArcGIS/ArcInfo. Grid surface data were then transferred into GMS for the 3D geologic modelling. Stratigraphic data were correlated, and in areas that lacked well data, GPR data was used to identify the distribution of sediments. The solid model of Quaternary deposit was constructed between topographic and bedrock surfaces. The model shows that the bedrock surface is highly undulated with low terrain bedrock in the east (zero to 10 m a.s.l. on average) and a buried bedrock valley in the north and NE-SW along the first Salpausselkä formation (-5 to < -25 m a.s.l. on average). Some kettle-hole features with maximum depth up to -65 m a.s.l. are found in this buried bedrock valley. The Quaternary deposit varies from less than one meter and up to 75 m thickness, with a mean thickness of approximately 13 m. The buried bedrock valley areas contain the thickest part of the aquifer with an average thickness of the aquifer of 21 m. The aquifer layers interval contain mainly sand, gravel but also till. The finer sediments, fine sand, silt and clay, are distributed throughout the area, and mainly in the distal slope in the eastern part above the bedrock terrain. Figure 11.10 presents the distributions of Quaternary deposit of the shallow aquifer in Hanko, viewed at an oblique angle northeast direction, showing the distribution of various geologic layers in the shallow aquifer in Hanko.

11.4.2 The 3D groundwater flow model

Model set up

Groundwater flow was simulated using the finite different MODFLOW2000 groundwater flow model (Harbaugh, Banta, Hill & McDonald, 2000), under the GMS graphic environment. The model comprises a rectangular grid of 186 rows and 269 columns and

covers a total area of 71 km^2. Grid cells are generally 50 m $\times$ 50 m, except for cells nearby the water intake plant area, where the grid cells are narrowed down to 5 m $\times$ 5 m. It is a one-layer model representing an unconfined shallow aquifer of the Quaternary sediments above the crystalline pre-Cambrian bedrock. The bedrock has a very low hydraulic conductivity and is regarded as an impermeable layer. Top of the model is the DEM surface and bottom of the model is the top of the bedrock surface.

The model boundaries were assigned based on the geological information of the study area including specific head boundary, General Head Boundary (GHB), seepage, drains and no flow boundary. The specific head boundaries are located along the interface of the groundwater aquifer and the coastline along the Baltic Sea. GHB is located in the north-east of the model domain, to describe the conditions between the model system and the external source/sink along the fractured bedrock area. Seepage flow boundary is located in the seepage flow areas along the coastline in the north-west and the east. The drains boundary is located in the east and south-east boundaries in the shallow bedrock surface area where the Quaternary sediment thickness is less than one meter, and the no flow boundary is located in the impervious bedrock area in the west of the model next to Hanko town.

Hydraulic parameters including hydraulic conductivity, specific yield and porosity were assigned to the model layer based on the hydrogeological data of the study area. The initial hydraulic conductivity (K) values for the model input were obtained from the soil analysis data and the slug test analysis. The estimates of K values correspond with the stratigraphy, for example, gravel and sand. Sand and gravel contain high K values of 7.6 to 12.48 m/d, fine sand the average K value is 6.12 m/d. In the high heterogeneous sediments such as interbedded sand and stone, and silt-fine sand-stone, the average K values are 4.79 and 2.79 m/d, respectively. Low estimated K values are found in silty sand, and silt and clay with the K values of 1.3 and 0.4 m/d, respectively. Since it is a one-layer model, the spatial distributions of K values assigned corresponded to the majority of the stratigraphy in the model domain. The K values were finally adjusted during the model calibration process. The specific yield value of 0.25 and the effective porosity values between 0.2 and 0.3 were assigned for the

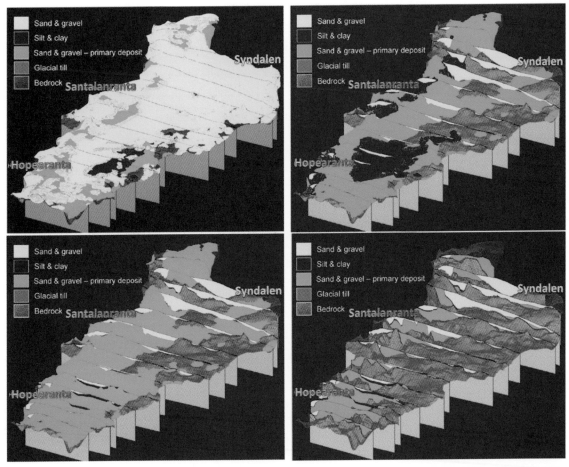

Figure 11.10 Geologic model of Quaternary deposit showing the distribution of various geologic layers of the shallow aquifer in Hanko (detailed location is in Figure 11.1).

model domain area. More details of the parameter assignment for the groundwater flow model can be found in Luoma (2012).

Groundwater recharge of the study area comes from rainfall and snowmelt (Figure 11.3). Plots of daily precipitation and hydrographs of daily groundwater level from the observation wells in the Santala area during 1 April 2009 and 6 April 2010 are presented in Figure 11.9. The hydrographs show the fluctuations of groundwater levels over time with high water levels occurring twice a year, during spring and late autumn–early winter, and low water levels occur during summer and late winter. This indicates that the main recharge processes in the Hanko area take place twice a year; during spring and late autumn–early winter.

During spring snow melts and with low temperature evapotranspiration is low. The snowmelt starts to fill up the soil pores and when the soil moisture reaches field capacity, the excess water passes through the vadose zone and reaches the groundwater table as a recharge. During summer and early autumn when evaporation takes place and plant growth resumes, the amount of recharge depends on the amount of precipitation and excess water over the field capacity of the soil. With no rain or insufficient amount of rain no recharge occurs in the aquifer area. The second phase of recharge occurs during late autumn – early winter, where evaporation and plant growth are diminishing. This can be seen from Figure 11.9 where the highest rainfalls took place on 3 October 2009 and continued

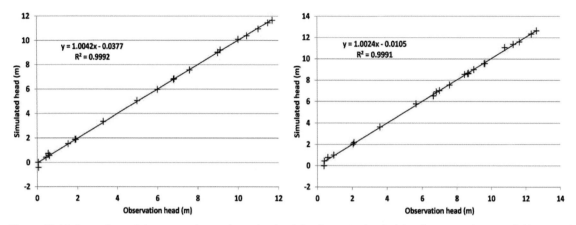

Figure 11.11 Comparisons of the computed groundwater head and the observation head of data from November 2009 (left) and April 2009 (right) conditions.

with small amounts of rainfall until the beginning of December. It took about 42 days from low groundwater level in autumn to reach its highest level at the beginning of winter.

In this study the snow-water equivalent values were received from SYKE. MODFLOW2000 requires an input of net recharge to a grid cell representing the groundwater table. The water lost due to evapotranspiration or run off are taken into account with the net recharge. As a percolation rate in this study was used the percentage of potential recharge based on the report from Lemmelä and Tattari (1986), which found that the mean cumulative percolations amount of water in the sandy aquifer in south Finland was about 95%, 60% and 20% of potential recharge during winter to spring, autumn and summer, respectively. These percentage values correspond to α (t, x). As mentioned in the previous section (Figure 11.3), recharge rate distribution depends on many factors including the amount of precipitation, land use, surface topography and soil moisture conditions. These cause the recharge rates to vary spatially and temporally. For the temporal variation as in winter to spring conditions, up to 95% of the potential recharge was used for the initial input of the maximum recharge, and in autumn 60% of the potential recharge was applied. For the spatial variation, the maximum recharge was assigned at 95% of the potential recharge in winter to spring in the areas of sand and gravel along the first Salpausselkä deposit in the north-west of the study area, and

10% of the potential recharge in clay deposit or urban areas where the percolation rates are very small.

Steady-state flow model – Current conditions
The steady-state flow models were simulated for April and November 2009, representing wet and dry conditions, respectively. Sea level values for the specific heads were assigned as zero and monthly mean pumping rates were used for the data input. Groundwater level data from 22 wells were used for manual model calibration. The initial K values were adjusted to fit the model with a reasonable range of data. There is a strong correlation between computed groundwater heads and observation heads from November and April 2009 (Figure 11.11). The head discrepancies overall are less than 10 cm, except for the well located next to the Santalanranta pumping station, where the computed heads were 35 and 38 cm (November and April 2009) lower than the observation head. This well is affected by the water drawdown from the pumping station, although the grid around the water intake plant was narrowed down to a cell size of 5 meters. More details of the study of the steady-state groundwater flow model can be found in Luoma (2012). A simulated groundwater table map of April 2009 (wet condition) from the results of a steady-state groundwater flow model in Hanko is shown in Figure 11.12.

Sensitivity analysis was performed to evaluate to which input parameters the groundwater level is most sensitive by varying (increasing or decreasing) by 10%

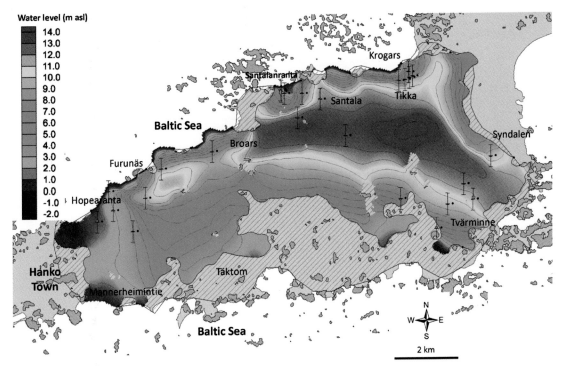

Figure 11.12 Simulated groundwater head map of April 2009 (wet condition) from the results of a steady-state groundwater flow model in Hanko.
Notes: Black dot represents observation data. Green bar indicates simulated head within 10 cm of the observed head and the colour bar above centre indicates simulated head greater than observed and vice versa. Contour interval is 2 m

the input parameters including K value, recharge, pumping rate, heads along the boundary condition areas, for example, specific head, GHB, seepage and drain. The changes of groundwater heads were compared. The sensitivity analysis shows that recharge and K values cause the greatest change in the simulation heads with a difference of 6.2 to 8.9% from the normal condition. The change of other parameters has only a small effect, except in the specific head boundary condition where the head difference depends also on the hydraulic conductivity of the aquifer materials in the boundary area. A high pumping rate from wells located in the high hydraulic conductivity materials along the coastal area contributes to a higher amount of sea water intrusion into the pumping well and the aquifer in the water intake plant area.

Transient flow model – Current conditions

The transient simulation of the current climate condition was run for one year during the abovemen-tioned investigation period with 20 stress periods and 75 steps varied based on the variations of precipitation data. Daily groundwater level data from 12 observation wells of the same year were used for the calibration process. For the input parameter, the calibrated K values were maintained the same as in the steady-state flow simulation. The other input parameters were varied during the year, for example, seepages, drains and GHB were assigned based on the calibrated data for April (wet) and November 2009 (dry) conditions, the daily mean sea level data of the investigation period were used as an input parameter for the specific head. The daily pumping rate of each pumping station was also available for the same period. Likewise, the recharge rate data was received from the daily potential recharge and its spatial and temporal factor (α (t, x)). The comparisons of transient simulated heads and observation heads of the selected wells is shown in Figure 11.13a for wells located inland in the Santala area, and Figure 11.13b for wells located

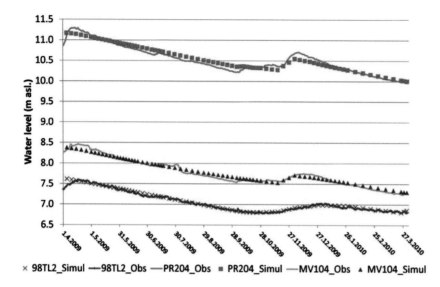

Figure 11.13a Transient simulation heads (_Simul) and observation heads (_Obs) of wells located in land in the Santala area.

along the coastline nearby the Santalanranta water intake plant area. Overall the simulated heads fit well with the observation heads, except for the wells along the coastline and close to the water intake plant area. The high fluctuations of groundwater level due to the influences from fluctuations of sea level and pumping rate make the calibration difficult.

Transient flow model – Climate scenarios

Transient flow models of the climate scenarios data were simulated in order to observe potential changes in groundwater head, and flow direction based on the scenarios data, relative to the current situations. Model domain and hydraulic parameters input for the scenario transient flow models were the same as in the transient flow model of the current condition (during the period 1.4.2009 – 31.3.2010), and the daily recharge rates were received from the percentages of the potential recharges which are the products of the daily potential recharges (pr (t)) and its spatial and time factors (α (t, x)). In this case, the daily potential recharges were based on the A1B 2071–2100 scenario

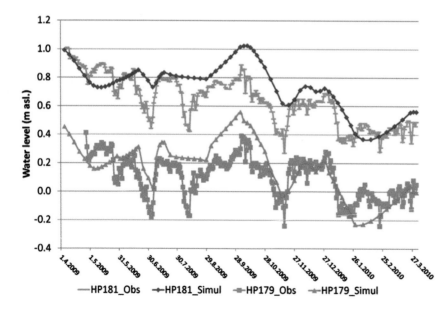

Figure 11.13b Transient simulation heads (_Simul) and observation heads (_Obs) of wells located along coastline nearby the Santalanranta water intake plant area.

data, and the spatial and time factors (α (t, x)) used the same values as in the current conditions. Sea level data contain high uncertainties. A sensitivity analysis was carried out for the scenario A1B 2071–2100 by simulating the transient groundwater flow model based on (1) the current sea level data, and (2) the sea level data at year 2100 (a 65-cm increase of sea level, based on the A1B regional high scenario). The groundwater heads from those different data sets were compared.

Water budgets

Simulated steady-state groundwater modelling of the current situation shows a good balance of the water budget in the Hanko area. Groundwater inflow from precipitations provides the main source to the groundwater system with the recharge of 41 947 and 28 251 m^3/d or about 100 to 99% of the total inflow for April (wet) and November (dry) 2009 conditions, respectively. While the main groundwater outflow derives from natural sources such as drains and seepage, specific head, and GHB with the outflow rates 39 589 m^3/d (94% of the total outflow), and 26 272 m^3/d (92%) for April and November, respectively. The outflow from all pumping wells in the model area is 2360 m^3/d and 2160 m^3/d (6% and 8% of the total outflow) for April and November 2009, respectively. This indicates that current pumping rates are still well below the recharge rates. However, a high pumping rate from a specific pumping well located in the high hydraulic conductivity area along the coastal area, will contribute more sea water intrusion into the pumping well and the aquifer area.

Impacts of climate change on a shallow aquifer in Hanko

Water budgets from the results of steady-state groundwater flow simulations both in the April (wet) and November (dry) 2009 conditions show good balance between inflow and outflow of the models. The natural outflow from drain and seepage and specific head contributes the most to the outflow, while the outflow from human activity from pumping wells at the current pumping rates contributes at a low level and is much lower than the inflow from natural recharge. Results of the steady-state flow model also imply that the natural processes have the strongest impacts on the groundwater budget of the shallow aquifer in Hanko. The sensitivity of the ground water model to

changing recharge and sea level rise was tested with two parameter set-ups. In Case 1 the daily potential recharges were based on the A1B 2071-2100 scenario data and the daily mean sea level data of the investigation period were used. In Case 2 potential recharge based on the A1B 2071–2100 scenario was used combined with a 65 cm increase of sea level. The results of the sensitivity analysis of the transient MODFLOW simulations based on the scenarios A1B 2071–2100 for the shallow aquifer in Hanko are summarized as follows:

Case 1: A1B 2071–2100 potential recharge data and current sea level conditions

Potential recharge increases approximately 15% on average (−2 to 55% in range; Figure 11.6) or 193 mm/yr compared to the current potential recharge (1971–2000) or about 310 mm/yr compared to the investigation period. These lead to an increase in groundwater levels by an average of 0.52 m (range of −0.01 to 1.47 m). The head differences vary between locations and seasons. In the southwestern part of the model area, where groundwater is currently about 7 meters below the surface, during the wet season with a high recharge (a 13 to 55% increase from the current condition), groundwater level can increase by 0.05 to 0.27 m. During the dry season with low recharge, the groundwater level changes between −0.01 m to 0.05 m. While in the central-eastern part of the model, the groundwater level is closer to the surface than in the south-western part (2 to 5 metres below the surface on average), the change of groundwater level maintains similar levels throughout the year from 0.93 to 1.47 m. This is the same as along the coastline where the groundwater level changes vary about 0.11 to 0.12 m from the current condition. Groundwater level and flow direction remain the same pattern as the current condition, but more flooding areas are observed in the west and eastern sides of the lake area, where the topographic elevation decreases to the east of the study area.

Case 2: A1B 2071–2100 potential recharge data and sea level conditions in 2100

In the flow model potential recharge based on the A1B 2071–2100 scenario combined with a 65 cm increase of sea level, causes an increase of groundwater level by an average of 0.88 m (range

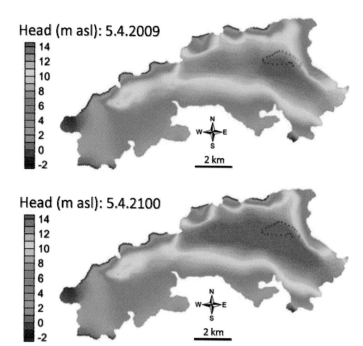

Head (m asl): 5.4.2009

Head (m asl): 5.4.2100

Figure 11.14 Distribution maps of simulated groundwater heads based on the current 2009–10 potential recharge data (above) and the scenario A1B 2071–2100 potential recharge data (below). Dashed line represents lake area.

of 0.33 to 1.70 m) from the current condition or about 0.36 m increase compared to Case 1. Groundwater flow direction and pattern remained the same as in the simulation result of Case 1, but more flooding areas occurred to the west and eastern sides of the lake area and also along the coastal areas. The surface elevation around the waterwork area along the coastline, for example, the Santalanranta pumping station is low and generally less than two metres. During spring time in the current condition, where groundwater reaches its highest level, groundwater levels in the observation wells along the coastline rise close to the surface. With an increase of a 65 cm sea level by the end of 2100, this area would be under water during the wet season and storm surges. Figure 11.14 presents the simulated groundwater head distributions based on the current potential recharge and the scenario A1B 2071–2100 potential recharge and sea level data at the end of 2100.

11.5 Discussion

The 3D finite difference groundwater MODFLOW models for the current condition of the shallow coastal aquifer in the Hanko area were constructed and calibrated with a satisfactory result from the comparison of simulation heads and observation heads. The same model domain and parameter input of the current groundwater flow conditions were applied to the climate scenario data A1B 2071–2100 by varying the input of the potential recharge and comparing the change in head between the current and scenario simulation heads. Scenario A1B for the period 2071–2100 was selected because it contained the highest change of climate data both in temperature and precipitation.

By the end of the 21st century, the mean temperature would increase 1.3 to 3.4 °C and the mean precipitation 5 to 12%. Seasonally, both temperature and precipitation would increase remarkably in autumn and winter (up to 4.6 °C and 26%). In summer, the temperature would slightly increase (2.6 to 2.9 °C), but the precipitation would decrease (−2 to −5%). Based on the scenario data, recharge is expected to increase because of a general increase of precipitation. However, in summer increasing temperatures can lead to an increase in evapotranspiration and can cause a slight decrease in precipitation low groundwater levels during summer.

Groundwater flow models for the current conditions were calculated with the recharge rates based on the percentages of the potential recharges which are the products of the potential recharges (pr (t)) and its spatial and time factors (α (t, x)). The same percentage values were applied for the transient flow models of the climate scenarios data. These percentage values can be used for the preliminary assessment of the groundwater flow models based on the scenario potential recharge data, in order to observe potential changes in groundwater conditions relative to the current situation, but they are probably not suitable for a detailed study of climate scenarios, where the temperature patterns differ from the current conditions.

Higher groundwater levels are expected in autumn and winter based on the groundwater monitoring data and the scenario data. This could affect the water quality negatively if the groundwater rises closer to the ground surface. The unsaturated zone and especially the pedological soil on the top part protects the underlying groundwater against microbe and other contaminants. Rain and melt water runs through this layer and the soil processes (chemical weathering, neutralization, cation exchange, sulphate adsorption) have sufficient time to treat the percolating water (Backman et al., 1999). When the groundwater table rises closer to the ground surface, the path through the unsaturated zone as well as the percolation time will be shorter and the contaminants might reach the groundwater.

The sea level rise scenarios based on the CLIMBER model indicate by 2100 a sea level rise between 0.22 and 0.81 cm in the Hanko area compared to 1995. The groundwater simulation results, based on the A1B regional high scenario, indicated that a rise of sea level of 65 cm would cause the groundwater rise for the whole area by an average of 88 cm.

Groundwater level of observation wells nearby the water intake plants along the coastline correlate well with the sea water level and also with pumping rates. Although the Baltic Sea water in the Gulf of Finland area contains low salinity and has little effect on groundwater resources, over-pumping or storm surges can cause risk to groundwater and water supply system e.g. pipeline corrosion. In addition, a comprehensive understanding of the geological background of the aquifer area aids the construction of the groundwater flow modelling close to real conditions. In this study the knowledge of sediments and the hydraulic properties of the aquifer along the coastal area are important to understand the interaction between seawater and groundwater levels. The groundwater simulation results indicate that the change of groundwater level in the coastal area that contains high K value materials depends directly on the change of sea water level. In addition, if this area is close to pumping wells, high pumping rates will cause more intrusion of the sea water into the wells and the aquifer.

11.6 Conclusion

The impacts of climate change on the groundwater resources of the shallow aquifer in Hanko have been assessed by utilizing a great variety of data including data from field investigation, climate data and 3D groundwater MODFLOW model, which was used as a tool to calculate groundwater flow behaviour under future climate conditions based on different climate scenarios. Based on the groundwater flow modelling simulation results, the main findings of the assessment of the impacts of climate change on groundwater are summarized as follows:

- Under current conditions groundwater levels in the Hanko area are high during spring and late autumn. During the summer and middle of winter the groundwater levels are lowest. The ranges of groundwater level during the year vary between 1 to 2 meters.
- Results of groundwater simulation models indicated that an increase in precipitation and potential recharge in the future would cause the groundwater level to increase and cause contamination risk for groundwater quality. High groundwater level also causes a risk to the infrastructure and water supply system.
- The mean temperature in summer would slightly increase from 2.6 to 2.9 °C compared to current conditions, this would cause more evapotranspiration during summer. In winter, higher temperatures reduce frost and cause earlier snow melt. The groundwater table would increase.
- The changes of groundwater levels along the coastline have a positive correlation with sea level, and in the water pumping area, the pumping rate has a negative correlation with the groundwater level.

Over-pumping in wells along the coastline causes a lower groundwater level, which will induce more sea water intrusion into the aquifer. This may cause problems for the groundwater quality. Groundwater flow simulation indicated that during the dry period in November 2009, the pumping rate of $1500 \text{ m}^3/\text{d}$ brought about $180 \text{ m}^3/\text{d}$ of sea water into the aquifer in the Santalanranta area.

- At the moment, the Santalanranta water intake plant is at risk from storm surges and sea water intrusion, because it is located at an elevation less than 2 metres a.s.l. and is very close to the seashore (20 to 60 m from the coastline). Based on the A1B regional high scenario, by the end of 2100, sea level would rise 0.65 m. The groundwater simulation results indicate that this would cause a groundwater rise by an average of 0.88 m. In addition, the Hanko area has experienced a storm surge of 1.24 m a.s.l. on 9 January 2005. A combination of over-pumping and storm surges can cause risk to the groundwater quality of the aquifer, and also to the water intake plant. In the future, based on the climate scenarios data, it is recommended that the water intake plant is moved to a higher elevation location.

- Climate change has impacts on recharge and groundwater resources. The shallow aquifer in Hanko is vulnerable to climate change. For future studies, an improved method for recharge estimation and better understanding of the recharge process are needed for a more accurate groundwater resources assessment and management.

Acknowledgements

We would like to thank Mr Kimmo Paakkonen and Hanko Water and Wastewater Works for providing us with data and background reports of the Hanko area, to Ms Heidi Sjöblom and SYKE for providing the snow water equivalent data.

References

Backman, B., Lahermo, P., Väisänen, U., Paukola, T., Juntunen, R., Karhu, J., Pullinen, A., Rainio, H. and Tanskanen, H., 1999. *Geologian ja ihmisen toiminnan vaikutus pohjaveteen. Seurantatutkimuksen tulokset vuosilta 1969–1996. (English: The effect of geological environment and human activities on groundwater in Finland. Results of monitoring in 1969–1996).* Report of Investigations 147, Espoo: Geological Survey of Finland.

Backman, B., Luoma, S., Schmidt-Thomé, P. and Laitinen, J., 2007. *Potential risks for shallow groundwater aquifers in coastal areas of the Baltic Sea, a case study in the Hanko area in south Finland.* CIVPRO Working Paper 2007 (2), Espoo: Geological Survey of Finland.

Breilin, O., Paalijärvi, M. and Valjus, T., 2004. *Hangon kaupunki – Pohjavesialueen geologisen rakenteen selvitys I Salpausselällä Hanko – Laoppohja alueella.* Tutkimusraportti. Espoo: Geological Survey of Finland.

Drebs, A., Nordlund, A., Karlsson, P., Helminen, J. and Rissanen, P., 2002. *Climatological Statistics of Finland 1971–2000.* Helsinki: Finnish Meteorological Institute.

Ekman, M., 1996. A consistent map of the postglacial uplift of Fennoscandia. *Terra Nova*, 8 (2), pp.158–165.

Fyfe, G.J., 1991. *The morphology and sedimentology of the Salpausselkä I Moraine in southwest Finland.* Fitzwilliam College, Cambridge University.

Ganopolski, A., Petoukhov, V., Rahmstorf, S., Brovkin, V., Claussen, M., Eliseev, A. and Kubatzki, C., 2001. CLIMBER-2: a climate system model of intermediate complexity. Part II: model sensitivity. *Climate Dynamics*, 17, pp.735–751.

Hänninen, P. and Äikää, O., 2006. *Hanko, Trollberget, DEMO-MNA hankkeen automaattiset seuranta-asemat (A field monitoring report for DEMO project, in Finnish).* Espoo: Geological Survey of Finland.

Harbaugh, A.W., Banta, E.R., Hill, M.C. and Mcdonald, M.G., 2000. *MODFLOW-2000, the U.S. Geological Survey modular ground-water model – User guide to modularization concepts and the Ground-Water Flow Process.* Reston, Virginia: U.S. Geological Survey.

IPCC, 2000. *Emissions Scenarios: Summary for Policymakers – A Special Report of IPCC Working Group III.* [online] Available at: <http://www.grida.no/publications/other/ipcc_msr/> [Accessed November 21, 2011].

Johansson, M.M., Boman, H., Kahman, K.K., and Launiainen, J., 2001. Trends in sea level variability in the Baltic Sea. *Boreal Environmental Research*, 6, pp.159–179.

Johansson, M.M., Kahma, K.K., and Boman, H., 2003. An improved estimate for the long-term mean sea level on the Finnish coast. *Geophysica*, 39 (1–2), pp.51–73.

Johansson, M.M., Kahma, K.K., Boman, H. and Launiainen, J., 2004. Scenarios for sea level on the Finnish coast. *Boreal Environmental Research*, 9, pp.153–166.

Jylhä, K., Tuomenvirta, H. and Ruosteenoja, K., 2004. Climate change projections for Finland during the 21st century. *Boreal Enviromental Research*, 9, pp.127–152.

Kakkuri, J., 1997. Postglacial deformation of the Fennoscandian crust. *Geophysica*, 33 (1), pp.99–109.

Kielosto, S., Kukkonen, M., Sten, C.G. and Backman, B., 1996. Hangon ja Perniön kartta-alueiden maaperä. (English: Quaternary deposits in the Hanko and Perniö map-sheet areas). *Geological map of Finland*, Sheets 2011/2012, 1:100000, Explanation to the maps of Quaternary deposits, Espoo: Geological Survey of Finland. 104 p.

Lemmelä, R. and Tattari, S., 1986. Infiltration and variation of soil moisture in a sandy aquifer. *Geophysica*, 22 (1–2), pp.59–70.

Luoma, S., 2012. *Steady-state groundwater flow model of the shallow aquifer in Hanko, south Finland*. Report 57/2012, Espoo: Geological Survey of Finland.

McGill, J.T., 1958. Map of coastal landforms of the world. *Geographical Review*, 48, pp.402–405.

Okkonen, J., 2011. *Groundwater and its response to climate variability and change in cold snow dominated regions in Finland: Methods and Estimations*. Ph.D. University of Oulu.

Peltier, W.R., 2000. Global glacial isostatic adjustment. In: B.C. Douglas, M.S. Kearney and S.P. Leatherman, eds. 2000. *Sea-Level Rise: History and Consequences*. San Diego: Academic Press, pp.65–95.

Petoukhov, V., Ganopolski, A., Brovkin, V., Claussen, M., Eliseev, A., Kubatzki, C. and Rahmstorf, S., 2000. CLIMBER-2: a climate system model of intermediate complexity. Part I: model description and performance for present climate. *Climate Dynamics*, 16, pp.1–17.

Rahmstorf, S., 2006. A semi-empirical approach to projecting future sea-level rise. *Science*, 315, pp.368–370.

Raupach, M.R., Marland, G.R., Ciais, P., Le Quéré, C., Canadell, J.G., Klepper, G. and Field, C.B., 2007. Global and regional drivers of accelerating CO_2 emissions. *PNAS*, 104 (24), pp.10288–10293.

Saarnisto, M. and Saarinen, T., 2001. Deglaciation chronology of the Scandinavian Ice Sheet from the Lake Onega basin to the Salpausselkä end moraines. In: J. Thiede, H. Bauch, C. Hjort and J. Mangerud, eds. 2001. *The Late Quaternary stratigraphy and environments of northern Eurasia and the adjacent Arctic seas – new contributions from QUEEN*. Global and Planetary Change, 31 (1–4), pp.387–405.

Sutinen, R., Hänninen, P. and Venäläinen, A., 2008. Effect of mild winter events on soil water content beneath snowpack. *Cold Regions Science and Technology*, 51, pp.56–67.

Veijalainen, N., Lotsari, E., Alho, B., Vehviläinen, B. and Käyhkö, J., 2010. National scale assessment of climate change impacts on flooding in Finland. *Journal of Hydrology*, 392, pp.333–350.

WBGU (Wissenschaftlicher Beirat der Bundesregierung Globale Umweltveränderungen), 2004. *The Future of the Seas: To warm, to high, to sour*. Berlin: German Advisory Council on Global Change to the Federal Government.

World Data Centre for Climate, 2011. CERA database. [online] Available at: <http://cera-www.dkrz.de/> [Accessed August 26, 2011].

12

Climate Change and Groundwater – From Modelling to some Adaptation Means in Example of Klaipėda Region, Lithuania

Jurga Arustienė[1], Jurgita Kriukaitė[1], Jonas Satkūnas[1] & Marius Gregorauskas[2]

[1] Lithuanian Geological Survey, Vilnius, Lithuania
[2] Vilniaus hidrogeologija Ltd, Vilnius, Lithuania

12.1 Introduction

Climate change belongs to one of the most recent emerging scientific endeavours to which society demands reliable scientific answers.

The concentration of large parts of its population and many larger cities in the coastal areas make the Baltic Sea Region sensitive to climate change. Among others, sea level rise and impacts on drinking water availability and quality (both surface and groundwater) can be expected to have important socio-economic impacts. The potential effects of climate change on natural hazards, as well as water and energy resources, are of great concern to geoscientists and stakeholders. Climate change has only recently been incorporated into decision-making processes and spatial planning (Schmidt-Thomé, Klein & Satkunas, 2010).

The investigations were carried out under the auspices of the BaltCICAproject. In the Klaipėda case study the BaltCICA project compiled a hydrodynamical model for the Klaipėda district and evaluated possible impacts of climate change on groundwater resources based on different scenarios. Adaptation measures on the use and protection of water resources were elaborated, including cost evaluations. The results of the project were incorporated into the Special Plan for the Water Supply of the Klaipėda District in January 2012.

12.2 Groundwater – the key geoenvironmental issue in Europe

At present, surface water plays a substantial role in public water supply in European countries (Table 12.1) and an increase in the share of groundwater to the public water supply is observed (Zektser & Everett, 2004). This is because groundwater has advantages over surface water. Groundwater generally contains micro and macro components needed for the human body; it does not require expensive treatment; and is much better protected from contamination (Zektser & Everett, 2004). The EU has a number of directives and regulations that aim to protect water resources (EU, 2008). These include the Urban Waste Water, Nitrates and Drinking Water Directives, and a Directive on Integrated Pollution Prevention and Control that requires licensing of discharges at sustainable levels. A coherent water policy was developed by 1995 but the measures focused

Climate Change Adaptation in Practice: From Strategy Development to Implementation, First Edition.
Edited by Philipp Schmidt-Thomé and Johannes Klein.
© 2013 John Wiley & Sons, Ltd. Published 2013 by John Wiley & Sons, Ltd.

Table 12.1 Groundwater percentage for potable water supply in European cities

CITY	Population (in million people)	Surface water (in %)	Groundwater (in %)
AMSTERDAM	1.3	52	48
BARCELONA	3.3	83	17
BERLIN	5.6	58	42
BRUSSELS	2.3	35	65
VIENNA	1.7	5	95
HAMBURG	3.6	–	100
GLASGOW	5.2	63	37
COPENHAGEN	1.0	16	84
LISBON	2.1	45	55
LONDON	6.7	86	14
MADRID	4.1	91	9
MOSCOW	8.5	98	2
MUNICH	1.6	–	100
PARIS	7.1	60	40
ROTTERDAM	1.4	90	10
ZURICH	0.5	70	30

Source: Zektser, 2000. Reproduced by permission of Taylor & Francis Group.

on preventing emissions leading to more pollution, rather than the improvement of water resources. The EU Water Framework Directive (2000/60/EC) was adopted in 2000 and establishes a framework for community action in the field of water policy. This Directive requires the establishment of technical specifications to complement the overall water (surface water, groundwater and coastal beaches) regulatory regime. These specifications cover a number of key elements that range from the characterization and analyses of pressures and impacts, towards monitoring of measures. All of these elements are linked to the development and implementation of river basin management plans aimed at achieving good ecological and chemical status of each body of water.

Taking into consideration the fact that groundwater is very important and in some countries is the only source of potable water (in general in Europe 65% of potable water is taken from groundwater aquifers) in 2006 the European Commission adopted a new Directive – the Groundwater Directive (2006/118/EC), which aims at protecting groundwater from pollution. Based on an EU-wide approach the Directive introduced for the first time the quality objectives which oblige member states to monitor and assess groundwater quality on the basis of common criteria, and identify and reverse trends in groundwater pollution.

The use of groundwater for public water supply is increasing in Europe, and especially in larger cities. However, it is estimated that 60% of European cities overexploit their groundwater resources (EU, 2008). European countries which use a high percentage of groundwater for public water supply can be listed in the following order: Lithuania – 100%, Denmark – 98%, Italy – 93%, Hungary – 90%, Poland – 70%, Estonia – 65%. This raises an important question of how climate change can potentially affect recharge, availability and quality of groundwater resources.

12.3 Groundwater in Lithuania

All potable water resources (100%) in Lithuania are extracted from groundwater aquifers. The amount of extracted groundwater resources is up to 2.2 million m^3 per day, or 25.5 thousand litres per second. Approximately one half of the evaluated groundwater resources are related to Quaternary deposits,

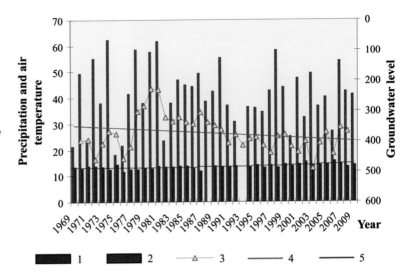

Figure 12.1 Shallow groundwater levels for the period 1962–2009 in Lithuania (exemplified by the Mikuziai station) indicating: the last 15 years are a period of decreasing levels, particularly low levels were observed in 2003–4 and 2006.
Source: Giedraitiene, 2009.
Legend: 1 – annual precipitation (cm); 2 – mean annual air temperature (°C); 3 – depth of groundwater level below ground surface (cm); 4 – linear trend of shallow groundwater level; 5 – linear trend of mean air temperature.

including alluvial sands. The other half is located in the Cretaceous, Permian and Devonian bedrock. The groundwater of the pre-Quaternary bedrock is artesian and generally of good quality. The chemical composition of some groundwater bodies is altered by natural saline water intrusions, which are in turn partly caused by groundwater abstraction. Despite generally sufficient resources and good quality, the drinking water supply is facing serious problems in rural areas. According to investigations of the State Public Health Service under the Ministry of Health, and the Lithuanian Geological Survey, over 50% of the examined shallow groundwater dug wells for rural water supply in Lithuania do not meet requirements of the bacteriological quality, while 48% of them are polluted by nitrogen compounds (Satkunas, Arustiene & Kanopiene, 2009).

Since 1962 a national groundwater monitoring programme has been carried out in Lithuania (Arustiene et al., 2009) and has shown that climate change is affecting shallow groundwater bodies (Figure 12.1). An increase in mean temperatures, together with lower precipitation amounts, and an increase in summer dry spells, were observed during the last 15 years, and have resulted in lower shallow groundwater levels. Therefore, this period is a time span of a general decrease of shallow groundwater resources, which in turn has most probably also led to a decrease in recharge rates of both deeper, and bedrock aquifers.

12.4 Case of the Klaipėda district – hydrogeological conditions

Klaipėda district is located in the western part of Lithuania on the Baltic Sea shore.

Groundwater is used for drinking water supply in all of Klaipėda district. Publically it is supplied in 72 settlements for ~ 58% of the population. Gargždai – main town of Klaipėda district – has the most developed infrastructure for public water supply. In the rural areas water supply infrastructure is weakly developed or old, installed before 1990. Most of the currently operated well fields do not fulfill environmental requirements – their resources are not approbated, sanitary protection zones are not established and they do not have water purification systems (for iron and manganese removal). People not connected to public water supply use water from individual drilled or shallow wells.

The Quaternary cover is ~120–160 m thick and consists of sandy and clayey layers of glacial and glaciofluvial origin. Shallow groundwater occurs close to the land surface in a depth of 2 to 3 metres in the biggest part of the district and in some areas even at 1 to 2 metres. In the valley of the river Minija groundwater occurs in sandy alluvial deposits and depth to groundwater level increases distinctly from 1.9 to 5.7 m. The groundwater quality of shallow aquifers is very sensitive to anthropogenic load. The best

Figure 12.2 Location of wells in the third well field of Klaipėda. The groundwater resources of this well field are very closely related with the surface water bodies.
Source: Underground Register, Lithuanian Geological Survey, 2012)

indicator of human impact is concentration of nitrates. In natural ambient average concentration of nitrates is limited to a few milligrams per litre, while in arable

land it increases to 10–20 mg/l and in urban areas to 30–40 mg/l (Arustienė, 2011).

The only well field, which extracts shallow groundwater from marine deposits, is located on the banks of the Wilhelm channel (Figure 12.2). It supplies drinking water for the southern part of Klaipėda and most of its resources (~70 000 m³/d) are surface water, hence the climate change impacts are quickly noticed.

Quaternary confined aquifers have limited distribution and use in the region. Sandy deposits are ~6-10 m thick, sometimes 20–30 m and specific yields of wells located into those aquifers are 0.2–0.3 l/s.

Fresh groundwater is contained in five different confined pre-Quaternary aquifers (Figure 12.3).

In the southern part of the district the most perspective for groundwater abstraction is the sandy part of Cretaceous aquifer (K2 cm + K1). The top of Cretaceous deposits lies in 62–135 m depth. Thickness of productive aquifer is around 5 to 30 m, values of hydraulic transmissivity ranges from 30–50 m²/d.

Jurassic aquifer is (J3-2) distributed in the whole district, but in the pre-Quaternary surface it appears only in the Northern part. Jurasssic aquifer is rather thin, 5 to 15 m consisting of sand, sandstone and marl. It is covered by a 5 to 30 m thick low permeable layer of aleurite and by Quaternary deposits. Hydraulic

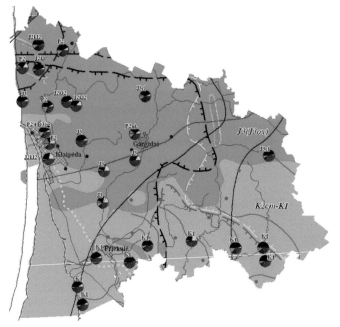

Figure 12.3 Hydrogeological map of Klaipėda district.
Source: Radzevičienė and Marcinkevičienė, 2007

properties of aquifer are moderate, values of transmissivity vary from 12–125 m²/d, but mostly between 20–40 m²/d. Water levels drop significantly in the Jurassic aquifer during groundwater abstraction, but extension of depression cones are limited due to poor hydraulic properties. Although the Jurassic aquifer is well covered, piezometric surface depends on landscape.

Beneath the Triassic aquitard 100-180 m thick, upper Permian and upper Devonian aquifers occur. Upper Permian (P2) aquifer is spread in the whole district. Its top is found in 220–300 m depth and its thickness varies from 6 to 70 m, on average 24 m. Groundwater is mostly contained in porous and fractured limestone. Piesometric level is found in different depths – from 253 m below ground surface to 9.4 m above ground surface. Hydraulic properties of the aquifer are quite uneven and strongly depend on fractures and karstic forms. Values of the specific well yields vary from 0.03–154 l/s (average 9.8 l/s).

In the biggest part of the Klaipėda district upper Permian and beneath upper Devonian layers form a joint aquifer. In the Klaipėda, Kretinga and Gargždai vicinities joint aquifer is the most productive. Values of conductivity in fractured limestone are up to 70 m/d. This aquifer is well confined and has very limited connection with upper aquifers.

The lowest laying upper Devonian carbonatic aquifer (D3fm) with fresh groundwater is found in 275–296 m depth. Productive layers are 10–20 m thick. The most productive areas are associated with fractured zones. In such a zone the well field of Klaipėda city is located, where values of hydraulic conductivity are 60–70 m/d, that is, 20 to 30 times more than background values. In the southeastern part of the district upper Devonian aquifer disappears.

Confined aquifers are better protected from anthropogenic impact, but with the increasing depth salinity of groundwater increases and because of confinement of aquifers iron, hydrogen sulphide and ammonium accumulate (Klimas, 2006).

The groundwater composition of Jurassic aquifer is distinct. Because of interaction of fresh ground water infiltrating from upper aquifers and glauconite particles present in the Jurassic sand, the fresh (total dissolved solids – 250–500 g/l) very soft (total hardness– 2–4 mg-eq/l) water of sodium bicarbonate type is formed. Such a type of water is rare in Lithuania.

The chemical composition of Upper Permian aquifer changes – from fresh calcium, magnesium bicarbonate type in the northern-central part of the district to saline (TDS 2–6 g/l) sodium calcium sulphate type in the southern part. Upper Permian aquifer was used for the Gargždai town supply, but because of saline water intrusion to the well field it was closed, and the well field was relocated and installed to Jurassic aquifer.

In general, confined groundwater in the Klaipėda district satisfies requirements of the drinking water standard (HN 24:2003). The concentrations of iron and ammonium are higher than permissible concentration for drinking water in the most of the well fields, so water treatment is needed. In some areas of Permian and Jurassic aquifers groundwater has increased concentration of fluorides (>1.5 mg/l).

12.5 Present and future groundwater resources in Klaipėda district

The main tasks of the research in the case study area were to evaluate the present groundwater resources and to estimate how climate change could affect them during the forthcoming 100 years. A mathematical flow model was applied to calculate the groundwater resources of different aquifers and to simulate the A1B and B1 climate change scenarios adapted to the Klaipėda region for the years 2025, 2050 and 2100. Potential future precipitation and evaporation changes were modelled by the Department of Hydrology and Climatology from Vilnius University.

Climate models foresee a rapid increase of air temperature in Lithuania for the forthcoming 100 years. The biggest changes of temperature are anticipated in winter time, while during the summer it should rise less intensely. All model runs result in an overall increase of precipitation as well. The highest increase is anticipated during the cold period, while during the summer the amount of precipitation could decrease.

A mathematical flow model was built for the Klaipėda region covering an area of 2078 km² inland and 695 km² undersea (Figure 12.4). To support the flow model a set of maps was compiled from available data – measurements of water levels in 170 dug wells and 400 hydraulically tested deep wells.

Hydrogeological conditions of the case study area were generalized and described in the model as seven

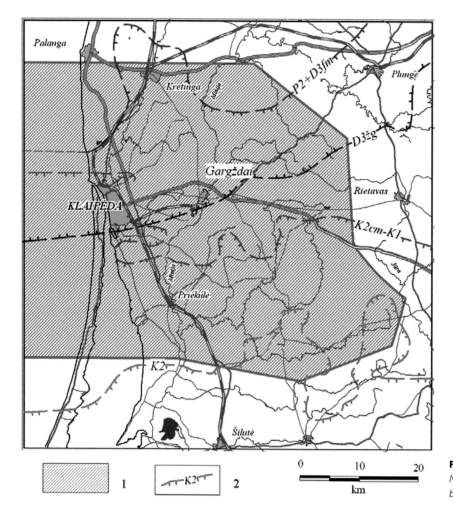

Figure 12.4 Case study area.
Notes: [a]1- modeled area; [b]2 - boundaries of aquifers.

structural units – shallow aquifer, Quaternary, Cretaceous and Jurassic confined aquifers with dividing layers of low permeability.

In the whole territory water exchange between shallow groundwater and surface waters as well as between groundwater and deeper confined aquifers is active, so interflow is the main source of dynamic groundwater resources and groundwater flow.

In the Klaipėda district calculated total shallow groundwater recharge, taking into account current groundwater extraction rate, is close to 350 000 m³/d (61.4 mm/year). The biggest part of it is formed by precipitation recharge (74.4%) (260100 m³/d or 45.7 mm/year) and by inflow from deeper confined aquifers (68 099 m³/d or 11.92 mm/year) (Figure 12.5).

The biggest part of dynamic groundwater resources (4.05 m³/s) is located in the shallow aquifer. The biggest part of shallow groundwater discharges to the rivers (246 800 m³/d or 2.9 m³/s). So dynamic resources of deeper Quaternary confined aquifers, which are mainly formed (75%) from shallow groundwater infiltrating in areas of groundwater recharge, are three times lower – 1.11 m³/s. The smallest dynamic groundwater resources are located in well confined Jurassic aquifer – 0.17 m³/s (Table 12.2).

Possible changes in groundwater resources were calculated using a mathematical flow model and based on correlations between precipitations, groundwater flow in the river basins and infiltration recharge of shallow groundwater (Figure 12.6).

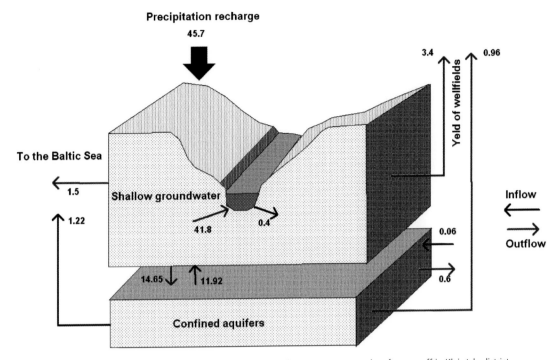

Figure 12.5 Modeled present balance (mm/year) of dynamic groundwater resources and surface runoff in Klaipėda district

Table 12.2 Modeled present (2010) balance of dynamic groundwater resources and their components

Dynamic resources		Modulus of dynamic resources; l/s*km²	Main sources	
thousand m³/d	m³/s		Source	Input; %
			Shallow groundwater	
349.56	4.05	1.95	Infiltration recharge	74.4
			Infiltration from the deeper confined aquifers	21.4
			Confined aquifers (agl III-II bl-žm)	
95.97	1.11	0.63	Infiltration from shallow aquifer	76.4
			Infiltration from the deeper confined aquifers	23.4
			Cretaceous (K_2cm-K_1) aquifer	
37.02	0.43	0.3	Infiltration from upper confined aquifers	86.1
			Infiltration from iš J $_{3-2}$ aquifer	13.6
			Jurassic (J_{3-2}) aquifer	
14.66	0.17	0.06	Infiltration from upper confined aquifers	99.7

Source: Gregorauskas, 2010.

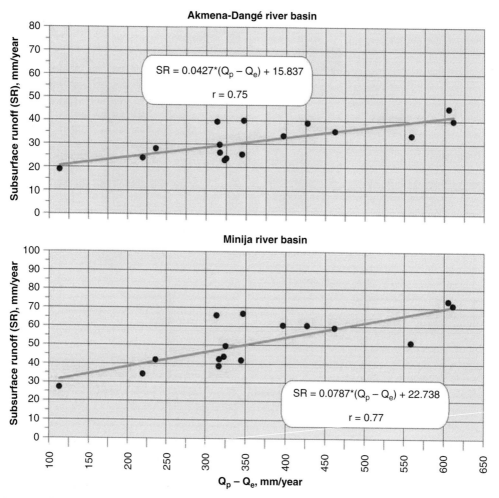

Figure 12.6 Correlation between groundwater runoff (mm/year) and part of precipitations left eliminating evaporation
Source: Gregorauskas, 2010.

Estimations were made using the following approach: the period of 2011–2100 was divided into 90 time steps, each of them equal to one year; using modeled values of precipitation and evaporation for each time step, values of infiltration recharge for each river basin were calculated; values of the specific storage and specific yield were put into the model; using maps of potential future groundwater level and after calculation of water balance, dynamic groundwater resources were assessed for all modeled aquifers for specified years.

Simulation results show that both climate change scenarios result in an increase of dynamic resources of groundwater in the area of investigation for spec-

ified years 2025, 2050 and 2100 years (Table 12.3). Since the main source of the groundwater resources and cause of its changes is infiltration recharge into shallow groundwater, therefore the highest increase of resources is anticipated in shallow groundwater aquifer. In the A1B case the highest groundwater level is expected in 2025 year. At that time the groundwater recharge will increase from 45.7 mm/year to 79.6 mm/year adding to dynamic resources from 4.05 m³/s to 5.99 m³/s. A similar growth of resources is anticipated in B1 case in year 2100.

Increase of groundwater resources declines as a function of depth. If the expected increase of dynamic resources in the shallow groundwater aquifer is

Table 12.3 Modeled future balance of dynamic groundwater resources and their components according to climate change scenario A1B

Year	Dynamic resources			Increase of dynamic resources*		
	thousand m³/d	m³/s	modulus, l/s * km²	thousand m³/d	m³/s	modulus, l/s * km²
Shallow groundwater						
2025	517.33	5.99	2.88	167.77	1.94	0.93
2050	426.64	4.94	2.38	77.08	0.89	0.43
2100	367.08	4.25	2.04	17.52	0.2	0.09
Quaternary confined (agl III-II bl-žm)						
2025	107.83	1.248	0.709	11.86	0.138	0.079
2050	101.74	1.178	0.669	5.77	0.068	0.039
2100	97.84	1.132	0.643	1.87	0.022	0.013
Cretaceous (K_2cm-K_1) aquifer						
2025	38.26	0.443	0.309	1.24	0.013	0.009
2050	37.63	0.436	0.304	0.61	0.006	0.004
2100	37.21	0.431	0.301	0.19	0.001	0.001
Jurassic (J_{3-2}) aquifer						
2025	14.688	0.17	0.06	–	–	–
2050	14.686	0.17	0.06	–	–	–
2100	14.685	0.17	0.06	–	–	–

*compared with present resources.
Source: Gregorauskas, 2010.

~1–2 m³/s (1.5 times higher than estimated present resources), then in the Quaternary aquifers, isolated from shallow groundwater by low permeable layers, reach only tenth or hundredth parts of m³/s. It means that resources in Quaternary aquifers will increase only by two to 12%. Deeper Cretaceous and Jurassic aquifers are almost not affected by climate change and the increase of water resources in later aquifers is almost negligible.

Changes of groundwater level in main aquifers are correlated with increase of groundwater recharge and dynamic resources. In the years 2025 and 2100 the highest increase of recharge is assumed for shallow groundwater, thus average annual groundwater level could rise in sandy lenses by 2.5–2.8 m, in till and peat between 0.2–0.5 m, in the remaining part of the territory in shallow groundwater aquifer by 1–1.5 m (Figure 12.7), in Quaternary aquifers between 0.3–0.5 m,

in Cretaceous aquifer by 10 cm and in Jurassic aquifers will remain practically unchanged.

Results show, that neither in the years 2025, 2050 and 2100, nor during all investigated time period is there any danger of groundwater resources decreasing. In times of decrease of dynamic resources, average annual groundwater level in shallow groundwater could fall to 10–15 cm, whereas in the Klaipėda third well field area and in Quaternary and Cretaceous aquifers just up to several centimetres.

The results of groundwater modelling indicate that the forthcoming century should be favourable for groundwater resources recharge. Modelled examples of groundwater resource changes are uneven – in some years significant increase is highly possible, while in other years it is changed by decrease. The estimation of groundwater resources strongly depends on input data reliability. No models could exactly

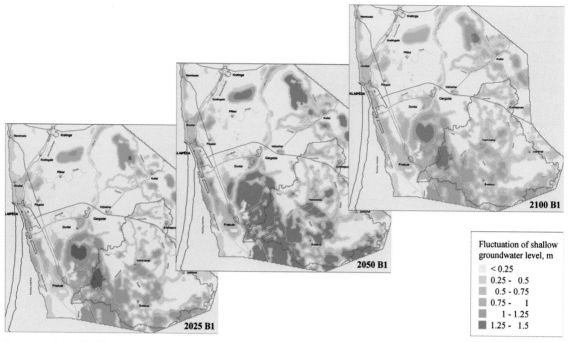

Figure 12.7 Modeled fluctuation of shallow groundwater level in the Klaipėda district according to climate change scenario B1 in 2025, 2050 and 2100.

foresee 'wet' or 'dry' years as well as exact climate values, but simulated trends allow the future situation to be assessed and to prepare for it. Increasing levels of shallow groundwater will increase the vulnerability of groundwater pollution and risk of inundation in a number of low-lying areas. Shallow groundwater is the most sensitive to climate changes, yet is still used as a main drinking water source in rural areas. The future drinking water supply should be based on deeper aquifers, less dependent on climate change.

12.6 The solutions of special water supply infrastructure development plan

The special plan of Klaipėda district drinking water supply and sewage system infrastructure development approved by Klaipėda district municipality in January 2012, foresees an expansion of public water supply system. The results of modelling of groundwater resources obtained within the BaltCICA project, as described in the previous chapter, were implemented

into the water supply plan. Two solutions from a special plan could be regarded as important for adaptation to climate changes and sustainable use of groundwater resources:

• Shift from individual water supply, based on shallow groundwater to public water supply, based on deeper confined aquifers.

• Planned measures for sustainable use and protection of water resources – development and reconstruction of well fields network (approbation of resources, establishment of sanitary protection zones).

According to the special plan, zones of public water supply should cover 115 settlements with a population of 44 400 (95.5% of inhabitants). The public water supply network will consist of 66 separate water supply systems (Urbanistika, 2011). Currently, 15 water supply systems are already established and fulfill all requirements, 11 have to be newly installed and the rest should be redeveloped (Figure 12.8).

The special plan includes schemes of future water supply system, but does not provide exact numbers – metres of pipelines or depth of the operational wells. This is done in detail at the technical planning stage. The total planned length of pipelines in Klaipėda

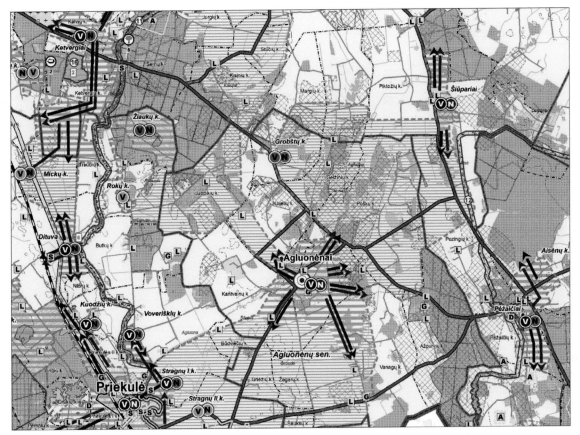

Figure 12.8 Solutions of water supply and sewage system infrastructure development in a special plan of Klaipėda district. *Source*: Urbanistika, 2011.

district is ~ 250 km. The biggest system of new pipelines is being installed around Gargždai town ~20 km; additionally ~30 km of pipelines will connect ~20 settlements to Klaipėda city water supply infrastructure. The length of separate water supply systems will be between 0.2 to 26 km, ~4.5 km in average. The total preliminary cost of new pipelines installation was calculated ~50 million lites (Lt).

Small well fields (<100 m³/d) will prevail (68%) in the new public water supply system (Figure 12.9). The bigger well fields extracting 100–300 m³/d will make 27%, and the biggest well fields extracting > 300 m³/d will account only for 5%.

The exact cost of each well field reconstruction/installation could be accessed only after close inspection and some investigations. Installation of new wells rather than reconstruction of old ones could be economically more effective. Based on currently used

rates, preliminary calculations indicate that reconstruction of a small well field could cost from 66 000–118 000 Lt and that of a bigger one from 117 000–201 000 Lt. The installation cost of a small well field per capita (360–650 Lt) is twice as high as of the bigger well field (150–260 Lt).

Concentrations of iron, manganese, ammonium and in some cases of fluorides in groundwater used for public water supply, exceeds values set in drinking water standard. This means that groundwater should be purified. Based on data derived from the special plan, only in six of currently operated well fields water purification is not necessary, in the rest 33 iron removal is necessary and in 12 well fields removal of fluorides. It is likely that for the newly installed well fields water purification should be foreseen as well. The cost of a water purification system, its installation and connection to the water supply system depend on

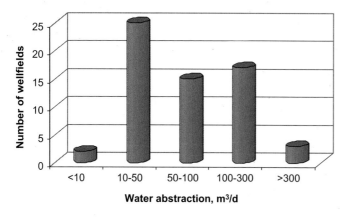

Figure 12.9 Distribution of well fields in Klaipėda district according to water abstraction.

volume and hydro chemical composition of groundwater as well as technical conditions and could be from 150 000–250 000 Lt. Taking into consideration that at least 45 water purification systems should be installed, the total price could be from 6 750 000–11 250 000 Lt. A more advanced technique is required for removal of fluorides from groundwater. Currently, installation of a water purification system with fluorides removal in one well field costs ~ 1 500 000 Lt. So, installation of 12 such systems could cost double that of 45 iron removal systems. The alternative water supply sources should be considered. In the small well fields the Quaternary confined aquifers could serve as drinking water source or at least 'fluoride' water dilution.

In general, development of the water supply system in Klaipėda district could cost from 71 million to 97 million Lt or 1600–2200 Lt. per capita. This figure could be used as a tentative cost of adaptation to climate change and its consequences related to securing safe drinking water. Installation of new pipelines is the costliest part of special plan implementation (50–70% of total cost). Installation of water purification systems

takes 13–20%, while development of well fields only 5–10% (Figure 12.10).

The process of water supply system development started in 12 territories. For the remaining territories technical documentation still needs to be prepared. Development of public water supply and sewage infrastructure in Lithuania is partly supported by EU funds. Usually implementation of separate projects continues for ~ two years, so full implementation of the special plan for Klaipėda district could take five to seven years.

12.7 Conclusions

Groundwater is the main drinking water source in Klaipėda district. During the last 15 years climate conditions have resulted in lower shallow groundwater levels in Lithuania. Therefore, this period is a time span of a general decrease of shallow groundwater resources, which in turn has most probably also led to a decrease in recharge rates of both deeper and bedrock aquifers.

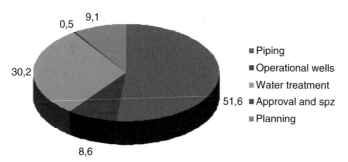

- Piping
- Operational wells
- Water treatment
- Approval and spz
- Planning

Figure 12.10 Proportions of costs for water supply infrastructure development.

Public water supply network currently is developed in 72 towns and settlements in Klaipėda district and is available for ~60% of population. In rural areas water supply infrastructure is poorly developed. A lot of people still use groundwater from individual shallow dug wells. During the last decade dense individual water supply systems (drilled wells together with local waste water treatment facilities) are rapidly developing in the new districts near Klaipėda city causing some threat to groundwater resources. In the area of investigation drinking water is located in shallow Quaternary, Cretaceous, Jurassic and Permian-famenian aquifers.

A mathematical flow model was applied to calculate groundwater resources of different aquifers and later used to simulate future scenarios, based on climate change data adapted to the Klaipėda region. The simulation results show that groundwater resources will increase during the forthcoming 100 years period. Though wet and dry periods repeat, the general trend of groundwater recharge points upward. Future groundwater resources are potentially up to 1.5 times higher than present resources. Since the main source of the groundwater resources and cause of its changes is infiltration recharge in shallow groundwater, therefore the highest increase of resources is anticipated in shallow groundwater aquifer. Increasing levels of shallow groundwater will increase the vulnerability of groundwater pollution and risk of inundation in a number of low-lying areas. So, current and future groundwater resources should be sufficient to sustain the shift from individual water supply based on shallow groundwater to public water supply, based on deeper confined aquifers, laid down in the special plan of Klaipėda district drinking water supply and sewage system infrastructure development. Planned measures – development and reconstruction of well fields network, approbation of resources, establishment of sanitary protection zones – will assure safe drinking water, sustainable use and protection of groundwater resources.

References

Arustiene, J., 2011. Groundwater quality and its variations: Groundwater monitoring in Lithuania 2005–2010: *Selec-* *tion of publications*. Vilnius: Lithuanian Geological Survey, pp.29–40.

Arustiene, J., Giedraitiene, J., Kriukaite, J. and Šimkovič, A., 2009. *Groundwater Monitoring in Lithuania 2008: Bulletin*. [pdf] Vilnius: Lithuanian Geological Survey. Available at: <http://www.lgt.lt/old/uploads/1254123084_2008_metu_MONITORINGAS.pdf> [Accessed 30 August 2012] (in Lithuanian with English summary).

Council Directive 2000/60/EC of 23 October 2000 establishing a framework for Community action in the field of water policy.

Council Directive 2006/118/EC of 12 December 2006 on the protection of groundwater against pollution and deterioration.

EU (European Union), 2008. *Water protection and management*. [online] Available at: <http://europa.eu/legislation_summaries/environment/water_protection_management/index_en.htm> [Accessed 6 September 2012].

Giedraitiene, J., 2009. Shallow groundwater resources: *Groundwater monitoring in Lithuania 2008: Bulletin*. Vilnius: Lithuanian Geological Survey, pp.10

Gregorauskas, M., 2010. *Evaluation of climate change impact on groundwater resources in Klaipėda district*. Vilnius: Vilniaus hidrogeologija Ltd. / Lithuanian Geological Survey.

Klimas, A., 2006. *Groundwater quality in Lithuania well fields*. Vilnius: LVTA.

Lithuanian higenic norm HN 24:2003, 2003. Requirements for drinking water safety and quality. *Published in Žin. 2003, No. 79–3606*.

Radzevičienė, D. and Marcinkevičienė, G., 2007. *The revision and renovation of the State hydrogeological maps at a scale of 1:200 000 (20–39/60–79 and 20–39/40–59 topographic lists)*. Vilnius: Lithuanian Geological Survey.

Satkunas, J., Arustiene, J. and Kanopiene, R., 2009. Environmental geology in Lithuania, state-of-art and European context: Special Report: Geo-pollution Science. *Medical Geology and Urban Geology*, 5 (1/2), pp.25–33.

Schmidt-Thomé P., Klein J. and Satkūnas J., 2010. Climate change, impacts and adaptation – some examples of geoscience applications for better environmental management in the Baltic Sea Region. *Episodes*, 33 (2), pp.102–108.

Urbanistika Ltd. 2011. *Concept of drinking water supply and sewage system infrastructure development. Special plan of Klaipėda district drinking water supply and sewage system infrastructure development.*

Zektser, I.S., 2000. *Groundwater and the Environment: Application for the Global Community*. Boca Raton: Lewis Publishers.

Zektser, I.S. and Everett, L.G. eds., 2004. *Groundwater Resources of the World and their Use*. [pdf] Paris: UNESCO. Available at: <http://unesdoc.unesco.org/images/0013/001344/134433e.pdf> [Accessed 30 August 2012].

13 Climate Change – A New Opportunity for Mussel Farming in the Southern Baltic?

Anna-Marie Klamt[1] & Gerald Schernewski[1,2]

[1] Leibniz-Institute for Baltic Sea Research Warnemünde, Rostock, Germany
[2] Coastal Research & Planning Institute, Klaipeda University, Klaipeda, Lithuania

13.1 Introduction

Aquaculture continues to be the fastest growing animal food-producing sector. It is set to overtake capture fisheries as a source of food fish. The production increased from less than one million tons in 1950 to 51.7 million tons in 2006 and the per capita supply from aquaculture increased from 0.7 kg in 1970 to 7.8 kg in 2006.

The Asia–Pacific region accounts for 89% of the production. Europe contributes only 4%. In 2001, mussel farming contributed 23% to the worldwide aquaculture production (Tacon, 2007) and the main cultivated species are the Blue mussel, *Mytilus edulis* (Europe and North America), *Mytilus galloprovincialis* (Mediterranean Sea) and the green-lipped mussel (*Perna canaliculus,* New Zealand). During 1950–2000 the production of mussels increased at an average rate of 5% per year. In 1999, the production reached nearly 1.7 million tons (Tacon and Metian, 2008). The United Nations Food and Agricultural Organization (FAO) estimated that the worldwide total value of mussels was roughly 645 million US$ in 2000. About 50% of the annual worldwide harvest of mussels is produced in Europe (Smaal, 2002). Along the Atlantic coast the main production areas are the Rias in Spain, the Dutch and German Wadden Sea, the Limfjord in Denmark, and the French coast (Smaal, 2002).

In Germany, Blue mussel production is located in the North Sea. So far, the production in the Baltic Sea is negligible. Neither mussel production nor its consumption has a tradition in the Baltic Sea Region. The commonly cultivated Blue mussel favours marine conditions with high salinity. The brackish Baltic Sea shows a decreasing salinity from southwest to northeast. Therefore, only the south-western parts are suitable habitats for Blue mussels. In salinities below five PSU, Zebra mussels occupy these habitats. Therefore, considerations on mussel farming in all southern Baltic coastal waters have to take into account both species. The lack of tradition certainly hampers the development of mussel farms in the Baltic but other aspects are important as well: drift ice in winter, lack of experience with mussel species in low salinity ecosystems, and a lacking economic profitability (Schernewski, Stybel & Neumann, 2012). Blue and Zebra mussels have many similarities. Therefore, knowledge and experiences can be transferred from existing mussel farms in other countries.

Mussel farming is not only an environmental friendly and sustainable way to produce high quality food, but an important measure to improve ecosystem quality. Baltic coastal waters are known for being oversupplied with nutrients. Large amounts of nitrogen and phosphor enter the Baltic Sea via rivers. The resulting eutrophication of coastal waters causes many problems and negatively affects ecosystem

Climate Change Adaptation in Practice: From Strategy Development to Implementation, First Edition.
Edited by Philipp Schmidt-Thomé and Johannes Klein.
© 2013 John Wiley & Sons, Ltd. Published 2013 by John Wiley & Sons, Ltd.

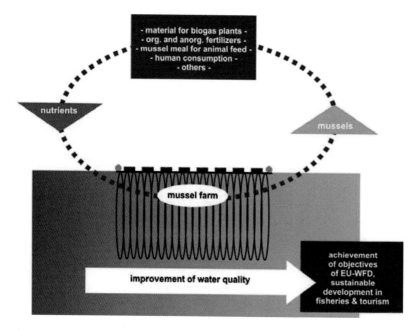

Figure 13.1 Schematic overview about the role of mussel farming for improvement of water quality, regional development and for a closed nutrient cycle, with possible products of mussels.

functions. The excess nutrients increase primary production and phytoplankton biomass. Heavy eutrophication has negative consequences on (bathing) tourism and affects fisheries. Measures to reduce nutrient loads in rivers are necessary. However, measures in the river basin alone may not be sufficient to reach a good water quality in lagoons (Schernewski, Behrendt, Neumann & Stybel, 2008). Additional internal measures like managed and enlarged natural mussel beds, mussel farms, algal farms, increased reed belts and extended submersed macrophyte areas and/or dredging of sediment and dumping on land are suggested (Schernewski, Stybel & Neumann, 2012). Mussel farming and harvesting is an important option to remove nutrients from water bodies. Mussels are effective filter organisms. They feed on plankton and hence fix considerable quantities of nutrients in their biomass. Harvesting the mussels closes the nutrient cycle, means a direct nutrient removal from the system and thus an improvement of water quality (Lindahl et al., 2005) (Figure 13.1). At the same time water transparency is improved with positive effects on bathing tourism.

The objectives of this chapter are (a) to document how climate change will affect the feasibility of mussel farming in the southern Baltic Sea; (b) to explore the options to cultivate the Blue mussel and especially the Zebra mussel, which is not cultivated yet; (c) to

reflect on possible farming and processing techniques, adapted to shallow coastal waters and finally (d) to briefly review possible products and the ecosystem services provided by mussel farms.

13.2 Baltic winters – a threat for mussel farms?

For instance, in winter 2010/2011 experimental Blue mussel farms for food mussel production in Kalmar Sound and Oresund have been largely destroyed by drift ice.

Ice is not a big problem for a mussel farm as long as the ice is not drifting. Drift ice, especially in Kalmar Sound where huge ice floes may sweep through the Sound, causes a lot of damage to bridges, constructions, buoys etc. A floating mussel farm must be designed so that the drift ice cannot grab any part of the farm. This is in practise very difficult to achieve and the only safe way is to lower the farm under the surface. To lower farms is technically not very difficult, but existing methods in e.g. Canada or New Zealand require some extra material and labour. [Consequently the lowering also increases the cost of farming the mussels. As long as mussels are produced as sea food, lowering the farm can be included in the price of the mussels.

However, environmental mussels which have a much lower value on a market may become too expensive as a measure to improve water quality if the farms have to be lowered by using exiting technique.] Therefore [. . .] I hope that the companies offering mussel farm equipment for use in the Baltic will development cheaper lowering systems. This work is in fact ongoing.

(Lindahl, 2005, pers. com.)

The alternative to lowering systems are farms located offshore. Offshore farms are subject to higher wave energy and require more stable farming systems, higher investment and maintenance costs, are less accessible and less flexible in daily management. Higher amounts of mussels are lost due to mechanical stress. Therefore, common recommendations for prospective mussel farmers suggest, not to choose sites where a risk of drift ice exists. From a practical and economic perspective, neither offshore nor lowering systems are a realistic option for Baltic coastal waters. Drifting ice is rare at the southern Baltic coast. In cold winters, the complex German Baltic coastline, with many embayments, is characterized by large trapped ice-sheets. However, but one cold winter with drifting ice could be enough to destroy a mussel farm and to cause a loss of the invested money. Mussel farms are not permanent, fixed installations but movable and floating systems. This is especially true for systems suitable for shallow waters. Mussel farms can be removed to sheltered areas, but this is laborious and reduces the profitability of farms. As long as ice-winters are the rare exception, mussel farms are an option in our shallow Baltic waters.

13.3 Climate change – creating new perspectives?

During the past century there was a general tendency towards warmer conditions in the Baltic Sea. According to the classification of Seinä and Palosuo (1996), all ice winters during the last ten years have been average, mild or extremely mild. For example, the German Federal Maritime and Hydrographic Agency (BSH) classified the ice winters in 2004/2005, 2006/2007, 2007/2008 and 2008/2009 as weak ice seasons (Schmelzer & Holfort, 2009). Ice occurred only in small harbours and in shallow, sheltered coastal lagoons. The outer coastal waters were ice-free. Along the Polish coast the length of the ice season has decreased by one to three days per decade during the period 1896–1993 (Sztobryn, 1994; Girjatowicz & Kozuchowski, 1995). Between 1720 and 2005 the ice extent in Szczecin Lagoon declined. Girjatowicz and Kożuchowski (1999) identified a statistically significant decreasing trend in the duration of the ice season in the period from 1888 to 1995.

These trends are representative for the southern Baltic and are supported by recent data. Figure 13.2 shows the number of days with ice cover and average winter (December, January and February) air temperatures exemplary for Ueckermuende station at Szczecin Lagoon, southern Baltic Sea. Between 1947 and 2011 winter temperatures increased by about 1 °C and on average are nowadays above 0 °C. During the last two decades only one winter had an average temperature below −3 °C, but four winters had average temperatures near or above 4 °C, which has not been observed before (4.1 °C in 1989/1990 and 4.4 °C in 2006/2007). Between 1947 and 2011, the number of days with ice cover on the lagoon decreased from about 70 to 40. Today, ice break-up happens much earlier compared to previously. In the Baltic, ice thickness has decreased during the last 20 years, as well (The BACC author team, 2008). However, with −3.9 °C and 141 days of ice, winter 1996 was extremely cold and had the highest number of ice cover days for the last 64 years. Extreme events always happen and model projections suggest that the inter-annual variability increases.

Water surface temperatures will increase by 1–4°C (Neumann, 2010) resp. 1.9–3.2°C (Meier, 2006). The already observed ongoing warming will not only continue, but speed up during the coming decades. Increasing water temperatures will have strong effects on sea ice conditions. The probability of ice free winters will increase significantly in the future (Stigebrandt, 2003), the duration of ice cover will be shorter because of later freezing dates and earlier ice break up (Jevrejeva et al., 2004). Today, on average approximately 50% of the Baltic Sea area is covered by ice in winter. According to model simulations the extent of total area of ice cover in the Baltic Sea will only be one third of today's expansion (Neumann, 2010) at the end of the 21st century. Other models suggest a reduction of annual maximal ice extent by

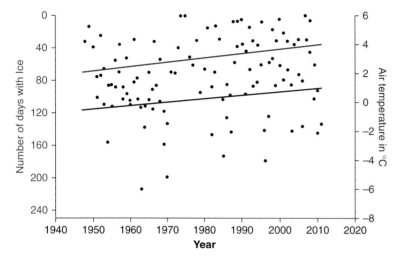

Figure 13.2 The trend for number of days with ice (blue) and trend for average winter (December, January and February) air temperatures (red) for Ueckermuende station at Szczecin Lagoon and photograph of ice floes at the coastline of the Baltic Sea (January 2010).
Sources: Ice data from German Federal Maritime and Hydrographic Agency (BSH), temperature data from German weather service (DWD).

46–77%, depending on the climate change scenario (Meier, Döscher & Halkka, 2004; Neumann, 2010) and 4% of the winter might be ice-free in the entire Baltic Sea in the period 2071–2100. At the same time these models suggest decreasing salinity in the Baltic Sea as a result of higher fresh water inputs. Salinity changes and other factors are of importance for ice-cover (Stigebrandt, 2003; Schrum & Schmelzer, 2007). Along the southern coast of the Baltic, winters with ice cover will be the exception in the future, but cannot be completely excluded because of increasing interannual temperature variability.

Salinity is a major controlling factor for aquatic ecosystems and determines the distribution of most organisms. River-runoff, evaporation and the exchange with the North Sea control salinity in the Baltic Sea (Meier, 2006, cf. Stigebrandt & Gustafsson, 2003). Due to increasing freshwater inflow salinity will decrease in the Baltic Sea (Neumann, 2010). Models predict a reduction of 8–50 % (Meier, 2006). The high volume of the Baltic Sea serves as a buffer and salinity changes are slow (Stigebrandt & Gustafsson, 2003). It takes about 30 years until a new steady-state has been reached after strong changes (Meier, 2006). The distribution of mussel species very much depends on salinity. *Dreissena* will benefit from a decreasing salinity and it is expected that Zebra mussels will extend their distribution area in the Baltic (The BACC author team, 2008). However, it is not likely that climate change will cause salinity changes that have serious consequences on mussel farming.

Climate change will prolong the growing season for many organisms. This in turn will have effects on food requirement and availability, on food chains and thus on species, especially phytoplankton and zooplankton (Neumann & Friedland, 2011). This is true for mussels, as well, and allows a higher production. In shallow waters increasing temperatures would enhance primary production and algae growth. This would increase the food supply for mussels. For mussel farming, the prolongation of the growing season, which could increase by 30 to 90 days, is much more important. Mussel farming in the Baltic Sea will benefit from climate change. There are certain risks and problems like enhanced cyanobacteria blooms or new competing invasive species. However this remains speculative and mussels are well able to deal with problems like algal toxins.

13.4 The Zebra mussel – a suitable farming species?

In the entire south-western Baltic Sea the salinity is high enough to allow the farming of Blue mussels (Figure 13.3). Blue mussels are larger than Zebra mussels and suitable for human consumption. From an economic, practical and food production perspective, Blue mussel cultivation is preferable to Zebra mussel cultivation. In several inner coastal waters, with very low salinities, only Zebra mussels are able to grow. The Szczecin lagoon is one of the largest Baltic lagoons,

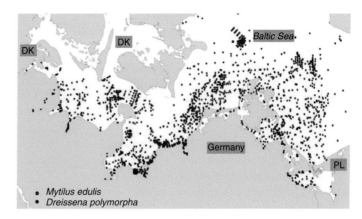

Figure 13.3 Proved occurrence of *Mytilus edulis* (blue points) and *Dreissena polymorpha* (red points) at the German Baltic coast (Abbreviations: DK = Denmark, PL = Poland).

with a high Zebra mussel biomass and a very high primary production (Fenske, 2003).

Szczecin lagoon is located at the border of Germany and Poland in the southern Baltic Sea Region and divided into two parts: Kleines Haff (German part) and Wielki Zalew (Polish part) (Figure 13.2). Three outlets link the inner coastal waters with the Baltic Sea: Peenestrom, Swina and Dziwna. The Odra is the most important river. It contributes 94% to the freshwater input and dominates the nutrient budget (Wielgat, 2002). The catchment area covers an area of about 120 000 km². The shallow lagoon (average depth 3.8 m) is in a polytrophic to hypertrophic state with low water transparency (usually below 50 cm). Salinity is about 1.4 PSU and this value increases from the southern to the northern parts. In summer months, cyanobacteria blooms appear regularly. In some years there are even anoxic conditions and fish deaths occur. However, the lagoon serves as converter, sink and sometimes as source of nutrients for the Baltic Sea (Schernewski, Hofstede & Neumann, 2011) and thus it is important for the nutrient loads to the Baltic.

Zebra mussels (*Dreissena polymorpha*) (Figure 13.4) can be regarded as a native species in Szczecin Lagoon. It was introduced in the 19th century (Brandt, 1896), but was once abundant before the last ice age throughout northern Germany (Meisenheimer, 1901). The species can tolerate low oxygen content in water for several days. The adults filter a wide range of size particles and feed on algae and zooplankton. Baker, Levinton, Kurdziel and Shumway (1998) showed that they were able to select: they prefer to feed on cyanobacteria than on diatoms. A shell length of 40 mm is pos-

sible, however, often the shells are well below this size. *Dreissena polymorpha* is a common species in many freshwater ecosystems, but also tolerates moderate permanent salinity between 2 and 5 PSU (Zettler & Röhner, 2004). A short-term salinity up to 12 PSU can be endured (Strayer & Smith, 1993). However, populations can be damaged by sudden changes (Wolff, 1969). In estuaries and brackish waters salinity is the most limiting factor (Strayer & Smith, 1993). Reproduction is possible below 7 PSU (Fong et al., 1995). In brackish parts of seas *Dreissena polymorpha* occurs from the lower shore to depths of 12 m.

Densities of *Dreissena polymorpha* populations are determined by mortality of planktonic veligers during settlement and in the post-veliger stage. Spawning

Figure 13.4 Experimental nets from Szczecin Lagoon grown over with one-year-old *Dreissena polymorpha* and shell of *Dreissena polymorpha* (approx. 2 cm length).

175

occurs in spring. The temperature threshold for synchronized spawning is 12 °C (Sprung, 1993). Mussels larger than 8 mm or two-year-old females are capable of reproducing. The fertilized eggs evolve into three different larval stages. The larvae tolerate 12–24 °C, the optimum being 18 °C (Sprung, 1993). The last larvae phase ends up with settling on suitable hard substrata and attaching firmly with byssal threads (Lewandowski, 1982a;b). Subsequently morphological changes take place. Most larvae (99%) die during the process of settlement (Stanczykowska, 1977). Reasons are a lack of suitable substrate or food, infections and predators. According to Sprung (1993) settling success depends on oxygen, bottom structures, water movements and substrata (e.g. plants, other shells, stones, artificial objects between 0.5 and 4.5 m). The natural life span of *Dreissena polymorpha* is variable and ranges from three to nine years. Adult mussels can detach and move to seek alternate habitats. Reasons for the fast and successful expansion of the Zebra mussel are their high rates of reproduction (r-strategist) and growth and their relative long planktonic veliger larvae phase which enables a spread by passive transport supported by current. Szczecin Lagoon is largely influenced by the Odra River. Hence it is likely that seed mussel supply is supported by river input.

Dreissena's high potential of dispersal became apparent in the US: the mussels were introduced there in 1985 (Hebert, Muncaster & Mackie, 1989). Claudi and Mackie calculated in 1994 a dispersal speed of 250 km/yr. Today *Dreissena* can be found in all Great Lakes, in many rivers and has been seen near the Gulf of Mexico. The high expansion and closed colonization caused lots of economic problems in the US, such as pipeline blockages or fouling of crafts. Today there is no known farming activity or large scale utilization of *Dreissena polymorpha*. Nevertheless, the mussels have already been used as a filter organism for restoration of eutrophied shallow lakes in the Netherlands (Pires, Ibelings & van Donk, 2010) and as a biofilter in waste water treatment (Kusserow et al., 2010).

Tests in the Szczecin lagoons (Fenske, 2003) and ongoing experiments by Dahlke (Figure 13.4) have shown that Zebra mussels can be grown on net and line systems, similar to those used for Blue mussel cultivation. Both species show many similarities, they require hard substrate, have a similar biomass per area mussel bed and a comparable filtration capacity per kg biomass (Table 13.3). The growth and productivity of both mussels are in a similar range and depend on similar environmental and food conditions. With respect to contents of raw protein, raw fat and the sum of unsaturated fatty acids both species are alike.

We can summarize that Zebra mussels seem to be well suited for farming in areas with low salinity, where Blue mussels are not able to survive. They can be farmed and processed in a similar way and the products are of a similar quality (Table 13.1).

13.5 Farming methods – the best choice for shallow waters

Worldwide there are many Blue mussel farming systems applied. They are adapted to very different environments and tidal waters are a special challenge for mussel cultivation. Baltic micro-tidal, relatively protected coastal waters allow the employment of commonly used systems. We compare the 'SmartFarm system' and two different 'longline systems' for potential use in the shallow Szczecin lagoon. The 'SmartFarm system' is a high-tech system compared to the longline systems. Investment costs are very high and a special vessel for care and harvesting is required (extra costs). The socking longline system needs intensive care and labour input (socking is done by hand). It has not yet been tested to see if *Dreissena* is a suitable species for this method. We think that an extensive loop-longline farm system most suits the purposes in Szczecin Lagoon. Settlement experiments showed highest densities of *Dreissena polymorpha* on PVC-nets compared to, for example, wood and stones (Fenske, 2003). Close-meshed nets proved their value for Zebra mussel colonization during experiments in Lake Constance or for development of mussel biofilters. The main advantages are: the system is stable, common, well approved and needs comparatively low investment costs. If the system is operated extensively the labour input can be reduced. Extensive farming means just bringing the growing lines into the water, letting them get colonized with larvae and letting the mussels grow. The mussel larvae are passively transported in the Szczecin Lagoon, which means that there are enough larvae available for settling. There is no special care required and lines do not have to be moved (from seeding to growing on areas). The harvesting is

Table 13.1 Comparison of *Dreissena polymorpha* and *Mytilus edulis*

Parameter	*Dreissena polymorpha*	*Mytilus edulis*
Temperature	−2–40 °C, best growth: 18–20 °C http://www.europe-aliens.org/pdf/Dreissena_polymorpha.pdf	10–22 °C (Bayne, 1965)
Salinity	opt. 0 PSU (freshwater species) max. 7 PSU for successful reproduction (Fong et al., 1995)	opt. 15–40 PSU (Bayne, 1965) min. 4–5 PSU for growth without distinct size reduction (Kautsky, 1982)
Substrata	hard (Zettler & Röhner, 2004), also on shells in sandy areas	hard, also on shells in sandy areas (Zettler & Röhner, 2004)
Raw protein*	8.4%	7.9%
Raw fat*	0.8%	1.5%
Sum unsaturated Fatty acids*	64.4%	70.8%
Filtration rate	10 to 100 ml/individual/hour (Claudi & Mackie, 1993)	300–600 ml/individual/hour (Abel, 1976)
Biomass	average 1.76 kg/m^2 in Skoszewska Cove (Woźniczka & Wolnomiejski, 2008) 2.98 kg/m^2 (max.) (Fenske, 2003)	3.0 kg/m^2 at northern point of Bornholm (Demel & Mulicki, 1958) 0.264 kg/m^2 in Pomeranian Bay (Piesik & Wawrzyniak-Wydrowska, 1997)

*data from mussel meat analysis, individuals from Kiel and Szczecin Lagoon.

by common fishery boats with some specialized equipment. The low effort reduces the costs.

Longline mussel farms are used in many countries. They consist of a series of horizontal parallel lines as backbone (headlines), which are fastened on concrete blocks or other anchors. On these lines hang looped growing lines (Figure 13.5). They are spaced at between approximately 0.5 m and 1 m intervals to minimize the chances of tangling. For the required buoyancy there are many floating bodies connected with the horizontal lines. Their number depends on mussel biomass and allows well adapted farming to different depths and other conditions.

Data about length of the lines vary and depend on site specific conditions. In deep waters growing loops can extend up to 10 m depth. They should not touch the bottom to avoid predators and mud intake. One unit is about 110 m long, but data differ from 70 m up to 200–400 m. The distance between the units is not specified as it depends on different conditions like local food availability and furthermore there should be enough space to easily use the (harvesting) vessel. For

harvesting the lines have to be raised by hoists and haul and brought on board. Afterwards the mussels are stripped of the lines and washed. A declumping machine divides the mussels and grades them if necessary according to size. Subsequently the outer side of the shells is cleaned to remove the byssel thread and other periphyton. After weighing and packing the mussels are ready for transport and selling. If necessary, mussels are subjected to purification and depuration process ashore.

13.6 Mussel products and ecosystem services

A lack of experience of how to farm them, as well as the small size and thin shells of *Dreissena polymorpha* reduce their potential for human consumption. A possibility means of making mussel farming more profitable in future is to use the mussel meat as a substitute for fishmeal as it has a high content of top quality proteins. Fishmeal contains up to 70% of

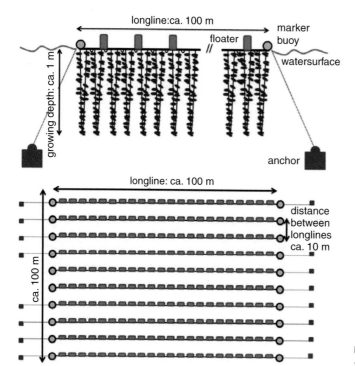

Figure 13.5 General build-up of a longline mussel farm, side view and from above.

proteins (with a well-balanced composition of amino acids), essential fatty acids (omega-3 fatty acids) and is energy rich (Schipp, 2008). Hence, it is a very common healthy and digestible feeding stuff. Today the largest share (68.2% in 2006) is used in aquaculture (Tacon & Metian, 2008). In particular, carnivorous fishes like salmon and trout need a certain share of animal proteins (Marine Harvest, 2011). Aquaculture production will grow during the next few years and thus fishmeal demand will increase further. Fishmeal production has been more or less stagnating or decreasing during the last few years. Consequently the price for fishmeal has increased drastically. In May 1999, the price for one ton was 363 € and in May 2010 had risen to 1520 €.

Figure 13.6 shows the correlation between worldwide fishmeal production and fishmeal price and the

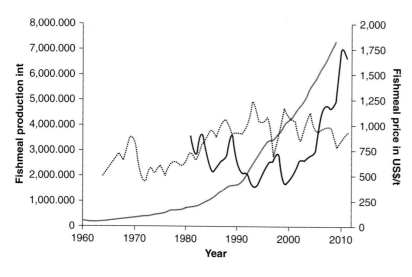

Figure 13.6 Correlation between worldwide fishmeal production in the top five fishmeal producing countries* (dotted line) and fishmeal price in US$/t (black line). The grey line shows the trend of development of aquaculture production (without dimension).
Sources: Data from www.indexmundi.com and www.fao.org
*=Peru, Chile, Thailand, U.S.A., Japan and additionally from 1999 on the European Union

Table 13.2 Comparison of 'SmartFarm' and two longline mussel farming systems for Blue mussels

Description	'SmartFarm'–system Vertical nets	Longline-socks Vertical growing lines, with socks for attaching support	Longline-loops Vertical growing lines
Common growing cycle	1. seeding of nets in harvest or spat fall regions	1. special farms with collector ropes in spat fall regions	1. seeding of lines in harvest or spat fall regions
	2. moving of nets to on-growing areas	2. seed-harvesting and 'retubing' (socking) on growing lines in on-growing areas (at water or land) by several employees	2. moving of lines to on-growing areas
	3. control (thinning and cleaning) by one special machine (1 person), submerged	3. control (thinning and cleaning) by several employees (lines out of water)	3. control (thinning and cleaning) by several employees, lines out of water
	4. harvest with one special machine (1 person)	4. harvest by boat and with the help of different machines to strip, clean, declump, grade and debyss mussels (several persons)	4. harvest by boat and with the help of different machines to strip, clean, declump, grade and debyss mussels (several persons)
Advantages	- not labour intensive - well elaborated system - highly mechanized system - numerous re-use of the nets, environmental friendly	- common and well approved system - low costs for the equipment - low-tech equipment, simple to use - breakdown-free material - numerous re-use of the ropes, environmental friendly	- common and well approved system - low costs of the equipment - low-tech equipment, simple to use - breakdown-free material - numerous re-use of the ropes, environmental friendly
Disadvantages	- high investment costs	- work and labour intensive	
Intensity	- very intensive	- intensive	- extensive till less intensive

trend of development of worldwide aquaculture production. It is generally approved that conventional fishmeal is not produced in a sustainable way, since the wild fishery catches add lots of pressure on the ecosystems (overfishing). For organic aquaculture it is mandatory to feed fishmeal or -oil which is produced in a sustainable way. From 2012 on it is in general no longer permitted to add synthetic amino acids to feed. For these reasons alternative protein sources have to be found. The experiments of Berge and Austreng (1989) showed that it is possible to use Blue mussels as an ingredient in diets for rainbow trout. Jönsson, Wall and Tauson (2011) demonstrated that the common Blue mussel is well accepted as feed for poultry. Shells could be and are used as fertilizers. Harvesting

extensive reared mussels provides a way of producing a sustainable product. Furthermore rising fishmeal prices and potentially paid subsidies could make mussel farming a cost-covering asset (Table 13.2).

The Blue mussel is a well known human food mussel. They can be sold fresh or frozen. Lack of experience in farming the Zebra mussel combined with its small size and thin shells prevent it being used for human consumption. To provide a storable food ingredient it is important to cook the mussel meat.

Zebra mussels can be processed into mussel meal according to these basic stages (Figure 13.7): After harvesting the fresh raw material it has to be cooked for reason of hygiene and to allow coagulation of

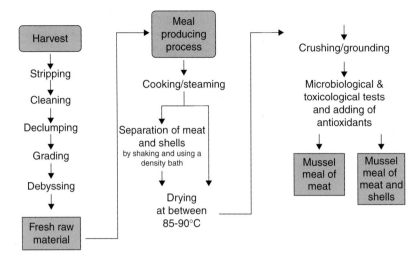

Figure 13.7 Overview of mussel meal processing steps.

proteins. Afterwards the stuff has to be dried at between 85–90 °C to a water content of maximal 10%. This is a critical process: under-dried meal may become a medium for bacteria, over-dried (max. 90 °C) meal may impair nutritional value. Afterwards everything has to be crushed. To fulfil the requirements demanded in animal feed production the mussel meal should be tested for microbiological (esp. *Salmonella* and *E. coli*) and toxicological (heavy metals, dioxins, PCBs) contaminants. To prevent oxidation it is necessary to add antioxidants. Afterwards the mussel meal can be weighed and packed. The finished product contains meat and shells. If it is necessary to separate shells from meat this can be easily done by shaking the cooked mussels and/or by using a density bath.

Dreissena meal can be used in a similar way to Blue mussel meal, however, at present we can find only one commercial *Dreissena* product: a German company sells fishing baits containing up to 15% Zebra mussel oil. These baits are used in particular for carp hunting. *Dreissena* is common in many rivers and lakes and thus the carps recognize their fragrance as natural food. Therefore *Dreissena* oil can be successfully used to attract and catch the fishes. As with all organic material the mussels can also be used for production of energy in biogas plants. No special processing after harvesting is necessary for it. This utilization is already being tested in Sweden.

If we try to to assume the production costs of producing mussel meal we see that they can be divided into costs for the raw material and costs for processing. Table 13.3 shows and compares the costs for the production of food and feed mussels (raw material) in different locations. The production costs may vary a lot. Of course there are differences, for example, in labour costs and material costs between the countries (cf. Europe and the US or northern and southern European countries). Furthermore the price depends on the different environmental conditions (mainly salinity) at the growing site, resulting in different farming of mussels (cf. examples from the Baltic Sea in Gren, Lindahl & Lindqvist, 2009). From an economical view it is evident that the production costs decrease with increasing farm size. Since there are different quality requirements for food and feed mussels the costs for their production are quite different. For instance the fodder mussels can be produced more extensively (with less care) and mussels of all sizes can be used. Although the range of the prices for fodder mussels (0.09–0.44 €/kg) is wide a typical price can be assumed to be approximately 0.25 €/kg raw material of feed mussels.

As there is so far no (large scale) mussel meal production, the costs can only be estimated from the known costs for the production of fishmeal: in 1984 the share of the production costs on the sales price came to 49.6%. For 2010 this would mean a processing price of approximately 0.58 €/kg. These low prices of course were only achieved with high amounts of produced meal. For a small scale production one has to assume higher costs. Lindahl (pers.

Table 13.3 Assumptions of production costs for Blue mussels (food and feed), in brackets: given raw data

Region	Note	Mussel species	Approx. production costs in €/kg	Source
mussels for industrial use				
Baltic Sea	Kattegat	*Mytilus edulis*	0.09 (0.08–0.10)	Gren et al., 2009
	The Sound		0.18 (0.16–0.19)	
	South Baltic Propper		0.20 (0.18–0.21)	
	North Baltic Propper		0.44 (0.33–0.54)	
	Kalmarsund		0.30 (0.20–0.40)	Lindahl (not published)
mussels for human consumption				
Baltic Sea	Kattegat	*Mytilus edulis*	0.30 (0.28–0.32)	Gren et al., 2009
	The Sound		0.68 (0.64–0.72)	
Greece	4 ha-Farm	*Mytilus galloprovincialis*	0.31*	Theodorou, 2010
	2 ha-Farm		0.40*	
	1 ha-Farm		0.59*	
USA		*Mytilus edulis*	0.36* (0.52 $/kg)	Langan, 2006
			0.38* (0.25 $/lb)	Hoagland, 2007

*calculated or translated from given data.

com.) assumed at least 2.50 €/kg for mussel meal processing. The sum of raw material and processing cost is at least 2.75 €/kg mussel meal. For comparison: actual fishmeal price is 1.59 US$/kg or 1.11 €/kg. There is a difference of 1.64 €/kg and thus no positive balance.

13.7 Conclusion

Climate change has positive effects on mussel cultivation. It extends the growing season and decreases the risk of farm damages due to icey winters. At the same

Table 13.4 Summary of positive and negative aspects of mussel-farming in the southern Baltic

Pros	Cons
• **Removal of nutrients** *via harvest and* **increased water transparency**	• **Lack of experience** *in Zebra mussel cultivation and processing methods*
• **Efficent WFD measure if water quality is indicated by transparency**	• **End-of-pipe solution for nutrient removal**
• **'Native' species** *used and* **knowledge transfer** *from Blue mussel cultivation*	• **Not profitable without subsidies and requires large scale investments**
• **High-quality protein and fatty acid source** *with increasing prices for products*	• **Un-settled legal situation**
• **Mussel meal as substitute for fish meal** *reduces pressure on wild fish stocks*	• **Uncertain effects on the ecosystem** (*denitrification, shifts in species composition, increased risk of hypoxia*)
• **Re-settlement of macrophytes** *due to improved water transparency (possibly regime shift)*	• **Accumulation of pollutants in mussels**
• **Synergy effects with local fisheries** *and alternative jobs for fishermen*	• **Damage of farms by drifting ice**
• **Increased number of tourists** *because of improved water transparency and a new local attraction (dishes, excursions)*	• **Losses due to bird and fish predation**
• *Reduction of summerly algal blooms*	• **Un-controlled settling of mussels** *on constructions and boats*
	• **Lack of tradition,** *poor acceptance of Zebra mussels and products as well as uncertain commercial use*

time the demand and price for high quality feed and food from an environmental friendly method of production is increasing. Already today, higher productivity and higher prices for mussel products might enable profitable Blue mussel farming in the Baltic. However, there are obstacles, and new innovative strategies, especially with respect to marketing are required. Low salinity in the Baltic favours the smaller Zebra mussel. The conditions for Zebra mussel cultivation on a larger scale are improving, but in the near future it will hardly be profitable. Mussel cultivation has the potential as a nutrient retention and water quality improvement measure. Table 13.4 gives an overview of the advantages and drawbacks of mussel farming.

Acknowledgements

The work has been partly supported by the project RADOST (BMBF 01LR0807B), the South Baltic Programme project ARTWEI (Action for the Reinforcement of the Transitional Waters' Environmental Integrity) as well as the Baltic Sea Region Programme projects BaltCICA (Climate Change: Impacts, Costs and Adaptation in the Baltic Sea Region) and AQUAFIMA (Integrating Aquaculture and Fisheries Management towards a sustainable regional development in the Baltic Sea Region).

References

Abel, P.D., 1976. Effect of some pollutants on the filtration rate of *Mytilus*. *Marine Pollution Bulletin*, 7 (12), pp.228–231.

Baker, S.M., Levinton, J.S., Kurdziel, J.P. and Shumway, S.E., 1998. Selective feeding and biodeposition by Zebra mussels and their relation to changes in phytoplankton composition and seston load. *Journal of Shellfish Research*, 17 (4–5), pp.1207–1213.

Bayne, B.L., 1965. Growth and the delay of metamorphosis of the larvae of *Mytilus edulis* (L.). *Ophelia*, 2 (1), pp.1–47.

Berge, G.M. and Austreng, E., 1989. Blue mussel feed for rainbow trout. *Aquaculture*, 81, pp.79–90.

BfG (Bundesanstalt für Gewässerkunde) ed., 2004. *Die Biodiversität in der deutschen Nord- und Ostsee*. Koblenz: BfG.

Brandt, K., 1896. Über das Stettiner Haff. *Wissenschaftliche Meeresuntersuchungen*, pp.105–141.

Claudi, R. and Mackie, G.L., 1993. *Practical Manual for Zebra Mussel Monitoring and Control*. Boca Raton: Lewis Publishers.

Demel, K. and Mulicki, Z., 1958. The zoobenthic biomass in the Southern Baltic. *J. Cons. int. Explor. Mer.*, 24 (1), pp.43–54.

Fenske, C., 2003. *Die Wandermuschel (Dreissena polymorpha) im Oderhaff und ihre Bedeutung für das Küstenzonenmanagement*. Ph. D. University of Greifswald.

Fong, P.P., Kyozuka, K., Duncan, J., Rynkowski, S., Mekasha, D. and Ram, J.L., 1995. The effect of salinity and temperature on spawning and fertilization in the Zebra mussel *Dreissena polymorpha* (Pallas) from North America. *The Biological Bulletin*, 189, pp.320–329.

Girjatowicz, K.P. and Kożuchowski, K., 1995. Contemporary changes of Baltic Sea ice. *Geographia Polonica*, 65, pp.43–50.

Girjatowicz, K.P. and Kożuchowski, K., 1999. Variations of thermic and ice conditions in the Szczecin Lagoon region. In: A. Järvet, ed. 1999. *Publications of second workshop on the Baltic Sea Ice Climate*. Tartu: University of Tartu, Department of Geography, pp.69–73.

Gosling, E.M. ed., 1992. *The mussel Mytilus: ecology, physiology, genetics and culture*. Amsterdam: Elsevier.

Gren, I.-M., Lindahl, O. and Lindqvist, M., 2009. Values of mussel farming for combating eutrophication: An application to the Baltic Sea. *Ecological Engineering*, 35, pp.935–945.

Hebert, P.D.N., Muncaster, B.W. and Mackie, G.L., 1989. Ecological and genetic studies on Dreissena polymorpha (Pallas): A new mollusc in the Great Lakes. *Canadian Journal of Fisheries and Aquatic Sciences*, 46, pp.1587–1591.

Hoagland, P., Kite-Powell, H.L. and Jin, D., 2003. *Business planning handbook for the ocean aquaculture of blue mussels*. Woods Hole: Marine Policy Center, Woods Hole Oceanographic Institution.

Järvet, A. ed., 1999. *Publications of second workshop on the Baltic Sea Ice Climate*. Tartu: University of Tartu, Department of Geography.

Jevrejeva, S., Drabkin, V.V., Kostjukov, J. et al., 2004. Baltic Sea ice seasons in the twentieth century. *Climate Research*, 25, pp.217–227.

Jönsson, L., Wall, H. and Tauson, R., 2011. Production and egg quality layers fed organic diets with mussel meal. *Animal*, 5, pp.387–393.

Kautsky, N., 1982. Growth and size structure in a Baltic *Mytilus edulis* population. *Marine Biology*, 68, pp.117–133.

Kusserow, R., Mörtl, M., Mählmann, J., Uhlmann, D. and Röske, I., 2010. The design of a Zebra-Mussel-Biofilter. In: G. van der Velde, S. Rajagopal and A.B. de Vaate, eds. 2010. *The Zebra Mussel in Europe.* Leiden: Backhuys Publishers, pp.323–330.

Langan, R., 2006. *Methods, economics and commercialization of offshore mussel production using submerged longlines.* [pdf] Available at: <https://www.was.org/Documents/MeetingPresentations/AQUA2006/WA2006-706.pdf> [Accessed 12 January 2012].

Lewandowski, K., 1982a. O zmiennej liczebnosci malza (*Dreissena polymorpha* (Pall.). *Wiadomosci Ekologiczne*, 28, pp.141–154.

Lewandowski, K., 1982b. The role of early developmental stages in the dynamics of Dreissena polymorpha (Pall.) populations in lakes. 1. Occurrence of larvae in the plankton. *Ekologia Polska*, 30, pp.81–109.

Lindahl, O., Hart, R., Hernroth, B. et al., 2005. Improving marine water quality by mussel farming: a profitable solution for Swedish society. *AMBIO*, 34 (2), pp.131–138.

Löser, N. and Sekścińska, A., 2005. *Integriertes Küste-Flusseinzugsgebiets-Management an der Oder/Odra: Hintergrundbericht.* IKZM-Oder Berichte 14, Rostock: EUCC.

LUNG (Landesamt für Umwelt, Naturschutz und Geologie), 2008. *Gewässergütebericht Mecklenburg-Vorpommern 2003/2004/2005/2006: Ergebnisse der Güteüberwachung der Fließ-, Stand- und Küstengewässer und des Grundwassers in Mecklenburg-Vorpommern.* Güstrow: LUNG.

Marine Harvest, 2011. *More farmed salmon using less wild fish as feed.* [online] Available at: <http://www.marineharvest.com/en/CorporateResponsibility/Salmon-feed/More-farmed-salmon-using-less-wild-fish-as-feed/#> [Accessed 11 November 2011].

Meier, M.H.E., 2006. Baltic Sea climate in the late twenty-first century: a dynamical downscaling approach using two global models and two emission scenarios. *Climate dynamics*, 27, pp.39–68.

Meier, M.H.E., Döscher, R. and Halkka, A., 2004. Simulated distributions of Baltic Sea-ice in warming climate and consequences for the winter habitat of the Baltic ringed seal. *AMBIO*, 33, pp.249–256.

Meisenheimer, J., 1901. Entwicklungsgeschichte von Dreissena polymorpha Pall. *Zeitschrift für wissenschaftliche Zoologie*, 69, pp.1–137.

Nalepa, T.F. and Schloesser, D.W. eds., 1993. *Zebra mussels: biology, impacts, and control.* Boca Raton: Lewis Publishers.

Neumann, T., 2010. Climate-change effects on the Baltic Sea ecosystem: A model study. *Journal of Marine Systems*, 81, pp.213–224.

Neumann, T. and Friedland, R., 2011. Climate change impacts on the Baltic Sea. In: G. Schernewski, J. Hofstede and T. Neumann, eds. 2011. *Global change and Baltic coastal zones.* Dordrecht: Springer, pp.23–32.

Piesik, Z. and Wawrzyniak-Wydrowska, B., 1997. Distribution and the role of Mytilus edulis (Linne) in the coastal zone of the Pomeranian Bay. *Baltic Coastal Zone*, 1, pp.45–53.

Pires, L.M.D., Ibelings, B.W. and van Donk, E., 2010. Zebra mussels as a potential tool in the restoration of eutrophic shallow lakes dominated by toxic cyanobacteria. In: G. van der Velde, S. Rajagopal and A.B. de Vaate, eds. 2010. *The Zebra Mussel in Europe.* Leiden: Backhuys Publishers, pp.331–341.

Schernewski, G. and Schiewer, U. eds., 2002. *Baltic coastal ecosystems: structure, function and coastal zone management.* Berlin: Springer.

Schernewski, G., Behrendt, H., Neumann, T. and Stybel, N., 2008. Managing the Baltic Sea: Lessons learnt from eutrophication history in a large river-coast-sea system. In: LITTORAL, *LITTORAL 2008: a changing coast: challenge for the environmental policies.* Venice, Italy November 2008. Warnemünde: IOW.

Schernewski, G., Hofstede, J. and Neumann, T. eds., 2011. *Global change and Baltic coastal zones.* Dordrecht: Springer.

Schernewski, G., Stybel, N. and Neumann, T., 2012. Zebra mussel farming in the Szczecin (Oder) Lagoon: water-quality objectives and cost-effectiveness. *Ecology and Society*, 17 (2), Art. 4.

Schipp, G., 2008. *Is the use of fishmeal and fish oil in aquaculture diets sustainable?* Darwin: Northern territory government.

Schmelzer, N. and Holfort, J., 2009. *Ice Winters 2004/05 to 2008/09 on the German North and Baltic Sea Coasts.* Hamburg, Rostock: BSH.

Schrum, C. and Schmelzer, N. eds., 2007. *Proceedings of the fifth workshop on Baltic sea ice climate.* Hamburg, Germany 31 August – 2 September 2005. Hamburg: BSH.

Seed, R. and Suchanek, T.H., 1992. Population and community ecology of Mytilus. In: E.M. Gosling, ed. 1992. *The mussel Mytilus: ecology, physiology, genetics and culture. Developments in aquaculture and fisheries Science.* Amsterdam: Elsevier, pp.87–170.

Seinä, A. and Palosuo, E., 1996. *The classification of the maximum annual extent of ice cover the Baltic Sea 1720–1995.* Helsinki: Finnish Institute of Marine Research.

Smaal, A.C., 2002. European mussel cultivation along the Atlantic coast: production status, problems and perspectives. *Hydrobiologia*, 484 (1–3), pp.89–98.

Sprung, M., 1993. The other life: An account of present knowledge of the larval phase of Dreissena polymorpha. In: T.F. Nalepa and D.W. Schloesser, eds. 1993. *Zebra mussels: biology, impacts, and control.* Boca Raton: Lewis Publishers, pp.39–53.

Stanczykowska, A., 1977. Ecology of Dreissena polymorpha (Pall.) (Bi- valvia) in Lakes. *Polskie Archiwum Hydrobiologii*, 42, pp.461–530.

Stigebrandt, A., 2003. Regulation of vertical stratification, length of stagnation periods and oxygen conditions in the deeper deepwater of the Baltic proper. *Meereswissenschaftliche Berichte 54*, Baltic Sea Research Institute Warnemünde, Germany, pp.69–80.

Stigebrandt, A. and Gustafsson, B.G., 2003. Response of the Baltic Sea to climate change – theory and observations. Proceedings of the 22nd Conference of the Baltic Oceanographers (CBO), Stockholm 2001. *Journal of Sea Research*, 49, pp.243–256.

Strayer, D.L. and Smith, L.C., 1993. Distribution of the Zebra mussel (Dreissena polymorpha) in estuaries and brackish waters. In: T.F. Nalepa and D.W. Schloesser, eds. 1993. *Zebra mussels: biology, impacts, and control*. Boca Raton: Lewis Publishers, pp.715–727.

Stybel, N., Fenske, C. and Schernewski, G., 2009. Mussel cultivation to improve water quality in the Szczecin Lagoon. *Journal of coastal research*, 56, pp.1459–1463.

Sztobryn, M., 1994. Long-term changes in ice conditions at the Polish coast of the Baltic Sea. *Proceedings IAHR Ice Symposium, Norwegian Institute of Technology*, pp.345–354.

Tacon, A.G.J., 2007. Fish meal and fish oil use in aquaculture: global overview and prospects for substitution. [online] Available at: <http://en.engormix.com/MA-aquaculture/articles/fish-meal-fish-oil-t323/p0.htm> [Accessed 13 June 2012].

Tacon, A.G.J. and Metian, M., 2008. Global overview on the use of fish meal and fish oil in industrially compounded aquafeeds: Trends and future prospects. *Aquaculture*, 285, pp.146–158.

The BACC author team, 2008. *Assessment of climate change for the Baltic Sea basin*. Berlin: Springer.

Theodorou, J.A., Sorgeloos, P., Adams, C.M., Viaene, J. and Tzovenis, I., 2010. Optimal farm size for the production of the mediterranean mussel (Mytilus galloprovincialis) in Greece. *Montpellier Proceedings*, pp.1–6.

Tidwell, J.H. and Allan, G.L., 2001. Fish as food: aquaculture's contribution. *EMBO reports*, 2, pp.958–963.

van der Velde, G., Rajagopal, S. and de Vaate, A.b. eds., 2010. *The Zebra Mussel in Europe*, Leiden: Backhuys Publishers.

Voss, M., Dippner, J.W., Humborg, C. et al., 2011. History and scenarios of future development of Baltic Sea eutrophication. *Estuarine, Coastal and Shelf Science*, 92, pp.307–322.

Walter, U. and de Leeuw, D., 2007. Miesmuschel-Langleinenkulturen – vom wissenschaftlichen Experiment zur wirtschaftlichen Umsetzung. *Informationen aus der Fischereiforschung*, 54, pp.34–39.

Wielgat, M., 2002. Compilation of the nutrient loads for the Szczecin Lagoon (Southern Baltic). In: G. Schernewski and U. Schiewer, eds. 2002. *Baltic coastal ecosystems: structure, function and coastal zone management*. Berlin: Springer, pp.75–92.

Wolff, W.J., 1969. The Mollusca of the estuarine region of the rivers Rhine, Meuse and Scheldt in relation to the hydrography of the area. II. The Dreissenidae. *Basteria*, 33, pp.93–103.

Woźniczka, A. and Wolnomiejski, N., 2008. A drastic reduction in abundance of Dreissena polymorpha in the Skoszecwska Cove: effects in the population and habitat. *Ecological Questions*, 9, pp.103–111.

Zaiko, A. and Olenin, S., 2006. *Dreissena polymorpha*. [pdf] Available at: <http://www.europe-aliens.org/pdf/Dreissena_polymorpha.pdf> [Accessed 20 June 2012].

Zettler, M.L. and Röhner, M., 2004. Verbreitung und Entwicklung des Makrozoobenthos der Ostsee zwischen Fehmarnbelt und Usedom – Daten von 1839 bis 2001. In: BfG, ed. 2004. *Die Biodiversität in der deutschen Nord- und Ostsee*. Koblenz: BfG.

14 Impacts of Sea Level Change to the West Estonian Coastal Zone towards the End of the 21st Century

Valter Petersell[1], Sten Suuroja[1], Tarmo All[2] & Mihkel Shtokalenko[1]

[1] Geological Survey of Estonia, Tallinn, Estonia
[2] Ministry of the Environment, Tallinn, Estonia

14.1 Introduction

Global warming is currently accepted as a fact by a large proportion of scientists and specialists. Together with the temperature rise, the level of the world ocean will rise, and the level of the Baltic Sea together with it. Thus it is most likely that shallow coastal areas will be submerged. An increase of storm-related floods, intensification of coastal abrasion and landslides, flooding and water regime changes caused by higher water level at river mouths, and so on are possible.

These issues triggered the evaluation of problems related to climatic changes in the Baltic Sea Region, for which the west Estonian coast was chosen as one case study area (Figure 14.1).

Publications of the Potsdam Institute for Climate Impact Research (PIK), the Intergovernmental Panel on Climate Change (IPCC) and the German Advisory Council on Global Change (WBGU), the databases of the Estonian Meteorological and Hydrological Institute (EMHI) and the Geological Survey of Estonia (EGK) as well as the field observations and land elevation measurements were an essential part of the work. The research was financed by the European Union (EU) and the Estonian Environmental Investment Centre (KIK).

According to Petoukhov et al. (2000), Ganopolski et al. (2001), IPCC (2007), Kropp (2007, 2010) and WBGU (2009), the temperature rise by the end of the century might be of 2–4 °C and the related sea level rise 0.6 to 1.4 m in the Baltic Sea Region. A large proportion of Estonians, including scientists and engineers, are not convinced of the reality of the predicted rise of temperature and sea level (Raukas, 2006a; Raukas, 2006b; Zirnask, 2008). They argue that the temperature rise will be short-lived and that the sea level rise will be compensated by a neotectonic uplift of western Estonia (Želnin, 1964; Potter, 2002). Because of this attitude, the tasks of the project turned into estimating the reality of sea level rise, predicting the circumstances caused in the coastal zone by climate change, analysis of possible means to mitigate and adjust, and how to pass information on possible hazards related to climate change on to the public.

14.2 The West Estonian Coastal Zone

The west Estonian coastal zone is almost 300 km long and is dominated principally by lowlands. The west Estonian coastal zone is located along the east coast of the Baltic Sea and is prone to south-western to north-western winds and storms. The coastal zone is bordered in the north by the shallow Väinameri strait rich

Climate Change Adaptation in Practice: From Strategy Development to Implementation, First Edition.
Edited by Philipp Schmidt-Thomé and Johannes Klein.

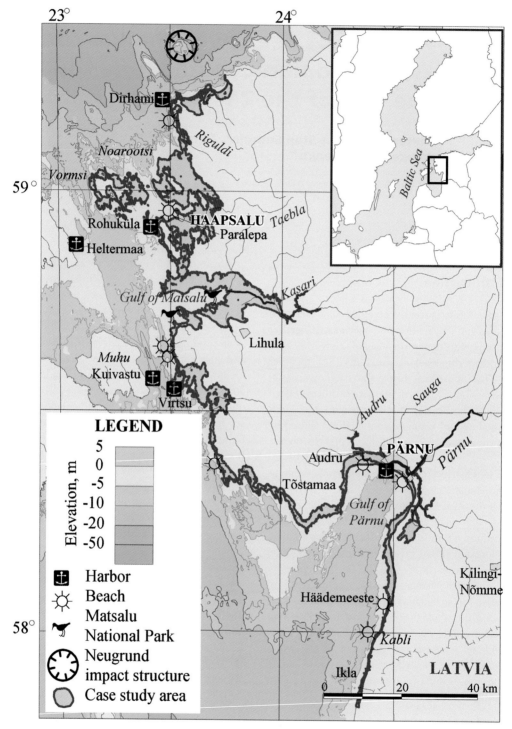

Figure 14.1 The location of the West Estonian coastal zone.

with islands and islets, and in the south by the Gulf of Riga. The coastline is cut by the bottleneck-formed Pärnu Bay and east-west elongated and narrow bays of Matsalu and Haapsalu. The eponymous cities, Pärnu and Haapsalu, are located on the coast, and there is an internationally renowned bird protection area in Matsalu Bay. The recently renovated harbours of Virtsu, Rohuküla and Dirhami are being developed in the coastal area, as well as the villages or settlements of Ikla, Häädemeeste, Audru, Tõstamaa, and so on (Roosaluste, Kask & Ektermann, 1998; Vunk, 2008). Recreational areas have spread along the beaches of the coastal zone. Their natural assets draw a great number of visitors.

The Väinameri strait is seldom more than 6 to 8 metres deep and it is rich with reefs near the Muhu and Vormsi islands. The Gulf of Riga is connected to the Gulf of Finland in a north-south direction. The principal sea-current flow is from the Gulf of Riga to the Gulf of Finland. The west Estonian islands are connected to the mainland via the harbours of Virtsu and Rohuküla.

The mouth of the Kasari river becomes the narrow Matsalu Bay with rambling coast and abundant islands. It forms the nucleus of the largest and richest protected area of migratory birds in northern Europe and the Baltic Sea Region (Figure 14.1). Haapsalu city is located on the south coast of Haapsalu Bay. The oldest part of the town is located on a peninsula that is formed of two eskers in a north-west direction. The eskers are low and form a chain of islets towards the mainland (Figure 14.2). The town is 10.6 km^2 in area and it started to form around a bishop's fort, founded in 1260 AD. Its ruins are one of the important historical relics of Estonia.

Haapsalu Bay is shallow and with modest water circulation. In the summer, the water warms up more than in other bays. The eskers divide the bay to the frontal bay (Eeslaht) in the south-west and the back bay (Tagalaht) in the north-east. The maximum depth of Eeslaht is 4.5 m and Tagalaht is less than 2 m deep. The partly underground Salajõgi river, which starts from a karst area almost 12 km to the east, flows into Haapsalu Bay. Curative mud is found in Haapsalu Bay, and has been a centre for mud spas since 1825.

The beaches of Pärnu and Valgeranna are internationally renowned (Photo 14.1). Up to 10 000 people visit these beaches every day during the summer holiday season. Various sporting attractions are being built on the beaches. The earlier ones were mainly destroyed by the 2005 January storm (Photo 14.1).

In Pärnu city there are also opportunities to observe and visit various architectural attractions.

14.3 Geology of the coastal zone

Coastal geology and hydrogeology are dominated on the one hand by the bedrock and the lithology of the Quaternary cover, and on the other hand by the sea level and the rate at which it changes. The gradient of the change can be estimated as the sum of sea level changes and through neotectonic movements of the ground.

The Ordovician and Silurian limestones, dolomites and marls in the coastal zone are with various clay content, frequency and thickness of clay interlayers. In the upper part of the cross-section (to a depth of approximately 50 m), the rock has varying porosity and amount of fractures. Fractured rocks are typically good aquifers and the sea water can penetrate far via cracks. The rock is quite resistant against erosion.

South of Pärnu and in the Tõstamaa peninsula there are Devonian terrigenous rocks in the form of weakly-cemented sandstones, siltstones and clays. Their grain size and porosity are varying. In places fracturing can be observed.

The thickness and lithology of the Quaternary cover varies considerably in the coastal zone (Eltermann, 1974; Kajak, 1972; Kajak & Liivrand, 1974). The thickness of the cover is dominantly 1 to 5 m, on alvars less than 1 m (e.g. Rohuküla and Virtsu regions). However, at locations where ancient valleys enter the sea, it is over 25 m thick (Haapsalu, Tõstamaa and Liu regions). On alvars, typically near the shoreline, the Quaternary cover is missing altogether, for example, Vormsi island (Photo 14.2), Rohuküla, and the sea bottom is formed by fractured carbonate rocks.

The most widely spread sedimentary cover is till with a varying content of clay and loose material. On Devonian sandstones, the cover is poor in clay particles but rich in sand. The undulating surface of the till, principally its lower part, is often covered by glacio-fluvial or glacio-lacustrine silt and sand, less often sand or gravel. The till and glacial sediments are covered in many locations by up to a few metres of thick sand and

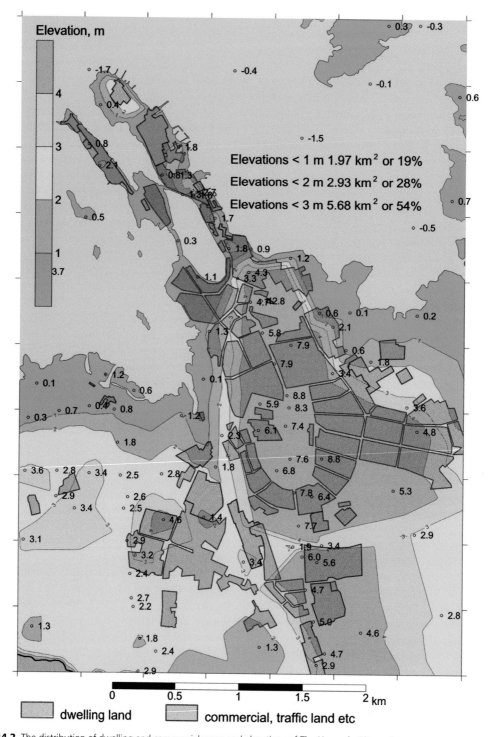

Figure 14.2 The distribution of dwelling and commercial areas and elevations of The Haapsalu City region.

Photo 14.1 The beach-cafe on Valgeranna before (on left) and after destroying by the storm in January 2005 (on right).

silt from different stages of the Baltic Sea. The latter also forms beach barriers and dunes. Low beach barriers are characteristic of the Pärnu Bay and Kabli shore to the south, as well as to the north-western Estonian shoreline. The areas between beach barriers and other depressions often become boggy or there are peat layers. To the south of Pärnu there is often a wide, shallow (<2 m) gently undulating belt between the old dune belt and the sea, where the till is covered with fine-grained sand. The belt has been developed by land uplift during the the last thousands of years. Within the belt, sand is absent only in the Tahkuranna region, where there is principally till by the coast.

The coastal morphology is dominated by the characteristics of the bedrock and the lithology of the Quaternary cover. The bedrock consists of Silurian and Ordovician carbonate rocks in the northern part of the coast that is often shallow, flat and rambling. There are few recreational areas and coastal sand often contains carbonate material. In the south, the bedrock is formed by Devonian weakly-cemented terrigenous sedimentary rocks, and the coastline is dominantly smooth and beaches more widespread. The beaches are typically separated from the forest by a terrace that has formed in the dune or beach barrier sands and that can be up to 4 to 6 m high.

The Dirhami area in the northern part of the coastal zone is particular in the sense that the beach is in places rich with erratic boulders, among which and of special interest are boulders (Photo 14.3) of impact

Photo 14.2 The limestone outcrop at Saxby lighthouse on the west coast of Vormsi Island.

Photo 14.3 The erratic boulders of the Neugrund breccia on the northern part of the Dirhami coastal zone.

breccia from the ring wall of the Cambrian Neugrund meteorite crater (Suuroja & Suuroja, 2010). These boulders are unique rocks and in them the clasts that are often cemented with a dark, typically aphanitic rock mass that melted during the meteorite impact.

The depth of the Väinameri strait varies and is mainly less than 5 m. Sediments are relocated along the shore. Clay particles are frequently transported to deeper depths or to coves with standing water. Patches of till (gravel) on the sea bottom are covered with rock fragments that resist water movement. Sedimentary rocks exposed in the sea bottom and on the shore maintain their infiltration properties, but may be prone to erosion by ice hummocks with boul-

ders or by large storms (at places such as Rohuküla and the west coast of Vormsi Island). In the south, the Gulf of Riga deepens relatively quickly and wave action causes only coastal abrasion. The beach barriers and developed dunes form the majority of beaches and recreational areas. The beaches and the terraces that often border them are easily prone to abrasion by stormy waves and can move forward up to 25 m towards the mainland during one single strong storm (Photo 14.4). Also, gently sloping beaches, and even shores with carbonate rocks, are prone to the action of stormy waves. Results of almost 15-year long monitoring show for example, that the coastline of the limestone western shore of Vormsi island is propagating inland almost 8–10 cm per year (Photo 14.2).

Photo 14.4 The traces of the coastal abrasion on the top of Dirhami Cape after storm in January 2005 (on left) and in 2010.

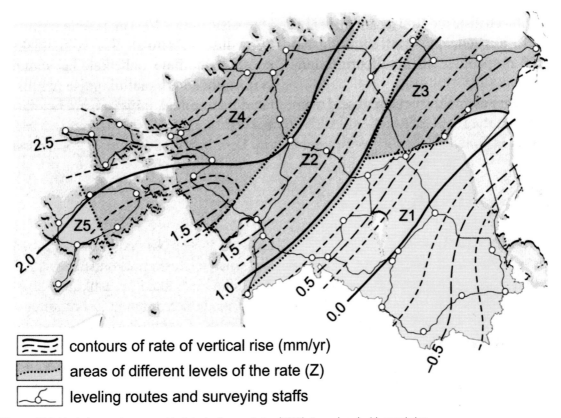

contours of rate of vertical rise (mm/yr)

areas of different levels of the rate (Z)

leveling routes and surveying staffs

Figure 14.3 Vertical ground movement in Estonia. *Source:* Potter (2002). Reproduced with permission.

A neotectonic map of Estonia, basing on precise levelling in the middle of the century, was widely used during the last century (Želnin, 1964; Torim, 1998). During this century, it has been replaced by a similar map, published as an official document (Figure 14.3; Potter, 2002). Based on this map, neotectonic uplift of the west Estonian coast continues. The uplift varies, having a minimum in the south (<1 mm/yr in Häädemeeste) and a maximum in the northwest (>2.5 mm/yr in Dirhami). Such a rise is at odds with observations by west Estonian marine stations in for example, Pärnu, Virtsu and Rohuküla. Based on observations of these stations, the rise has currently been replaced by subsidence (Figure 14.4; Garetsky, Ludwig, Schwab & Stackebrandt, 2001; Jevrejeva, Rüdja & Mäkinen, 2001; Liibusk, 2007), in Pärnu up to 1.3 mm/yr) and in Rohuküla up to 0.3 mm/yr). The elevation of beach barriers from different stages of the Baltic Sea is proof that

the land uplift with regard to the sea level has been continuous during the latest thousand years (Raukas & Kajak, 1997). From information on the database of EMHI, it is apparent that the uplift has been replaced by subsidence, which is difficult to explain with only the sea level rise, as the difference is too large (Figure 14.5). The postglacial isostatic uplift has ceased and has been or is being replaced by progressive subsidence. This presumption definitely needs to be specified, as it is a factor that increases the relative sea level rise.

14.4 Mean and extreme sea levels

Temporal development of sea level rise can be determined in west Estonia based on data from EMHI marine stations. The rising trend is not continuous, but changes periodically). The variation of annual

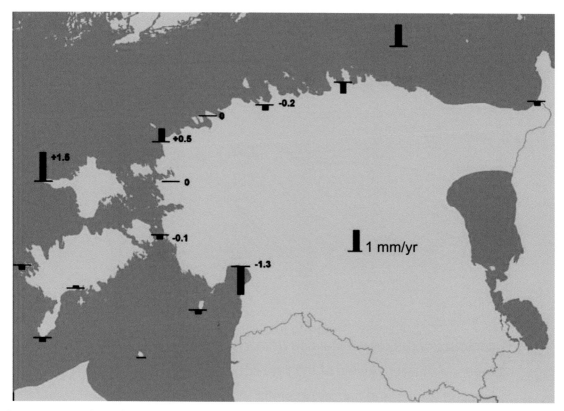

Figure 14.4 Vertical ground movement in Estonia in regard with sea level (Jevrejeva et al., 2001).

averages in Pärnu Bay is up to 37 cm and variation of short-term maximal changes is within 360 cm (Figure 14.5). As a result of stormy winds, sea level in the Pärnu bay has risen up to 275 cm and in Rohuküla up to 171 cm (Figure 14.6). The absolute minimum of the annual average in the 60-year period can be seen in 1996 and the maximum in 1990. These extremes are regional and can be observed all around west Estonian territorial waters, and also in observatories in south and south-west Finland (Johansson, Kahma, Boman & Launiainen, 2004). On top of the general background, there are periodic 10–15-year cycles of rise and fall (Figure 14.6) and against this background also several months-long periods of rise and fall (Figure 14.7). In Pärnu Bay, the rise exceeds the general linear trend by up to 8 cm during periods of rise of annual averages, and up to 55 cm during short-term periods of rise. High sea levels lasting a few months are principally observed during autumn and

winter months, simultaneously also in the Haapsalu region and similarly through the whole west Estonian coastal zone.

There are also years when against a background of generally low water (−3 cm in 2002) the wintertime sea level per month is high (+40 cm). Storms during these periods cause maximal erosion and the most hazardous floods. It is not beyond possibility that landslides become more frequent during rapid lowering of sea level (e.g. 2002 Pärnu landslide). Causes of this sort of sea level behaviour require additional research. While short-term rise and subsidence are very likely related to meteorological causes, long-term changes along the west coast of Estonia are almost definitely synchronized with sea level changes in the North Sea (west coast of Denmark, Figure 14.6). It is possible that sea level rise in the North Sea forms a blockage to the water flowing from the Baltic Sea. It can also cause sea level rise in the Baltic Sea by water flowing from

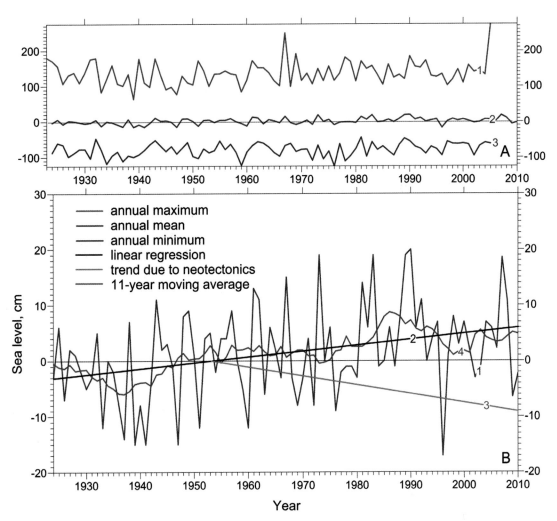

Figure 14.5 The Pärnu tide gauge measurements A 1-annual maximum, 2-annual mean, 3-annual minimum, B 1-annual mean, 2-linear regression 0,12 cm/yr, 3-calculated linear regression relation to uplift of land −0,15 cm/yr (Potter, 2002), 4–11-year moving average.

land areas and decrease of salinity, or inflow of salty water from the North Sea causing an increase of salinity. Other causes cannot be excluded.

14.5 Hydrology and hydrogeology of the coastal zone

The hydrological and hydrogeological conditions of the west Estonian coast and the impact the sea has on them depend on the elevation, thickness of Quaternary cover, its lithology and on the charac-

teristics of the underlying bedrock to some 50 m in depth.

Water flows to the sea via rivers, streams and numerous draining channels. The largest river is the Kasari river flowing into Matsalu Bay, the Kasari river with its tributaries (Suitsu, Peni), and the partly subterraneous Salajõgi river that flows into Haapsalu Bay. Average discharge of the Kasari river is 25–30 m³/s, the maximum can exceed 750 m³/s during the spring thaw and rainy season, and the minimum can be less than 1 m³/s during the dry season. Near the mouth, almost two kilometres upstream, a hydrological

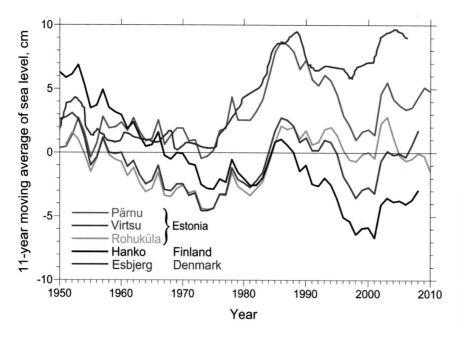

Figure 14.6 11-year moving average of sea level in W Estonia, SW Finland (Johansson, et al., 2004) and W Denmark marine stations 2011 (http://www.klimatilpasning.dk/DA-DK/KYST/PÅVIRKNING/Sider/Forside.aspx).

station has observed river surface variation of 2.38 to 4.84 m. The river flows in the west Estonian lowland. Thawing snow and the rainy season cause wide flooding in the middle and lower sections of the river. The area around the river mouth is embogged. In the region of both Matsalu Bay and the lower section of the Kasari river there are wide lowland areas (< 3 m a.s.l.; Figure 14.8). The discharge of the Salajõgi river flowing to Haapsalu Bay is modest, mainly less than 3 m³/s, but its lower section runs several kilometres in subterranean karst caverns and shows that the sea water can affect mainland areas to a great extent via karst formations.

There are problems related with the Ordovician-Cambrian water layer, which deepens from 100 m in the north to 550 m in the south (EGK, 2011). The water layer is confined and from Haapsalu towards the south the piezometric surface in the coastal zone is higher than the elevation of ground surface. Mineralization of the water increases towards the south, mainly because of an increasing content of Cl^- and Na^+. While in Haapsalu mineralization of the water approaches 0.4 g/l, in Häädemeeste it is already 4.7 g/l and the water is counted as mineral water. Because of the high level of pressurized water, water with high mineral content migrates upwards, towards the free-surface groundwater.

More than 1000 wells, down to a depth of 50 m, have been drilled in the coastal zone and further inland, up to 15 m elevation, during the last 60 years. The content of Cl^-, $Na^+ + K^+$ and other macroelements has been determined during various times in the free-surface water of wells that subsist from the groundwater. The map (Figure 14.9) shows the concentration and distribution of Cl^- in the water. The map is quite intriguing and very likely represents the intensity of subsistence (pollution) of the free-surface groundwater from the sea water.

The picture is essentially more complicated in the areas of Silurian and Ordovician carbonate rocks than in the areas of terrigenous Devonian rocks. The Cl^- content in the water is often higher than the conditional natural background level (60 mg/l) and also higher than the maximum amount of Cl^- (250 mg/l) in drinking water according to the EU recommendation (Council Directive 98/83/EC). The high Cl^- content varies mainly between 220 and 500 mg/l and in some cases is up to 2000 mg/l. High contents of $Na^+ + K^+$ are typically at the same levels and correlate very well with the Cl^- content. Mineralization of the free-surface groundwater layer decreases unevenly with distance from the sea. The trend of Cl^- content change in relation to depth in drilled wells between Haapsalu and Pärnu Bays shows that

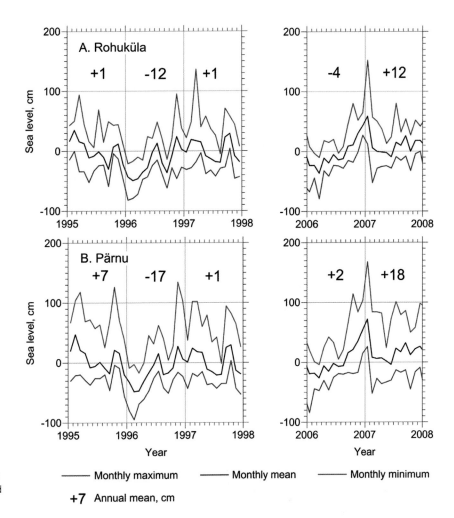

Figure 14.7 Montly variation of sea level in Rohuküla (A) and Pärnu (B) marine stations.

———— Monthly maximum ———— Monthly mean ———— Monthly minimum

+7 Annual mean, cm

Cl⁻ content decreases to the depth of 50–100 m and starts to increase at greater depths. This relationship leads to the assumption that the aquitard between the high Cl^--content Ordovician-Cambrian water layer and Ordovician-Silurian water system is not continuous but passes confined water in places.

Causes for increased mineralization in the coastal free-surface groundwater (subsoil water) are varying. The coast is bordered by shallow sea, mainly less than 20 m deep and with exposed sedimentary rocks at its bottom and shores. Sea water forms a permanent reservoir that permanently affects the free-surface fresh water of the coastal zone. Pollution starts and spreads mainly by infiltration through cracks in the limestone to areas where the dynamic water level is below sea level because of a cone of depression caused by water consumption. This phenomenon has been proved with single wells and water intakes (e.g. in the cities of Haapsalu and Pärnu).

The Quaternary cover forms a conditional aquitard with varying protective qualities (Perens, 2001). It is not continuous (Figure 14.9) and water in the cover and underlying sedimentary rocks typically forms a uniform layer of subsoil water or free-surface groundwater. This is the case principally in the areas of Devonian sedimentary rocks where the filtration coefficient is often more than 0.1 m/day.

Subsoil water subsists mainly from precipitation. Thus the water in sedimentary rocks directly below the Quaternary cover should be principally originated from precipitation, with low or moderate mineralization (< 400 mg/l). The real picture is essentially more

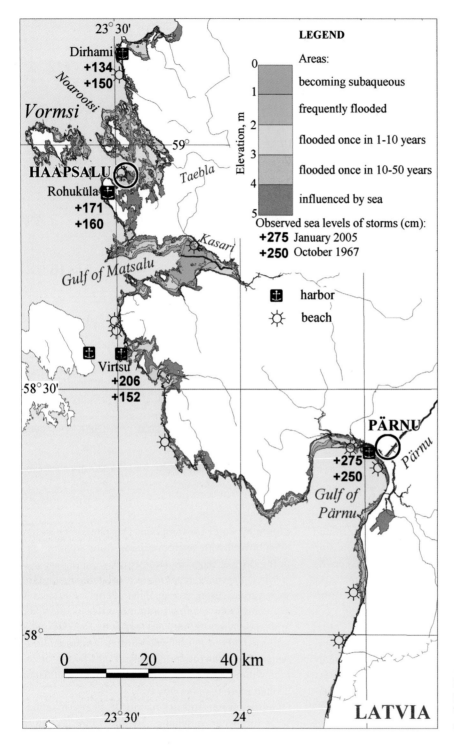

Figure 14.8 Map of predicted sea impact to the W Estonian coastal zone by the end of the 21st century.

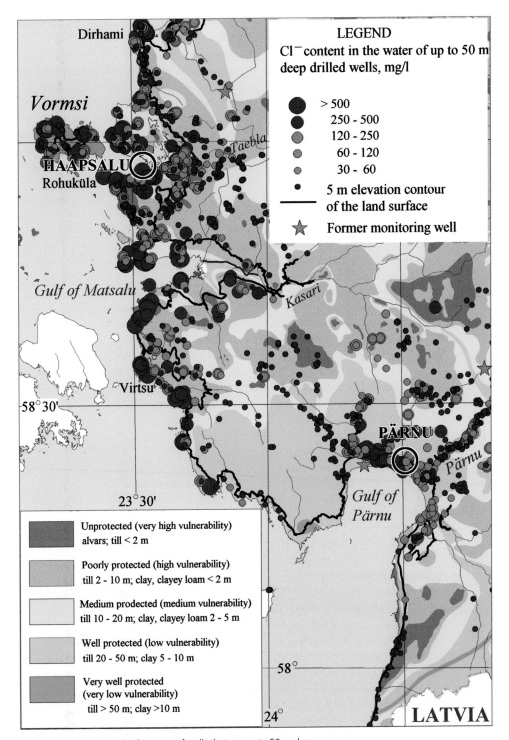

Figure 14.9 Chlorine ion content in the water of wells that are up to 50 m deep.

complicated. Water mineralization varies between 0.1 and 2 g/l. As earlier measurements during short-term monitoring have shown, Cl^- concentration has increased in drilled wells that are located at considerable distance from the sea. Although the observations have been conducted during a relatively short period, Cl^- content increases in all three wells in a linear manner – 0.5 to 1.3 mg/l per year.

In addition to water directly infiltrating from the sea, free-surface groundwater pollution occurs also by flooding in areas where an aquitard is conditional or absent. The uppermost water-free soil layer is filled with sea water as a result of flooding, and the water infiltrates more deeply. Based on studies by Väärsi et al. (1969) added pollution is caused by sea water sprayed in the atmosphere in storms. The extent of this phenomenon can be up to several kilometres. This kind of pollution has been observed in Malda village (Audru municipality) from the shoreline up to 10 km inland and up to an altitude of almost 6 m. As a result, water mineralization in a well increased to up to 2000 mg/l, mainly because of increased Na^+ and Cl^-.

The situation is less striking in the areas of Devonian terrigenous rocks. If the Pärnu water intake is excluded, free-surface water with high mineralization (> 600 mg/l) has not been observed. However, there are drilled wells, where the Cl^- and Na^+ contents are elevated. This shows that the process of increased salt content, however slow, also occurs in the areas of Devonian sedimentary rocks and definitely in the areas of ancient valleys extending to the sea (e.g. Tõstamaa).

14.6 Climate change impacts

Based on observations of EMHI Pärnu and Tallinn (Harku) stations, the average annual temperature rise in the region has been 0.022 to 0.026 °C during the last 50 years. Thus during the years from 1955 to 2005, the annual mean temperature has risen almost 1.1 °C and 1.3 °C, respectively. Also the amount of precipitation has an increasing trend and increased during the abovementioned period 1.6–2.6 mm. It is likely that these trends will continue, and by the end of the century the annual mean temperature in Estonia could be 7.8 to 8.8 °C.

It has been observed during recent years that in the shallow Haapsalu Bay the sea surface temperature has periodically increased to over 26 °C during the summer season due to inefficient water exchange. This has increased the mortality rate of fish. If temperature rise continues, fish mortality and water purification will intensify. Increase of temperature causes an increase of evaporation. At the same time, periods of rich precipitation and drought will develop and the contrast between them become more pronounced. More frequent occurrence of intensive rain periods has already been observed during the last decade. During such periods, precipitation across a few days exceeds the monthly average by two or more times, accompanied with floods. For instance in August 2003, precipitation in north-eastern Estonia was over 3-fold larger than the monthly average (86 mm) and the precipitation in August 2008 in the Kuusiku-Kohila area was over 2.5-fold greater than the monthly average (83 mm).

During recent decades, progressive development of a wide east-west anticyclone has been observed around Moscow during the summer season. The western edge of this anticyclone extends sporadically to Estonia. Thus at times during summer, the maximum temperatures in Estonia can increase significantly above 31–34 °C and may reach 35–38 °C by the end of the century.

14.7 Sea level rise

Satellite measurements indicate that the sea surface height increases by three to four millimetres per year (Kropp, 2007; WBGU, 2009) and is progressing. Based on the studies by WBGU, IPCC, PIK and EMHI, and known information (Conway & Tans, 2011) and taking into account the neotectonic movements, it is realistic that the sea level rise related to the temperature rise can be up to 0.9–1.1 m in the west Estonian coastal zone by the end of this century.

The critical sea level in west Estonia is close to 150 cm in the Pärnu region and 100 cm in the Haapsalu region. According to the marine station observations, storm-related sea level during the last 60 years has exceeded these values 19 and 14 times and extended in January 2005 to 275 cm and 171 cm, respectively (Figure 14.8). During later measurements of absolute ground elevation, it was found that in

Audru almost three kilometres inland, the sea level rose to 298 ± 5 cm and almost five kilometres upstream from the mouth of the Pärnu river 316 ± 5 cm. Although the water level observations were made by the local inhabitants and can be somewhat inaccurate, it can be presumed that the elevated water level in Audru was caused by stormy winds, but in the lower reaches of the Pärnu river by additional river water.

It is realistic that in the circumstances of sea level rise the critical sea level remains similar. Based on a ground elevation estimation with the aid of basic Estonian maps and RTK-GPS measurements data, it is possible that by the end of the century mainland areas of west Estonian coast currently up to 1 m a.s.l., forming an area up to 146 km^2 or 0.3% of the mainland will be submerged. Areas up to 2 m a.s.l., with a summed up area of 253 km^2, up to 3 m a.s.l. and 449 km^2 in area, and up to 5 m a.s.l. and 760 km^2 in area will be affected by the sea in various intensities (Figure 14.8).

As a result of sea level rise, houses that are currently located by the shoreline or at lower than 2 m elevation will become uninhabitable. This may also happen in lowland areas where houses at higher elevations may be confined to islets. Also harbour piers need to be reconstructed, as their current elevation is 1.9–2.5 m and there is the risk that they will be frequently flooded by stormy waves by the end of the century.

Such coastal areas are most sensitive to sea level rise, where there are wide areas of fine sand and silt with glacio-fluvial, glacio-lacustrine or various Baltic Sea stage origin, or varved clay. Existing drainage systems are at risk in such areas, and their reconstruction may be costly or unrealistic. These areas are also prone to intense erosion and material transport may be extensive in the event of sea level rise. The sea may extend far further than the absolute ground elevation suggests. At the same time, favourable conditions will develop for 'drifting' sand spits, beach barriers and other formations, as well as those areas that experience accumulation of sea mud and seaweed or development of mires. This coast type is widespread in the Matsalu and Haapsalu bays and in the area of the Noarootsi peninsula.

Such coasts are more stable where the limestone is covered by till of various thicknesses. As a result of abrasion by the sea, these areas will be covered by material washed out from the till and become resistant against abrasion and erosion. Such areas will be partly covered by sea, or form low islands, but the outlines of land morphology will be preserved. This coast type is frequent in Tõstamaa and Tahkuranna, as well as in the area between Haapsalu and Matsalu bays.

Carbonatite rock cliffs are resistant against the impact of sea level rise. These are found in the region of the Rohuküla harbour and along the west coast of Vormsi island (Photo 14.2). These coasts are also resistant to storm waves and abrasion works slowly according to monitoring data, 8–10 cm/yr on average. It is also true that in the event of sea level rise, abrasion will not increase.

It is probable that abrasion of sandy terraced beaches such as Dirhami, Kabli and Valgeranna will intensify considerably. Although they will not be universally prone to storm-related floods by the end of the century, there will be relative sea level rise with regard to the terraces, and material transport per surface unit will decrease. As a result, resistance of terraces against waves may decrease and drifting of terraces inland may be measurable in hundreds of metres by the end of the century. As mentioned earlier, even in the current circumstances the shore of Valgeranna moved up to 25 m inland during the 2005 January storm (Photo 14.1).

There is no doubt that pollution of subsoil water and groundwater will intensify under conditions of sea level rise. Cl$^-$ content in wells will very likely exceed 250 mg/l by the end of the century at locations where it is now 120 to 250 mg/l, and sea water pollution will spread considerably further inland.

Intensification of free-surface groundwater pollution by sea water occurs as a result of widening of flooding, via bedrock outcrops created by sea water rise and by sea water sprayed in the atmosphere.

Although the picture is less severe in the areas of Devonian terrigenous rocks, there also the free-surface groundwater pollution from sea water will intensify. The higher the sea level rises, the more serious problems there will be with household and drinking water.

As a result of sea level rise, the water level will rise in the lower reaches of rivers or brooks, and in channels or ditches draining into the sea, and the salt content in subsoil water will increase further. Although downpours related to storms or intense snowmelt increase flow rates by several magnitudes, they may also widen considerably the effect of floods mixed with sea water. In this sense, the most endangered area are the lower reaches of the Kasari river. Water level increase in the lower reaches of river beds favours the development of landslides.

It is realistic that the described processes will intensify in the progressing circumstances of sea level rise, and become more complicated due to the geological characteristics of the coastal zone. More and more areas will become submerged or become influenced by the sea level rise, causing additional problems. The most pertinent of the problems are:

• pollution (e.g. P, N, Pb, and petroleum products) may be transported to the sea from the soil;

• warming up of the sea water and pollution from the mainland favour growth of vegetation and accumulation of its residues in half-closed bays or shores with slow water exchange, which has a negative impact on the sea environment.

14.8 Prediction of damage caused by sea level rise

As a result of sea level rise, an increasing amount of land, buildings and constructions will be submerged or prone to the negative impact to a varying extent. There will be economic and societal losses with varying degrees.

The methodology for estimating economic losses is not uniquely clear, as the value of estimated targets changes and in general increases. This is an argument for why the estimates here are based on current price level and direct real damage. It is simple to adjust this when the price level changes, as has been tested in the US (Michael, Sides & Sullivan, 2003) and as is recommended by Estonian economists (Levald, 2009). As the related economic losses are formed over 90 years more or less equally, discounting is not used in calculating the losses.

By the end of the century mainland areas, by the end of the century mainland areas outside the cities of Pärnu and Haapsalu, up to 146 km², will be permanently submerged. Because of frequent flooding (below the critical level) an area of 107 km² will become useless, of which plot area is 2.4 km² and 11.2 km² is fields. An additional 196 km² will be prone

to frequent (10-year event) floods, and 90 km² to rare (50-year event) floods. Additionally, almost 220 km² will be affected by the indirect effects of future storm surges.

In estimating the losses it was presumed that the value of continuously submerged land will become 0 and it will be replaced by mainland areas below the critical level. Losses related to frequent floods will form 20% and to rare floods 4% of the average value of one hectare of lost (below critical level) land. Areas of frequent flooding will be those that have an absolute elevation of less than 3 m and areas of rare flooding those that have an elevation of less than 4 m. These thresholds are almost 10% higher in northwest Estonia and 10% lower in the Pärnu area.

A harvest of two tons of wheat per hectare was taken as the basis for estimating the value of fields. When arable, forest and grazing land were estimated, the method was differential rent, where in the conditions of continuously rising value of land, the normative efficiency of capital is $E = 0.1$ and the duration 45 years. In the case of residential land, regional market value of state-owned land (0.64 €/m²) and housing stock cost of 1 m², mainly similar to the average monthly salary in west Estonia (577 €/m²), were considered. Costs of communal amenities were equalized with housing stock costs.

Without doubt the estimated economic losses are only approximate. The state (cadastre) price of a hectare and m² cost of housing stock in Estonia are considerably lower than the average in EU and they are constantly approaching this value. Thus, the estimates represent minimal values. Defining the housing stock caused big problems, for two reasons mainly. On the one hand, inhabitants leave their residence and on the other hand, new homes are being built in the coastal zone. In order to solve this dilemma, an approximate number of inhabitants who need to find a new residence was taken into account, with the estimate of almost 10 m² per person, if the sea level rises up to 100 cm. The number of inhabitants was estimated to be one third of the towns, villages and separate residences and one out of four of inhabitants of Haapsalu town. The calculated number of people is almost 5000.

Based on the presented information, the economic losses related to sea level rise are presented in Table 14.1:

Table 14.1 Economic losses related to sea level rise. Mainland that will be submerged (below critical level)

Arable land	2.19 million €
Forest and grazing land	5.09 million €
Residential areas	1.54 million €
Housing stock with services and maintenance	57.69 million €
Total	66.51 million €

The values do not reflect societal losses. Also, they do not reflect the losses caused to recently renovated harbours, piers and roads that connect the islands with the mainland (Rohuküla, Virtsu, Dirhami).

14.9 Possibilities for mitigation of losses that are related to sea level rise

The expected rise of global temperatures and sea levels are alarming. As can be seen in satellite images the rate of sea level rise is progressive and is currently already 3–4 mm/yr (Conway & Tans, 2011). The majority of research on causes of temperature rise shows that the rate of the trend is directly related to the amount of greenhouse gases that are released into the atmosphere from caustobiolites.

On the basis of the Kyoto protocol, the trend of using caustobiolites should have decreased or at least stabilized during the last decade. In fact, it shows a continuous growth. Increase of the CO_2 content in the atmosphere continues at an increasing rate by almost 2% per year (Kropp, 2009; Conway & Tans, 2011) and the global temperature increases (Drange, 2010). It seems more realistic that during the coming decades it is not possible to stabilize the consumption of caustobiolites, and it is less possible to diminish their usage to the level of 1990. As proof, decisions and recommendations of the 1997 Kyoto protocol have been not fulfilled to a wide extent. And if the view of the majority of scientists, that the principal cause for global temperature rise is man-made release of solar energy that has been accumulated in caustobiolites during hundreds of millions of years, becomes reality, the temperature rise and related sea level rise are unavoidable.

Estonian inhabitants cannot stop or change negative phenomena related to the global rise of temperature and sea level. However, they can minimize and mitigate their impact, even though the actions are long term and costly, and may also be temporary. With this background, it is relevant to define the possible extent of negative impact on the most essential constructions in the coastal zone, as well as the costs and relevance of possible actions. It needs to be taken into account that the sea level rise may not halt by the end of the century but may continue into the future.

The Geological Survey of Estonia decided to resume or start with the monitoring of free-surface groundwater and compilation of geological and hydrogeological maps of the coastal zone on the scale 1:50 000 to provide a basis for defining adaptation measures. Complex analysis of measurements and mapping, together with databases of marine and meteorological observations will provide a means for monitoring environmental changes in the coastal zone and, based on geological and hydrogeological conditions, provide a means for mitigation of impacts. They should give an answer to the characteristics of intensification and extent of increase of salt content in groundwater, and to the nature of changes of physiochemical properties of the soil while subsoil water level rises.

It is likely that the causes of rising global temperature and sea level, new trends and effect of state-level actions will become apparent during the next 20 years, in order to deal with the impact of temperature rise. However, it should be mentioned again that the actions by states will likely be rather modest. As proof of this, there is little response in the form of actions to the recommendations of the 1997 Kyoto protocol and 2011 Durban protocol.

Based to a large extent on the long duration of the predicted losses, it is recommended that during the next 20 years:

• until the prediction of sea level rise is elaborated, construction of buildings that are expensive or meant to be in long-term use should be halted if they are in areas that are less than 3 m a.s.l. in the Pärnu region, and less than 2.5 m a.s.l. in the Haapsalu region, respectively;

• monitoring of salt content in the free-surface groundwater in the coastal zone should be resumed and geological and hydrogeological mapping of the coastal zone on the scale 1:50 000 should be conducted;

- explain to Haapsalu and Pärnu cities the possibility of minimizing the impact of sea level rise by constructing protective dams;
- study the possibility of rising piers in harbours and take action over it.

In the case of continuing sea level rise, the collected information makes it possible to plan and either incorporate or discard methods in order to minimize the effect.

14.10 Public outreach and conclusion

As mentioned, a major portion of Estonian inhabitants, including scientists and engineers, do not believe in sea level rise and long duration of global temperature rise. Accordingly, the principal task in the project was to assess current changes in climate and model potential future effects and impacts of climate change, to explain related risks to the people and to analyse minimizing actions and their efficiency. When the people become convinced of the realness of temperature and sea level rise, existence and expansion of risks, local and central governments start to analyse the circumstances and apply measures for risk mitigation.

During the project, the results have been published by printed media (Petersell, 2009; Petersell, All & Suuroja, 2010; Petersell, Suuroja, All & Shtokalenko, 2012), discussions and presentations to local stakeholders and inhabitants, including local government representatives, architects, entrepreneurs, officials of the Ministry of the Environment and environmentalists. Natural and man-made causes of climatic change, views of those who deny the man-made warming, and related environmental, economic and societal losses were dealt with.

The architects both in Audru and in Haapsalu region promised to ban the future building of expensive and possibly environmentally unfriendly constructions at lower elevations than 3 and 2.5 m a.s.l., respectively. It is worth noting that this has been acted upon by the Audru municipal council.

Acknowledgements

We are grateful to the European Union and the Estonian Environmental Investment Centre (KIK) who considered this study necessary and provided the funding. We thank the Estonian Meteorological and Hydrological Institute (EMHI) for agreeing to let us use the sample meteorological data of their database and ETF grants 8266, JD172 for support. Particular thanks go to the organizers and methodological leaders of this research, the geologists of the Geological Survey of Finland Dr Philipp Schmidt-Thomé and senior geologist Johannes Klein.

References

Conway, T. and Tans, P., 2011. *Trends in Atmospheric Carbon Dioxide*. [online] Available at: <www.esrl.noaa.gov/gmd/ccgg/trends> [Accessed 23 May 2012].

Council Directive 1998/83/EC of 3 November 1998 on the quality of water intended for human consumption.

EGK (Eesti Geoloogiakeskus), 2011. *Põhjavee kadaster, seisuga 2010*. Tallinn: EGK.

Elterman, G.J., 1974. *NSVL riiklik geoloogiline kaart*. Leht O-34-XII, (Haapsalu), 1:200000 (Kvaternaari setted), Moskva: Aerogeoizdat.

Ganopolski, A., Petoukhov, V., Rahmstorf, S., Brovkin, V., Claussen, M., Eliseev, A. and Kubatzki, C., 2001. CLIMBER-2: climate system model of intermediate complexity. Part II: model sensitivity. *Climate Dynamics*, 17, pp.735–751.

Garetsky, R.G., Ludwig, A.O., Schwab, G. and Stackebrandt, W. eds., 2001. Neogeodynamics of the Baltic Sea Depression and adjacent areas. Results of IGCP project 346. *Brandenburger Geowissenschaftliche Beiträge* 1, pp.1–48.

IPCC, 2007. *Climate Change 2007: Synthesis Report. Summary for Policymakers*. Geneva, Switzerland: IPCC.

Jevrejeva, S., Rüdja, A. and Mäkinen, J., 2001. Postglacial rebound in Fennoscandia: new results from Estonian tide gauges. In: M. Sideris, ed., *Gravity, Geoid and Geodynamics. IAG International Symposium*. Banff, Alberta, Canada 31 July – 5 August 2000. Heidelberg: Springer-Verlag, pp.193–198.

Johansson, M.M., Kahma, K.K., Boman, H. and Launiainen, J., 2004. Scenarios for sea level on the Finnish coast. *Boreal Environment Research*, 9, pp.153–166.

Kajak, K.F., 1972. *NSVL riiklik geoloogiline kaart*. Leht O-34-XVIII (Virtsu), 1:200000 (Kvaternaari setted), Moskva: Aerogeoizdat.

Kajak, K.F. and Liivrand, H.I., 1974. NSVL riiklik geoloogiline kaart. Leht O-35-XIII (Pärnu), 1:200000 (Kvaternaari setted), Moskva: Aerogeoizdat.

Kropp, J.P., 2007. *Climate Change Scenarios and costs of Sea-Level Rise in the Baltic Sea Region*. [pdf] Available at: <http://www.astra-project.org/sites/download/Tampere_kropp.pdf> [Accessed 23 May 2012].

Kropp, J.P., 2010. *The conclusions drawn from Copenhagen: The need for sound climate information and solution options. Climate Change in the Baltic Sea Region Conference*. Kalundborg, Denmark 27 January 2010. Potsdam: Potsdam Institute for Climate Impact Research.

Levald, H., 2009. *Tehnoökoloogia*. Tallinn: Euroülikool.

Liibusk, A., 2007. *Ülevaade maakoore vertikaalliikumistest Eesti aladel*. Tartu, Maaülikool: EGF.

Michael, J.A., Sides, D.A. and Sullivan, T.E., 2003. *The Economic Cost of Sea Level Rise to Three Chesapeake Bay Communities*. [pdf] Available at: <http://www.dnr.state.md.us/irc/docs/00015534.pdf> [Accessed 23 May 2012].

Perens, R., 2001. *Groundwater Vulnerability Map of Estonia*. 1:500000, Tallinn: Geological Survey of Estonia, EGF.

Petersell, V., 2009. Eesti rannikualad on ohus. *Eesti Päevaleht*, 19. Nov., p.1.

Petersell, V., All, T. and Suuroja, S., 2010. Kliima ja 21. sajandi ohud. *Keskkonnatehnika*, 1 (10), pp.8–11.

Petersell, V., Suuroja, S., All, T. and Shtokalenko, M., 2012. Meremõju prognoos Lääne-Eesti rannavööndile XXI sajandi lõpuks. *Eesti Geoloogiakeskuse Toimetised*, trükis.

Petoukhov, V., Ganopolski, A., Brovkin, V., Claussen, M., Eliseev, A., Kubatzki, C. and Rahmstorf, S., 2000. CLIMBER-2: climate system model of intermediate complexity. Part I: model description and performance for present climate. *Climate Dynamics*, 16, pp.1–17.

Potter, H., 2002. Kõrgussuhted ja gravimeetria. EE 11. Eesti Entsüklopeedia, Eesti Üld. Tallinn, Eesti Entsüklopeediakirjastus, 88–90 lk.

Raukas, A. and Kajak, K., 1997. Quaternary Cover. In: A. Raukas and A. Teedumäe, eds. 1997. *Geology and Mineral Resources of Estonia*. Tallinn: Estonian Academy Publishers, pp.125–136.

Raukas, A., 2006a. Milles seisneb Science'i artikli tuum?. *Postimehe ajakiri*, 1 (317), 8p.

Raukas, a., 2006b. Kliima ja teadusmüüdid meie ümber. *Horisont*, 4, pp.34–40.

Roosaluste, E., Kask, J. and Ektermann, M., 1998. Läänemaa Loodus, 1988. Haapsalu. Schmidt-Thomé, P., 2006. *Integration of natural hazards, risk and climate change into spatial planning practices*. PhD University of Helsinki.

Suuroja, K. and Suuroja, S., 2010. The Neogrund meteorite crater on the seafloor of the Gulf of Finland, Estonia. *BALTICA*, 23 (1), pp.47–58.

Torim, A., 1998. *Renovation of the Estonian Levelling Network*. Tartu: Estonian Land Board Development Centre.

Väärsi, A., Kajak, K., Kajak, H., Kirss, J. and Liivrand, H., 1969. Otčet Južno- Èstonskogo otrjada o kompleksnoi geologo-gidrogeologičeskoi s'emke m: 1:200 000 jugo-zapadnoi časti Èstonii za 1966–1968 gody. EGF.

Vallner, L., Sildvee, H. and Torim, A., 1988. Recent crustal movements in Estonia. *Journal of Geodynamics*, 9 (2-4), pp.215–223.

Vunk, A. ed., 2008. Pärnumaa. Loodus aeg inimene, 2008. Eesti Entsüklopeedia Kirjastus.

WBGU (German Advisory Council on Global Change), 2009. *Solving the climate dilemma: The budget approach - Special Report*. Berlin: WBGU.

Želnin, G.A., 1964. Totčnost i vozmožnosti povtornogo nivelirovanija. In: Gudelis V.K., ed. 1964. *Covremennõje i noveišie dviženija zemnoi kõrõ v Pribaltike*. Vilnjus, pp.17–24.

Zirnask, M., 2008. Akadeemik lippmaa: globaalne soojenemine on jama! Ostke rahumeeli suur maastur. *Eesti ekspress*, 16 Oct. p.1.

15

Geodynamical Conditions of the Karklė Beach (Lithuania) and Adaptation to Sea Level Change

Jonas Satkūnas[1], Darius Jarmalavičius[2], Aldona Damušytė[1] & Gintautas Žilinskas[2]

[1]Lithuanian Geological Survey, Vilnius, Lithuania
[2]Nature Research Centre, Institute of Geology and Geography, Vilnius, Lithuania

15.1 Introduction

Changes of coastlines and beaches have significant socio-economic impacts on the population concentrated in coastal areas. Beaches in particular play an important role in coastal protection and recreation. The length of the Lithuanian Baltic Sea coast is only 90.6 km (Žilinskas, 1997): 51.03 km on the Curonian Spit and 38.49 km on the mainland. These are separated by the 1.08 km wide Klaipeda strait (Figure 15.1).

The Baltic mainland coast of Lithuania is geomorphologically rather diverse – artificial foredunes as coastal protection measure occupy 28.1 km, or 74.4% of the coastline, morainic landscapes and sand cliffs occupy 5.62 km, or 14.9%, and natural coastal dunes 3.73 km, or 9.87% of the continental coastline (Jarmalavičius, Satkūnas, Žilinskas & Pupienis, 2012). The Klaipeda City municipality has initiated feasibility and planning studies for determination of conditions for the establishment of recreational infrastructure in the vicinitiy of Karklė village (10 km north of Klaipeda) (Figure 15.2). The studies are partly carried out within the BaltCICA Project. Klaipeda is the only larger city in Lithuania located at the coastline and as

a result, its environs are especially sensitive to climate change.

Though the coastline of Karklė village is only 1.4 km long (Figure 15.2), complex geological processes take place. Beside the climate change driven factors, Karklė's shoreline is constantly exposed to increasing human impacts (e.g. dredging in Klaipeda harbour) which affects shore-formation processes independently (Žilinskas & Jarmalavičius, 2000). Lately, problems affecting the Lithuanian coast of the Baltic Sea have become particularly acute (Žilinskas, Pupienis & Jarmalavičius, 2010).

According to in-house observations, the tourist figures for Karklė are evidently increasing – their number sextupled in the summer peak season (July) for the period 2001–11. In response to this trend, the Klaipeda municipality initiated some regional development plans. Nevertheless, the Karklė beach is located in the Seaside Regional Park – with the protected landscapes of Olando Kepure (The Dutch Cap) south of Karklė, and the Karklė talasological reserve in the north (Figure 15.2). The Karklė village itself belongs to the ethno cultural reserve. Therefore, development of recreational infrastructure is constrained by the occurence of these protected areas.

Climate Change Adaptation in Practice: From Strategy Development to Implementation, First Edition.
Edited by Philipp Schmidt-Thomé and Johannes Klein.
© 2013 John Wiley & Sons, Ltd. Published 2013 by John Wiley & Sons, Ltd.

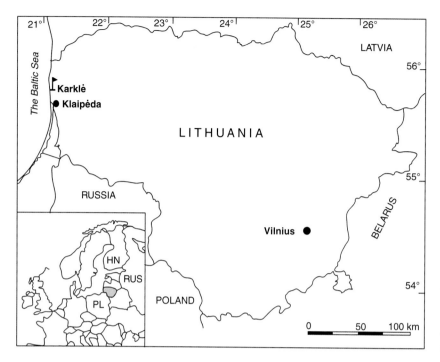

Figure 15.1 Location of the case sudy area. Karklė is located in the north of Klaipėda and marked by flag.
Source: Lithuanian Geological Survey.

15.2 Methods of investigations

The study area includes the Karklė beach and cliff and reaches isobaths of 15 to 20 m in the offshore region. Investigations were accomplished according to coastal environment monitoring methods. The geological-geomorphological map (scale 1:5000) covers the beach and the cliff, that is, an onshore belt 100–300 m wide, and the underwater slope of the offshore zone up to the 10–20 m isobaths. It was compiled based on interpretation of aerial photographs (scale 1:18 000), drilling and sampling data. The geological-geomorphological map was illustrated by offshore morpholithological profiles and two geological cross-sections on the onshore zone. These cartographic data were used for identification of the natural conditions to serve as a basis for planning of beach infrastructure. Due to shallow occurrence of till the beach is rather characterized by an abundance of boulders, cobbles and coarse gravel. Therefore, it is very important to identify the offshore places with sand accumulations that would serve as a resource for beach nourishment.

In order to evaluate the change of beach profile in sectors dominated by different dynamic processes – accretion and erosion – those variations of morphometric parameters that occurred during the last 18 years (period 1993–2011) were analysed. Levelling of cross profiles of the coast was performed by using the electronic tachometer TOPCON GTS 229 in two measuring stations installed along the Karklė coast since 1993. The following morphometric indicators were chosen for analysis of morphological changes of the beaches: beach width (distance between the shoreline and the dune ridge foot or cliff) and beach height (difference between the altitude of beach surface at the dune ridge foot and its altitude at the shoreline in view of the average long-term Baltic Sea level). These parameters were selected as being the best to represent the features of beach profile (Thom & Hall, 1991; Jarmalavičius, Satkūnas, Žilinskas & Pupienis, 2012). The obtained field data served as a basis for interpretation and evaluation of the shoreline dynamics between 1993 and 2011.

In the present chapter, sea level measurement data from the Klaipeda hydrological station (period of

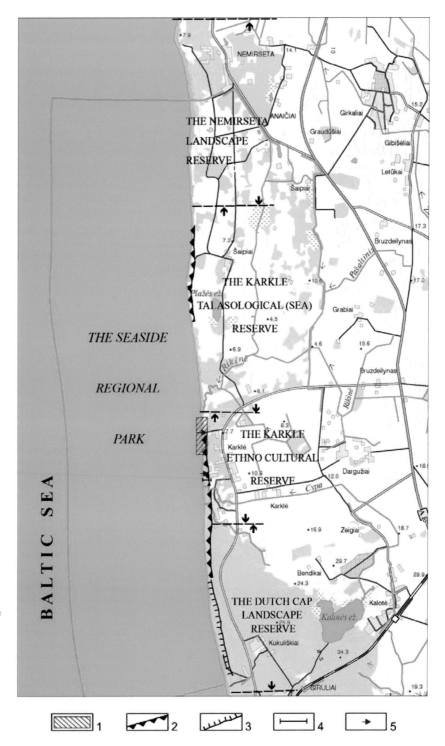

Figure 15.2 Situation of the of the Karklė beach.
Legend: 1 – proposed territory of the Karklė beach; 2 – active cliff; 3 – inactive cliff; 4 – line of geological cross-section; 5 – line of profile of measurements of coastal dynamics (see Figures 15.4–15.5).

observations 1898–2011) were used. These datasets were obtained from the Department of Marine Research, Klaipeda. The Klaipeda sea level trends were determined for two different time periods, 1898–2011 and 1976–2011.

15.3 Geomorphological and geological features of the Karklė beach

The Lithuanian coast is formed exclusively of Quaternary deposits (Bitinas et al., 2005). Glacial deposits formed during the last glaciation are distinctive of the central part of the continental coast and are exposed

in the cliffs south of Karklė. The sandy sediments that were mainly formed in the Littorina and Post-Littorina Seas as well as their lagoons prevail in the coastal structure of the northern part of Karklė. Due to the fact that the Baltic Sea is tideless the main beach-forming factors are wind-generated waves and currents along the coast.

Karklė's coast is in particular characterized by morainic and sandy cliffs of 4.5 to 6.5 m height (Figure 15.3). Gravel, cobble and boulders cover up to 70–90% of the beach surface at Karklė and close to Olando Kepure (Photos 15.1–15.4). The beach is 18–42 m wide and only the northern part of it is sandy (Photo 15.5). The littoral zone is very uneven. Its

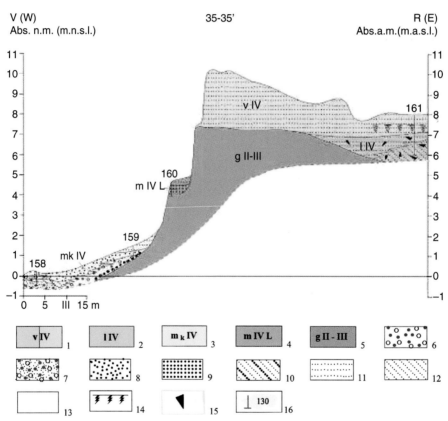

Figure 15.3 Typical geological cross-section of the southern part of the Karklė beach. Brown colour – till outcroping at the cliffs of Karklė village.
Source: Compiled by A. Damusyte.
Legend: *Stratigraphic and genetic; 1 – Holocene, aeolian deposits, 2 – Holocene, limnic sediments, 3 – Holocene, Baltic sea beach sediments, 4 – Holocene, Litorina Sea marine sediments,*

5 – Pleistocene, glacial deposits of Middle and Upper Pleistocene (unseparated); Lithological 6 – Mixture of shingle and gravel, 7 – Mixture of gravel and sand, 8 – Course sand (Ø 0,5 – 1,0 mm), 9 – Medium sand (Ø 0,25 – 0,5 mm), 10 – Fine and medium sand, 11 – Fine sand (Ø 0,1 – 0,25 mm), 12 – Extra–fine and fine sand, 13 – Glacial loam, till; Other signs 14 – Burried soil, 15 – Fine grained organic matter, 16 – Borehole and its number.

Photo 15.1 Cobble and boulders in the southern part of Karklė beach.
Source: Photo by J. Satkunas.

surface is composed of till with pits and boulders. Accordingly, bathing conditions there are quite complicated.

15.4 Geodynamical conditions and sea level rise

Coastal change is a complex result of the interaction of climate-driven eustatic sea level change, vertical crustal movements, hydrodynamic and aeolian processes as well as human activities. Coastal change may be a relatively slow or very rapid process. Often, during storms, beach morphology can change substantially in a few hours. Hydrodynamic and aeolian processes on the beach act as the main driving forces in sand movement along and across the shore (dune ridge – sea – beach – dune ridge) changing not only the morphometric characteristics of the beach but also the composition of beach sediments (Musielak, 1989).

Measurements of coastal morphology since 1993 demonstrate that until 2011 the coastline north of

Photo 15.2 Central part of the Karklė beach, composed exclusively of gravel and cobble.
Source: Photo by J.Satkunas.

Photo 15.3 Big single boulders create a specific attractive feature of the beach.
Source: Photo by J. Satkunas.

Photo 15.4 Narrow beach at the Olando Kepure – the protected area south of the Karklė beach.
Source: Photo by A. Damusyte.

Photo 15.5 Sandy beach at the mouth of Rikine rivulet. This part is most favourable for the traditional summer recreation activities.
Source: Photo by A. Damusyte.

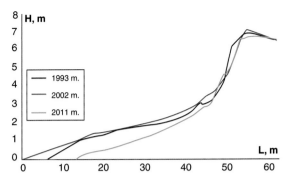

Figure 15.4 Dynamics of coastal profile in 1993–2011 south of Rikine rivulet; L – width of the beach, H – height of the beach.

Photo 15.6 Cliff at the Karklė cemetary after the storm 'Ervin'. *Source*: Photo by D. Jarmalavičius.

Rikine mouth retreated approximately by 6 m (Figure 15.4), in contrast, at the Karklė graveyard it proceeded by 2 m towards the sea (Figure 15.5). During the period 1993–2002 the coastline proceeded towards the sea by 3 m in the south of the study area to 6 m in the north of the study area. It has to be concluded that morphology of the beach changed rather insignificantly during the period of investigations (18 years), even the impacts of storms 'Anatolij' (4 December 1999) and 'Ervin' (8/9 January 2005) (Photo 15.6) were not particularly hazardous.

Analysis of historic cartographic material revealed that during the period 1910–1947 the coastline between Rikine and Cypa rivulets proceeded approximately by 15 m towards the sea and accumulation of sand on the beach prevailed. However, during 1947 and 1984 erosion predominated so that the coastline retreated by 10 m. It has to be noted, that during

this period, in 1967, the strongest storm of the 20th century impacted the coastline. Further strong storms occured in 1981 and 1983. Therefore, the mentioned period is characterized by prevailing coastal erosion. Later, up until the end of the 20th century accretion processes dominated. Generalized results of analysis indicate that the coastline of Karklė was developing in a cyclical manner – erosion periods are followed by accretion periods. But it has to be noted that the state of the beach was better (wider and more sandy) at the beginning of the 21st century compared with that of the beginning of the 20th century.

In spite of that historical trend, the development of Karklė morphology has to be modelled according to predicted sea level change in the 21st century in order to establish the infrastructure of the Karklė beach and to increase its attractiveness for tourism.

Based on data of the Klaipeda hydrological station the Baltic Sea water level rose about 14.9 cm at Klaipeda during the 20th century. Even more dramatic sea level changes are expected for the 21st century. According to observations, at least once per year sea levels reached 50 cm which is higher than the long-term average level. The highest value ever measured was 186 cm (17 October 1967).

It should be taken into accout that sea level could change in an oscillation manner. There were periods of increased rates of sea level rise (e.g. end of 19th / beginning of 20th century and end of 20th century) and periods of stabilization (ca. 1920–1950 and 1990–2005), so that 100 years of record could be too short for a reliable future scenario.

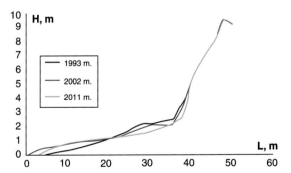

Figure 15.5 Dynamics of coastal profile in 1993–2011 near the Karklė cementary; L – width of the beach, H – height of the beach.

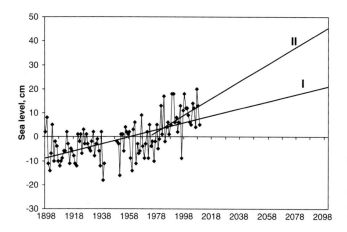

Figure 15.6 Forecast of the Baltic sea level rise in Lithuania in the 21st century: I – if the trend of the 20th century remained, II – if trend of last 35 years remained. *Source*: based on data obtained from the Department of Marine Research, Klaipeda.

Nevertheless, it is assumed that there are two possible scenarios for sea level rise. If the rate of sea level rise of the 20th century remained (1.5 mm/yr), sea level would rise by 15 cm (scenario I). But if the rate of sea level rise of the last 35 years remained (3.5 mm/yr), sea level would even rise by 35 cm (scenario II, Figure 15.6). These scenarios are based on observational data of sea level change during the 20th century and linear extrapolation of the observed trends without consideration of possible climate change acceleration. But, forecasted in this way the sea level rise corresponds to the IPCC (2007) scenario B1 (lowest level of greenhouse gas (GHG) emissions), according to which global sea level rise during the 21st century will be 19–37 cm. Similar forecasts of future sea level rise were proposed in Poland (Pruszak & Zawadzka, 2005). However, taking into account the highest GHG emissions scenario (A1FI), the global sea level would rise by 43–58 cm (IPCC, 2007). Current observations at the beginning of the 21st century do not show an acceleration of sea level rise. Even more, its rate diminishes. According to data obtained from the archive of the Department of Marine Research, Klaipeda, the sea level trend in 2000–11 was only 2.8 mm/yr. Therefore, an accelerated sea level rise could not be proven. Similar conclusions could be drawn from the satellites TOPEX, Jason-1 and Jason-2 data, according to which the rate of global sea level rise since 1993 is averaging 2.8–3.2 mm/yr, however, without traces of acceleration (Nerem, Chambers, Choe & Mitchum, 2010). Keeping in mind these observation data the scenario of moderate sea level was taken for modeling of the sea level change in the Karklė beach.

But, other factors also affect the peculiarities of coastal dynamics. With average sea level rise, maximum levels are also increasing. If during the period from 1910 until 1955 sea level only three times exceeded a height of 100 cm above long-term average level, in period 1956–2010 this level was exceeded more than 20 times (Archive of the Department of Marine Research, Klaipėda). As a consequence, in the latter period an increased shore erosion was observed as such high water level events have come along with storm surges.

The annual amplitude of water level at Lithuanian coasts varies from 90 to 240 cm (average 140 cm). Therefore erosional impact respectively varies during one year. With increasing peak values, coastal erosion also increases.

During the stormy autumn-winter period, the beach is suffering the strongest erosion impact and, therefore, it becomes narrow and flat. During the calm weather of the spring-summer period, sand accumulation usually prevails on the beach. Due to this regularity the beach of Karklė keeps more or less the same width during summer season and, thus, it is suitable for recreation. Severe storms constitute an exception, as restoration of the beach afterwards takes up more time.

Taking into account beach inclination in Karklė (aspect ratio: 1:11 – 1:14), its coastline retreated by 1.6–2.1 m during the 20th century. If sea level rises around 15 cm in the 21st century, projections reveal a coastline retreat of between 1.6 to 2.1 meters. If sea level rises by 35 cm, the coastline will retreat by 3.9 to 4.9 m.

However, greater coastal retreat values during the 21st century could be calculated using Bruun Rule (Bruun, 1962):

$$R = -Hw/h + B$$

where: R – change in shoreline location; H – change in water level; h – water depth at the seaward limit of sediment motion (closure depth); B – beach height; w – width of the active profile (distance from the shoreline to the seawards limit of sediment motion).

According to this rule and taking into acount the morphological pecularities of nearshore, the coastline of Karklė could retreat between 6.5 m (scenario I) and 15.5 m (scenario II). In the worst case scenario beaches would reach widths similar to those at the cliff of Olando Kepure, where they are only 15 – 25 m wide in summer time.

15.5 Conclusions

The aim of the current research was to identify the future climatic conditions at Karklė beach in order to address the need for the development of beach infrastructure in Karklė due to rapidly rising numbers of holiday makers. At first, the analysis of observational data revealed that the coastline at Karklė beach was developing in a cyclical manner in the past – erosion periods were followed by accretion periods. During 1993 to 2011, the coastline north of Rikine mouth retreated by aproximately 6 m, but proceeded towards the sea by about 2 m at the Karklė graveyard. Different coastal areas or individual profiles may vary considerably due to local geological conditions, morpho-lithological diversity and anthropogenic loads. The geodynamic stability of beaches can be revealed only by an analysis of long-term measurements of its morphometric indicators.

The modelling of different possible scenarios of future sea level rise showed a wide variety for the development of Karklė beach. Assuming that most likely the sea level will rise by 35 cm (IPCC scenario B1), the coastline will retreat only between 3.9 to 4.9 m, allowing the Karklė beach to function properly as a recreation area in future. In the worst case scenario with coastal recession up to 15.5 m, recreational space on the Karklė beach will be rather limited. Moreover, maintenance of the beach infrastructure will be complicated due to increasing storm damage.

On the other hand, neither the observational data nor the forecasted values show alarmingly high rates of coastal erosion. That is why, according to our modelling, there is no need to implement coastal protection measures yet. However, in the case of an acceleration of future sea level rise, a strategy for coastal protection would be urgent. The Karklė settlement and especially the cemetary would be critically threatened by accelerated erosion as they are located closest to the erosion-prone cliff. To protect the Karklė cemetary use of protective gabions would be recommended. Besides, as beaches tighten, beach nourishment will be necessary in order to maintain recreational space for an increasing number of beach visitors.

The conclusions and recommendations were presented to spatial planners to encourage the elaboration of a detailed spatial development plan for Karklė beach that also contains sea level change adaption measures. The overall conclusion based on research of the specific geodynamic conditions is that the beach could be further developed as a recreational area and for tourism, but its proper functioning in the future will require observations and maintenance (e.g. beach nourishment).

Acknowledgements

Thanks are extended to Professor Szymon Uscinowicz, Dr Philipp Schmidt-Thomé, Anika Nockert and unknown reviewers for valuable comments and remarks that served for improvement of the chapter.

References

Bitinas, A., Žaromskis, R., Gulbinskas, S., Damušytė, A., Žilinskas, G. and Jarmalavičius, D., 2005. The results of integrated investigations of the Lithuanian coast of the Baltic Sea: geology, geomorphology, dynamics and human impact. *Geological Quarterly*, 49 (4), pp.355–362.

Bruun, P., 1962. Sea-level rise as a cause of shore erosion. *Journal of the Water and Harbors Division, Proceedings of the American Society of Civil Egineering*, 88 (WW 1), pp.117–130.

IPCC (Intergovernmental Panel on Climate Change), 2007. Climate and sea-level scenarios. In: IPCC, 2007. *IPCC Fourth Assessment Report: Climate Change 2007 – Contribution of Working Group II to the Fourth Assessment Report of the Intergovernmental Panel on Climate Change*. [online] Available at: <www.ipcc.ch/publications_and_data/ar4/wg2/en/ch6s6-3-2.html> [Accessed 01 June 2011]. Ch. 6.3.2.

Jarmalavičius, D., Žilinskas, G. and Kulviciene, G., 2001. Peculiarities of long-term water level fluctuations on the Lithuanian coast. *Acta Zoologica Lituanica*, 11 (2), pp.132–140.

Jarmalavičius, D., Satkūnas, J., Žilinskas, G. and Pupienis, D., 2012. Dynamics of beaches of the Lithuanian coast (the Baltic Sea) for period 1993–2008 based on morphometric indicators. *Environmental Earth Sciences*, 65 (6), pp.1727–1736.

Musielak, S., 1989. Morpholithodynamics of the sandy sea beaches. *Studia i materialy oceanologiczne, brzeg morski*, 1, pp.67–78 (In Polish).

Nerem, R.S., Chambers, D., Choe, C. and Mitchum, G.T., 2010. Estimating Mean Sea Level Change from the TOPEX and Jason Altimeter Missions. [online] *Marine Geodesy*, 33 (1), p.435. Available at: <http://sealevel.colorado.edu/wizard.php?dlon=21&dlat=56&map=t&fit=n&smooth=n&days=1> [Accessed 1 June 2011].

Pruszak, Z. and Zawadzka, E., 2005. Vulnerability of Poland's coast to sea-level rise. *Coastal Engineering Journal*, 47 (2–3), pp.131–155.

Thom, B.G. and Hall, W., 1991. Behavior of beach profiles during accreation and erosion dominated periods. *Earth Surface Processes and Landforms*, 16 (2), pp.113–127.

Žilinskas, G., 1997. The length of the Lithuanian shore of the Baltic Sea. *Geografijos metraštis*, 30, pp.63–71 (In Lithuanian).

Žilinskas, G. and Jarmalavičius, D., 2000. Coast condition and dynamics. In: L.L. Lazauskienė, G. Vaitonis, A. Draugelis, eds. 2000. *Klaipėdos uostas: ekonomika ir ekologija*. Vilnius: Baltic ECO, pp.55–67 (In Lithuanian).

Žilinskas, G., Pupienis, D. and Jarmalavičius, D., 2010. Possibilities of regeneration of Palanga coastal zone. *Journal of Environmental Engineering and Landscape Management*, 18 (2), pp.95–101.

16 Consequences of Climate Change and Environmental Policy for Macroalgae Accumulations on Beaches along the German Baltic Coastline

Matthias Mossbauer[1,3], Sven Dahlke[2], René Friedland[1] & Gerald Schernewski[1,4]

[1] Leibniz Institute for Baltic Sea Research, Warnemünde, Rostock, Germany
[2] University of Greifswald, Kloster/Hiddensee, Germany
[3] EUCC – The Coastal Union Germany, Rostock, Germany
[4] Coastal Research & Planning Institute, Klaipeda University, Klaipeda, Lithuania

16.1 Introduction

In the coming decades changes in natural processes and conditions in coastal waters will have an impact on the use of Baltic beaches for recreation. With regard to the natural conditions for beach management, the impacts of these changes have not previously been assessed. We define beach management as activities to maintain or improve a beach as a recreational resource while recognizing ecosystem links.

Climate change is an ongoing trend which will continue in the future. Global warming will affect the structure and functioning of Baltic benthic ecosystems (BACC, 2008). A second source for change is the EU policy seeking to achieve a good environmental status of Baltic coastal waters. To achieve this, eutrophication is tackled by the BSAP. This action plan was prepared by the Helsinki Commission in 2007. The programme is designed to reduction 18% of waterborne nitrogen and 42% of phosphorus until the year 2021 (HELCOM, 2007). Changes will have an impact on the production capacity of underwater eelgrass and macroalgae habitats on the foreshore. During storms

or at the end of the vegetation period, eelgrass and macroalgae are uprooted, transported by currents and accumulated in the surf zone of beaches. On many beaches along the German Baltic coastline detached seaweed or 'beach wrack' is a major aspect of beach management (see Figure 16.1).

The biomass represents a nuisance for beach tourism resulting from the foul smell of decomposing plant material and an affected accessibility of the water. As a consequence, beach wrack accumulation on the beaches of seaside resorts leads to decreasing beach tourism resulting in falling income of coastal communities. Good management of beach wrack with a strong focus on keeping beaches clean from natural debris is essential. This is expensive (Mossbauer, Haller, Dahlke & Schernewski, 2012). Knowledge of a change in state is vital to limit specific impacts (Williams & Micallef, 2009). In this respect, knowledge of future amounts of beach wrack is crucial for effective adaptation measures in beach management and provides a basis for a proactive future plan, which aims to save costs.

We confined ourselves to the investigation of hard substrates which are habitats for permanent

Climate Change Adaptation in Practice: From Strategy Development to Implementation, First Edition.
Edited by Philipp Schmidt-Thomé and Johannes Klein.
© 2013 John Wiley & Sons, Ltd. Published 2013 by John Wiley & Sons, Ltd.

Figure 16.1 Beach wrack accumulation on Warnemünde beach (Germany) in summer 2010.

macroalgae communities (Duphorn, 1995) because rotting macroalgae accumulations spread an intensive smell resulting from the degradation of dimethylsulphoniumpropionate to dimethylsulphide (Wiesemeier et al., 2007). We limited our research on perennial algae because our methods provide scenarios on changes over decades. Short term conditions are important for ephemeral algae, something which is not modelled in our simulations.

It is estimated that two thirds of beach wrack along the German Baltic coast is composed of uprooted eelgrass (*Zostera marina*) (Mossbauer, Haller, Dahlke & Schernewski, 2012). However, we excluded eelgrass in our study because the plants respond rather slowly to better growing conditions resulting from a large component of vegetative reproduction by rhizome branching which allows a maximum horizontal spread of 50 cm per year (Palacios & Zimmerman, 2007). A large scale spread of eelgrass in newly available habitats is a very slow process and unlikely to occur (Meyer & Nehring, 2006).

Near shore habitats are the sources of algae biomass ending up as beach wrack (see Figure 16.2). Approx-

imately 10% of the biomass produced by perennial macroalgae communities ends up as beach wrack (Grave & Möller, 1982). Red-, green- and brown algae grow on almost every hard material (Wahl & Mark, 1999). Pleistocene lag sediments (Duphorn, 1995) and artificial structures provide hard substratum in the near shore coastal zone all along the German Baltic coast. An artificial hard bottom consists of coastal protection structures, breakwaters, jetties and artificial reefs.

In recent years, different types of artificial reefs were installed in water depths between two and 20 m for ecological compensation measures, applied research and dive tourism promotion (Karez & Schories, 2005). A second source of artificial hard substrates is coastal protection. To counter an intensified erosion of the sandy beaches induced by sea level rise and climate change, more coastal protection measures will be necessary (MLUV, 2009).

Breakwaters, which are installed around 100 m in front of the water line, are one possible scenario to face increasing coastal erosion rates. Today offshore breakwaters are already installed along some sections of the

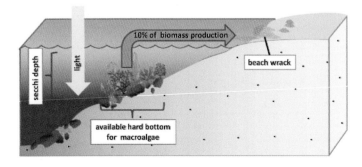

Figure 16.2 Simplified interactive structure of subaqueous geology, bathymetry, Secchi depth and beach wrack accumulations which form the structure and research questions.

Figure 16.3 Breakwaters for coastal protection along the German Baltic coastline. Besides preventing erosion the artificial structures provide hard substrate in the euphotic zone for macroalgae.
Source: Pictures by Stefanie Maack and EUCC – The Coastal Union Germany.

coastline with severe erosion (see Figure 16.3). This hard substrate is both coastal protection measure and artificial habitat for macroalgae. The effect of artificial hard substratum on the quantities of beach wrack has not yet been quantified. We expect an increase of artificial hard substratum surface in the euphotic zone available as habitat for macroalgae.

In addition to hard substratum, macroalgae need light for proper growth (Fürhaupter, Grage, Wilken & Meyer, 2008) and the lower distribution of macroalgae is limited by underwater light conditions (Lüning, 1990). Water dissolved nutrients promote the phytoplankton production that lowers the water transparency. It is assumed that the lower distribution limit of macroalgae correlates directly with the underwater light availability. Numerous studies have demonstrated the validity of that concept (Krause-Jensen, Sagert, Schubert & Boström, 2008). Our study reveals to what extent growing conditions for macroalgae and therefore beach wrack accumulations are affected by climate change and the BSAP on a regional scale. We hypothesize that the implementation of the BSAP could lead to an increased water transparency resulting from a decreased primary production on the basis of a decreased availability of water-dissolved nutrients. As a result, macroalgae could spread to greater depths. Climate change will have significant impacts on benthic communities in the Baltic Sea as well (HELCOM, 2007). Our ecosystem modelling will provide scenarios on what extent the water turbidity is affected. The results are synthesized with the bathymetry and geological data to assess the impacts on potential macroalgae habitats.

16.2 Methods and materials

The study area covers 550 km of the 720 km of the German Baltic outer coast but with a very different availability of near shore hard bottom along that length. To point out geological differences in the study area we subdivided it into three macro regional sections (see Figure 16.4). All coastal sections are characterized by long sandy beaches with a curved platform and are often bounded by cliffs. Section A has a coastal length of 270 km, section B has a coastal length of 130 km and section C comprises 140 km of outer coast. The seaward limit of the study area is the border of the German Baltic exclusive economic zone (EEZ).

To study the future development of the Baltic Sea with regard to mean water temperature and water transparency (Secchi depth), several simulations for the period 1960–2070 with the dynamic ecosystem model ERGOM (Neumann, Fennel & Kremp, 2002; Neumann & Schernewski, 2008), were conducted. The ecosystem model consists of three phytoplankton groups (diatoms, flagellates and cyanobacteria), on which grazing pressure is provided by a bulk zooplankton variable. Primary production only takes place as long as enough nutrients (ammonium, nitrate and phosphate) are available. Cyanobacteria are able to fix airborne nitrogen, so that their growth only depends on the phosphate availability. A part of the dead organic matter is mineralized back to ammonium and phosphate, while another portion accumulates in the sediment, where it can again be discharged depending on the oxygen conditions.

The physical part of the model is based on the circulation model MOM (Pacanowski & Griffies, 2000), which was adapted to the Baltic Sea. The atmospherical forcing was provided by the regional weather model CLM of the German Weather Service. For the simulations the two greenhouse gas scenarios A1B and B1, proposed by the Intergovernmental Panel on Climate Change (IPCC) (2007), as well as two nutrient scenarios were used. In scenario one, which has the high input of the late 1990s (called 'business as usual' or 'BAU '), and in the second case the nutrient input was reduced according to the BSAP (HELCOM, 2007).

For the calculations of the distribution limits depending on the light availability the Secchi depth was used as a lead parameter. At the Secchi depth the irradiance falls to nearly 10% of the surface intensity.

From the phytoplankton and detritus densities (averaged over the upper 15 m of the water column) the Secchi depth was computed after the

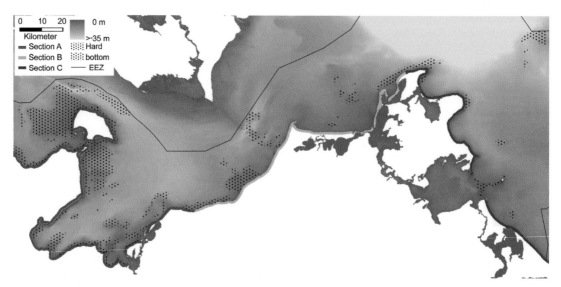

Figure 16.4 Study areas A, B and C along the German Baltic coastline water depths, exclusive economic zone (EEZ) and the hard bottom on the sea floor.

Beer-Lambert Law as depth, where 10% of the surface light is still available. The equation was calibrated with measurements at the Mecklenburger Bay. High wind speed with wave turbulence and low solar radiation do not change this value significantly (Mur & Visser, 1996; Paulson & Simpson, 1977). The value of 10% irradiance represents the lower distribution limit for the majority of macroalgae along the German Baltic coastline (Blümel et al., 2002).

Geological data (provided by HELCOM) in combination with bathymetry data generated information about the sum of the area of hard substrate surface in different water depths.

In addition we evaluated the percentage of artificial hard bottom as a habitat for macroalgae. With the help of georeferenced aerial pictures, taken from Google Maps (www.google.de/maps), we measured the surface of all breakwaters and jetties in the study area. Measurement data were validated with construction data, available for 30% of all breakwaters along the German Baltic coast.

Breakwaters in the study area are built in a mean water depth of 3 m and have an aspect ratio of 2 to 1 (MLUV, 2009). With regard to this data, 1 m length of a breakwater provides 13.4 m^2 of underwater hard bottom.

16.3 Results

The Secchi depth was computed by using the phytoplankton and dead organic matter densities. For the reduction scenario its annual mean significantly increases in contrast to the high input case (BAU) (Figure 16.5).

The simulations show that climate change will increase the temperature in the western Baltic Sea by up to 2 K (at A1B) and 1.5 K at the B1-simulations until 2070 respectively (Figure 16.6). Because of the water warming the sea ice cover will be reduced, so that it is probably likely that one winter with ice creation occurs every decade after 2040 (not shown).

A synthesis of the bathymetry and the geology of the nearshore sea bottom to a depth of 20 m is shown in Figure 16.7. Lag sediments in study area A are particularly common in depths between 6 and 15 m. In study area B lag sediments are concentrated in water depths between 6 and 10 m. The distribution of lag sediments in study area C varies widely up to a depth of 20 m (see Figure 16.7).

The synthesis of bathymetry, geology and the modelling of water transparency shows the future availability of hard substrate. The hard substrate can serve as a potential habitat for macroalgae. Changes in nutrient inputs have a stronger influence on underwater light conditions than changes in climate. Thus, the implementation of the BSAP will lead to much better underwater light conditions resulting in a macroalgae growth on the hard bottom in greater depths than today. Altogether more hard bottom will be available as habitat for macroalgae in the future (see Figure 16.8).

On the basis of the regional distribution of hard substratum at different depths (Figure 16.7) nutrient reduction measures will have a major impact on hard bottom availability in study area B. Here, a doubling of hard bottom surface is possible. In study area A an increased Secchi depth will have a comparable effect. In contrast, future impacts on hard bottom availability

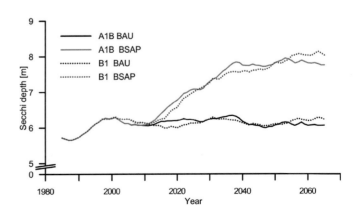

Figure 16.5 Ten-year-running-mean of the Secchi depth of the southern Baltic Sea.

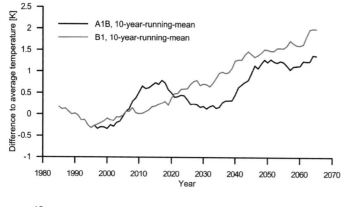

Figure 16.6 Ten-year-running-mean of water temperature differences in relation to the average water temperature between 1980 and 2000.

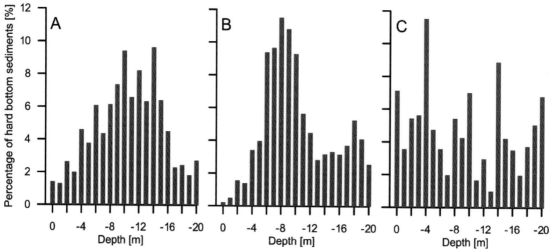

Figure 16.7 Distribution of hard bottom sediments in relation to the water depth for the study areas A, B and C.

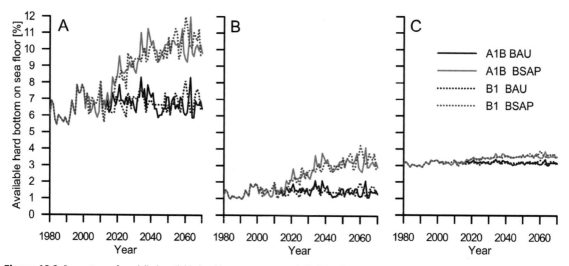

Figure 16.8 Percentage of modelled available hard bottom as a potential habitat for macroalgae for study area A, B and C (1980–2070).

in study area C can be expected to be comparatively small (see Figure 16.8).

Aerial surveying and verification with available construction data showed that breakwaters with a total length of 19.7 km and a total underwater surface of 0.27 km^2 are installed in the study areas A, B and C.

16.4 Discussion

In this study, data and information originate from various sources with a different degree of validity. Geodata analysis suffers from the classification compromises of input data used in our study (HELCOM, 2011). For the study area only sediment data is available with a lack of information on boulder reefs and their probable area of formation (Al-Hamdani & Reker, 2007).

The scenario simulation model ERGOM we used in our study is linked with some limitations. Firstly, regional climate change effects are not taken into account. Secondly, the ERGOM simulation is simplified to the most important factors and interactions of the Baltic Sea ecosystem. A further source of uncertainty is the allocation of nutrient reductions in the study area because concrete reductions in the future are still under discussion. Lastly, the grid resolution of

three nautical miles does not allow model ecosystem processes in front of the waterline.

In addition, there is increasing evidence that the correlation of water transparency and distribution depth is not generally true. Along the German Baltic coastline significant differences of the lower distribution limit of different species with identical minimal light demand are known (Schories, Selig & Schubert, 2004).

To summarize different causes for the changes in the amounts of beach wrack we created a hypothetical model (see Figure 16.9).

Against the background of future nutrient input and climate change, the hypothetical model takes influencing factors on habitats of macroalgae and consequences for biomass growth into account (see Figure 16.9). Simulation results show that nutrient reduction leads to an increased light availability resulting from a decreased primary production of phytoplankton due to a decreased nutrient availability. A comparison of higher mean water temperatures of the southern Baltic Sea (climate change), and the nutrient reduction scenarios (implementation of the BSAP) shows that for the Secchi depth the nutrient budget is more important. The increased Secchi depth provides more underwater light and therefore better growing conditions for stable macroalgae communities (consisting of for example *Fucus spp.*). We estimate that the

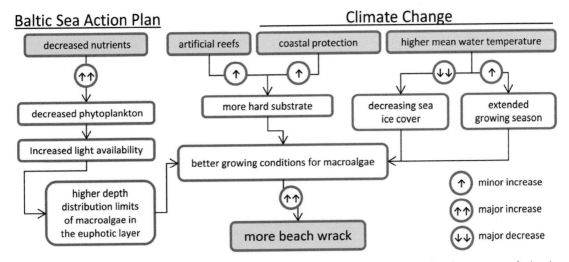

Figure 16.9 Hypothetical model of beach wrack influencing factors, consequences for biomass growth and consequences for beach wrack along the German Baltic coastline.

distribution limits of macroalgae species will spread to a greater depth and hence, more biomass will be produced. This estimation is based on original growing conditions of macroalgae in the past. Before eutrophication of the Baltic Sea took place in the middle of the last century, the maximum growing depth of *Fucus ssp.* was 10 m (today 3 m). Macrophyte biomass production was 95% higher compared to today (Vogt & Schramm, 1991). The impact of increased hard substratum availability on the amount of beach wrack can be expected to be relatively high.

Model results from ERGOM show an increase of the mean water temperature for the Climate Change scenarios A1B and B1 by the end of the 21st century. Higher water temperatures will lead to a prolonged vegetation period and a reduced sea ice cover. Sea ice is a limitation factor for perennial macroalgae in shallow waters (Fürhaupter, Wilken & Meyer, 2006) and can be responsible for the destruction of underwater macrophytes in very low depths (Vogt & Schramm, 1991). Accordingly winters without sea ice lead to better growing conditions for macroalgae during the vegetation period.

With regard to an upper strong limitation of perennial macroalgae habitats caused by dehydration, which is in turn generated due to fluctuating water level and wave turbulences (Schories, Selig & Schubert, 2004), the overall impact of decreased sea ice is expected to be low.

The biomass productions caused by macroalgae will likely increase in the shallow waters of the study area, if an extended vegetation period, also based on higher water temperatures, occurs. We assume that this effect has only a minor impact on future beach wrack accumulations.

Artificial reefs and structures in the study area provide habitats for macroalgae as well. The total effect of artificial habitats is small. Artificial reefs cover a total hard substratum surface area of 0.05 km^2 (Karez & Schories, 2005) corresponding to 0.03% of the total natural available hard bottom surface in all three study areas. Recent artificial breakwaters and other coastal protection measures provide only a total share of 1.6% of the total natural available hard bottom. Even if breakwaters with a total length of 720 km were able to protect the complete German Baltic coastline, only 5.8% of the total hard bottom available as habitat for macroalgae would be manmade. Due to this small per-

centage the effect of artificial macroalgae growth on future beach wrack is marginal.

Due to the complexity of ecosystem interactions and the abovementioned errors of the used methods, our conclusion remains speculative to a certain degree. Better physical growing conditions do not necessarily lead to a spread of macroalgae communities. Interspecific competitions and disturbed intraspecific interactions can be a hindering factor for the spread of macroalgae. Hard bottom habitats which are partially covered with blue mussels possibly prevent new populations of *Fucus spp.* Certainly, the number of species and the coverage of red algae are significantly lower when blue mussels occur (Schories, Selig & Schubert, 2004). On unpopulated hard substratum germ buds of macroalgae (for example *Fucus serratus*) do not find any shelter from parental plants. As a consequence, wave and current-induced mechanical stress prevents the succession of macroalgae (Isæus et al., 2004).

The trends and natural interactions described above outline the complex interactions of biophysical coastal processes leading to beach wrack accumulations. For the development of reliable scenarios of future macroalgae growth which can be used as basic criteria for future beach wrack management, a better understanding of these processes is necessary.

As for the dimension of future beach wrack it can be summarized that the full implementation of the BSAP will lead to a significant increase of the availability of hard substrate in the underwater euphotic zone. This will allow macroalgae to cover larger areas. As a result we expect a doubling of beach wrack in the study area. This trend will put pressure on recreational beach management. Effects could lead to an increased pressure on the handling and management of beach wrack and increased costs for beach cleaning.

Acknowledgement

This work was funded by the project BaltCICA (Part-financed by Baltic Sea Region Programme of the European Union) and the BMBF-Project RADOST. HLRN (Norddeutscher Verbund für Hoch- und Höchstleistungsrechnen) has provided the supercomputing power. Construction data of breakwaters were kindly provided by the National State Agency for Coastal Protection, Agriculture and Conservation (StALU).

Thanks to Tina Medenwald for her comments on the text.

References

Al-Hamdani, Z. and Reker, J. 2007. *Towards marine landscapes in the Baltic Sea – BALANCE Interim Report No. 10.* [pdf] Available at: <http://balance-eu.org/xpdf/balance-interim-report-no-10.pdf> [Accessed 15 June 2011].

Blümel, C., Domin, A., Krause, J.C., Schubert, M., Schiewer, U. and Schubert, H., 2002. *Der historische Makrophytenbewuchs der inneren Gewässer der deutschen Ostseeküste.* Rostocker Meeresbiologische Beiträge, 10, pp.5–111.

Duphorn, K., 1995. *Die deutsche Ostseeküste.* Stuttgart: Borntraeger.

Fürhaupter, K., Wilken, H. AND Meyer, T. 2006. *WRRL-Makrophytenmonitoring in den Küstengewässern Mecklenburg Vorpommerns – Teil B: Äußere Küstengewässer.* Güstrow: MariLim, Abschlussbericht für das LUNG-MV (final report).

Fürhaupter, K., Grage, A., Wilken, H. and Meyer, T. 2008. *Kartierung mariner Pflanzenbestände im Flachwasser der Ostseeküste. Schwerpunkt Fucus und Zostera – Außenküste der schleswig-holsteinischen Ostsee und Schlei.* Flintbek: Landesamt für Natur und Umwelt.

Grave, H. and Möller, H. 1982. Quantifizierung des pflanzlichen Strandanwurfs an der westdeutschen Ostseeküste. *Helgoland Marine Research,* 35 (4), pp.517–519.

HELCOM (Helsinki Commission) 2007. Climate change in the Baltic Sea area. [pdf] *Baltic Sea Environment Proceedings No. 111.* Available at: <http://www.helcom.fi/stc/files/Publications/Proceedings/bsep111.pdf> [Accessed 15 June 2011].

HELCOM 2011. *HELCOM Map and Data Service.* [online] Available at: <http://maps.helcom.fi/website/mapservice/index.html> [Accessed 15 June 2011].

IPCC 2007. Climate Change 2007: Synthesis Report Available at: <http://www.ipcc.ch/pdf/assessment-report/ar4/syr/ar4_syr.pdf > [Accessed 15 June 2011].

Isæus, M., Malm, T., Persson, P.S. and Svensson, A. 2004. Effects of filamentous algae and sediment on recruitment and survival of Fucus serratus (Phaeophyceae) juveniles in the eutrophic Baltic Sea. *European Journal of Phycology.* 39 (3), pp.301–307.

Karez, R. and Schories, D., 2005. Stone extraction and its importance for the re-establishment of Fucus versiculosus along its historical reported depths. *Rostocker Meeresbiologische Beiträge,* 14, pp.95–107.

Krause-jensen, D., Sagert, S., Schubert, H. and Boström, C. 2008. Empirical relationships linking distribution and abundance of marine vegetation to eutrophication. *Ecological Indicators,* 8 (5), pp.515–529.

Lüning, K., Yarish, C. and Kirkman, H. 1990. *Seaweeds – Their environment, biogeography, and ecophysiology.* New York: Wiley.

Meyer, T. and Nehring, S. 2006. Plantation of seagrass beds (Zostera marina L.) as internal measure for restoration of the Baltic Sea. *Rostocker Meeresbiologische Beiträge,* 15, pp.105–119.

Mluv (Ministerium für Landwirtschaft, Umwelt und Verbraucherschutz Mecklenburg-Vorpommern), 2009. *Regelwerk Küstenschutz Mecklenburg Vorpommern.* [pdf] Available at: <www.kfki.de/asset/kfki/pdf/uebersichtsheft.pdf> [Accessed 15 June 2011].

Mossbauer, M., Haller, I., Dahlke, S. and Schernewski, G. 2012. Management of stranded eelgrass and macroalgae along the German Baltic coastline. *Ocean & Coastal Management,* 57, pp.1–9.

Mur, L.R. and Visser, P.M. 1996. *Aquatische Milieubiologie.* Amsterdam: Deel I.

Neumann, T., Fennel, W. and Kremp, C. 2002. Experimental simulations with an ecosystem model of the Baltic Sea: a nutrient load reduction experiment. *Global Biogeochemical Cycles,* 16 (3), pp.1033–1054.

Neumann, T. and Schernewski, G. 2008. Eutrophication in the Baltic Sea and shifts in nitrogen fixation analyzed with a 3D ecosystem model. *Journal of Marine Systems,* 74 (1–2), pp.592–602.

Paulson, C.A. and Simpson, J.J. 1977. Irradiance measurements in the Upper Ocean. *Journal of Physical Oceanography,* 7 (6), pp.952–956.

Pacanowski, R.C. and Griffies, S.M. 2000. MOM 3.0 manual. Geophysical Fluid Dynamics Laboratory. (technical report)

Palacios, S.L. and Zimmerman, R.C. 2007. Response of eelgrass Zostera marina to CO2 enrichment: possible impacts of climate change and potential for remediation of coastal habitats. *Marine Ecology Progress Series,* 344, pp.1–13.

Schories, D., Selig, U. and Schubert, H. 2004. Testung des Klassifizierungsansatzes Mecklenburg-Vorpommern (innere Küstengewässer) unter den Bedingungen Schleswig-Holsteins und Ausdehnung des Ansatzes auf die Außenküste, Küstengewässer-Klassifizierung deutsche Ostsee nach EU-WRRL Teil B: Innere Küstengewässer Schleswig-Holstein. (research report).

The Bacc Author Team 2008. *Assessment of Climate Change for the Baltic Sea Basin.* Berlin, Heidelberg: Springer, Regional Climate Studies.

Vogt, H. and Schramm, W. 1991. Conspicuous decline of Fucus in Kiel Bay (Western Baltic): what are the causes? *Marine Ecology. Progress Series*, 69 (1–2), pp.189–194.

Wahl, M. and Mark, O. 1999. The predominantly facultative nature of epibiosis: experimental and observational evidence. *Marine Ecology. Progress Series*, 187, pp.59–66.

Wiesemeier, T., Hay, M. and Pohnert, G. 2007. The potential role of wound-activated volatile release in the chemical defence of the brown alga Dictyota dichotoma: Blend recognition by marine herbivores. *Aquatic Science – Research Across Boundaries*, 69 (3), pp.403–412.

Williams, A.T. and Micallef, A. 2009. *Beach management – Principles and practice*. London: Earthscan.

17 Climate Change Impacts on Baltic Coastal Tourism and the Complexity of Sectoral Adaptation

Christian Filies & Susanne Schumacher

EUCC – The Coastal Union Germany, Rostock, Germany

17.1 Introduction

Coastal tourism in moderate marine climate zones is often seen as a potential winner of climate change. The prediction of a moderate temperature rise and changes in precipitation from summer to winter seem to favour conditions for coastal tourism at first glance. But upon closer examination the risks for coastal tourism are significant and could exceed the chances by far if no adaptation strategies are established. Sea level rise, changes in ecosystems and the potential dangers of extreme weather events are only a few of the changes that come to mind quickly. But the analysis of touristic climate impacts needs to go further. Tourism as an interdisciplinary field with close connections to other economic sectors and highly dependent on social and economic external factors is probably one of the most vulnerable fields climate change will have impacts on. Therefore far-reaching, interdisciplinary adaptation strategies are critical and will not only need to imply reactions to direct ecological changes, but will also have to deal with altered social and economic realities. These indirect and induced effects are making the prediction of the influence of climate change on tourism very complex. In this chapter all three types of changes (direct, indirect and induced) will be introduced, applied to the tourism sector in the Baltic Sea Region (BSR) and compared to other global touristic destinations.

The research on economic and sociological impacts of climate change and following on from this the interpretation of nature–science-based scenarios is a relatively new approach in climate sciences. Dealing not only with the question of what exactly will happen, but to picture the consequences of these changes for an economic sector like coastal tourism is challenging; whereas some problems occur for the whole touristic sector, touristic climate adaptation needs to be broken down to fit the individual needs and special characteristics of tourism destinations. An overall approach of adaptation strategies for the tourism sector is therefore not a desirable attempt.

This chapter will depict global challenges, but will focus on the BSR and analyse the risks and chances of climate adaptation strategies on the example of the German Baltic Coast. To give an account of the needs and requirements, difficulties and opportunities necessary to establish touristic climate adaptation, the results of exemplary expert interviews with touristic stakeholders will be considered and compared to the requirements of adaptation that climate sciences suggest.

17.2 The challenges of climate change for coastal destinations

Climate change in coastal areas is a wide field. The coast is not 'just' a spatial area, but a complex system

Climate Change Adaptation in Practice: From Strategy Development to Implementation, First Edition.
Edited by Philipp Schmidt-Thomé and Johannes Klein.
© 2013 John Wiley & Sons, Ltd. Published 2013 by John Wiley & Sons, Ltd.

of natural, sociological and cultural subsystems, influencing each other. It is therefore not surprising that defining coastal tourism is challenging. While nearly everyone has a – probably even first handed – idea on what it is, from a scientific point it is not clearly defined.

Where does the coast start, where does it end? Is it the pure shoreline? Definitions vary from a few meters to some kilometres inland. Distances for the water side of the coast vary as well and it is not clearly determined on which 'point' in the water the coast ends and the open sea begins.

Struggling with a lot of different definitions for coastal tourism it is essential to agree on a specific one first. Therefore coastal tourism in this chapter will be understood as

> All infrastructure and activities with the main focus on the shoreline and the beach as a contact-zone between sea and land and with the main purpose to accommodate, to supply or entertain local, domestic and international day- and overnight tourists. The coast is a functional area which can't be limited by terms of space but of function and is mainly characterized by ecologic, economic and social interdependencies of marine and terrestrial processes.
>
> (Gee, Kannen and Licht-Eggert, 2006)

17.2.1 Vulnerability of the coastal tourism sector

Interpreting the chances and risks determined by climate change for coastal tourism presumes an estimation of its vulnerability. Tourism itself is known as a highly vulnerable sector, due to its dependency on intact landscapes and other vulnerable industries such as forestry and agriculture (Zebisch et al., 2005). Tourism development is closely linked with economical and sociological developments; key framework conditions which are likely to be affected by climate change. The European Travel Commission rates climate change as one of the most important long-term challenges for tourism, and a direct threat for many destinations (ETC, 2006). Still, vulnerability to sub-segments of tourism has not been classified yet. It is obvious that tourism activities with a strong context to nature (such as coastal tourism) will be more affected than others, such as cultural or city tourism (Ehmer

& Heymann, 2008). To define the vulnerability of coastal tourism it is essential not only to understand the effects of climate change, but to analyse the 'inherent ability of a system to respond to and recover from climate change' (O'Hare, Sweeney & Wilby, 2005). Following this definition, the vulnerability of coastal tourism can only be estimated by knowing the effects *and* possible adaptation strategies.

As stated above, the effects of climate change can basically be classified into three types: direct, indirect and induced changes (Simpson et al., 2008). The most important direct effects include a possible rise of temperature that may be as high as up to 5.5 °C in Europe (Alcamo et al., 2007) until 2100 – even though the proximity to large marine water bodies will prevent the rise from being this severe in most coastal areas. In the Baltic Sea region, for example, due to specific climatic conditions the predicted increase will be less, probably up to 3.3 °C (Werner & Gerstengarbe, 2007). Shifts in precipitation patterns from summer to winter and an increasing number of extreme weather events are two other major challenges that are expected as a consequence of climate change. For coastal tourism all three of these developments mean extensive challenges. But even though these developments are likely to happen all over the world, affecting every coastal tourism destination, their consequences for different coastal areas will vary. An increase of temperature will lead to heat-related problems in some already warm tourism spots – on the other hand it is expected to favour marine destinations in higher altitudes and latitudes, providing longer seasons and warmer summers (Simpson et al., 2008). For European tourism an increase of tourism numbers in western and northern Europe and a loss for southern Europe is expected – already within the next few years (Alcamo et al., 2007). Global warming may also lead to an extension of the touristic season into spring and autumn. Predicted changes in precipitation, likely from summer to winter, will bring either risks or chances to destinations, depending on the ecosystem they are situated in. Although the changes will be similar everywhere, the effects on individual coastal tourism destinations will differ – destinations with now 'wet' summers may become more attractive for beach and bathing tourism, already dry spots may face desert-like peak seasons, narrowing their attractiveness to tourists.

The number of indirect climatic effects on coastal tourism is much higher. Indirect effects include all changes and developments that are a consequence of the direct effects. Probably the best known effect is the melting of polar ice shields as a result of warmer air and water temperatures, leading to sea level rise of up to 1 m (Dow, Downing & Schellnhuber, 2007), endangering one of the most important resources for coastal tourism: beaches and coastlines. Other indirect effects which are important to coastal tourism are changes in terrestrial and marine ecosystems, including possible impairment of bathing water quality (Matzarakis & Tinz, 2008), accelerated erosion of beaches and coastlines, shortage of water during the summer months (Ehmer & Heyman, 2008), leading to problems for agriculture and forestry, including crop failure and increasing risks of forest fires (Alcamo et al., 2007). These are only a few examples of a variety of possible challenges. Coastal tourism may also benefit from indirect changes on the other hand: Increasing air temperatures and less cloudiness in summer lead to warmer water temperatures and offer the potential to extend the number of days suitable for sea bathing in moderate coastal destinations. These few examples underline the need for an individual regional approach when dealing with indirect climatic developments. They have to be analysed in the context of regional, individual characteristics and cannot be evaluated for coastal tourism in general.

Induced climate changes present as variations in social, cultural or economic systems. This approach is relatively new in climate science and has become possible through the participation of social sciences and the humanities in discussions about climate change within the past few years. Whereas direct and indirect climate impacts are mostly a matter of research by natural sciences, induced climatic changes are the interpretation of these changes (Welzer, Soeffner & Giesecke, 2010). They vary in levels and must be analysed carefully. For coastal tourism the most important induced climate changes can be shifts of tourism patterns, due to shifts of the perceived attractiveness of competing destinations, alterations in the financial situation of tourists due to costs of climate change and increase of water and energy price levels. These few examples of induced changes demonstrate that they cannot be predicted easily and can only be understood in the context of the system they are imbedded

in. An overall attempt to analyse the risk of induced climate changes for coastal tourism is therefore not possible.

Understanding these three dimensions of direct, indirect and induced changes for coastal tourism makes it possible to evaluate the vulnerability of coastal tourism, but it also shows that a general approach is too vague and therefore not fruitful. Direct and indirect changes require downscaling to a regional, spatial attempt. An analysis of induced climate change implies a detailed insight in all the involved social, cultural and economic subsystems.

Obviously, evaluating the chances and risks of climate change for the BSR cannot be done by a standardized risk analysis of global coastal tourism but needs a detailed analysis of all the unique characteristics of the region.

17.3 Baltic Sea tourism: characteristics and challenges

17.3.1 Economic relevance

Determining the exact amounts of the tourism share of GDP is a challenge facing tourism's multidimensional character. It is difficult to narrow down the amounts solely resulting from tourism, especially when dealing with indirect or induced economic effects. These effects describing the economic revenue of sectors which benefit from tourism expenditures, even though they are not basic touristic services or institutions, such as the food trade or building industry (indirect effects) or economic developments as a consequence of the rise of GDP by for example, strengthening the domestic purchasing power and creating jobs (induced effects). But even an interpretation of direct effects only indicates the importance of tourism for the BSR. Figures 17.1 and 17.2 compare the relevance of tourism for Baltic Sea countries in matters of direct and relative economic effects. It is obvious that relevance for the GDP and the job sector varies, but will remain significant for all countries. The lack of an explicit definition of coastal tourism, different statistical methods or simply non-existing explicit data in the littoral states and overlapping of direct, indirect and induced economic effects based on equivocal definitions poses difficulties for determining the exact economic relevance of tourism for coastal regions only.

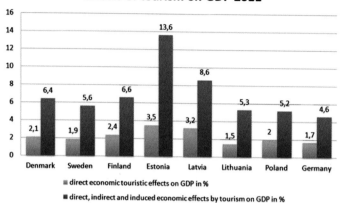

Figure 17.1 Direct effects of tourism on GDP 2011 in BSR countries.
Source: compiled by the author, based on data of World Travel and Tourism Council 2012, www.wttc.org

Therefore, numbers for whole countries were used. It is very likely though that for many BSR countries these numbers are a valuable indicator of the tourism situation on their coastlines, since coastal areas are among their major tourism attractions. In terms of regional distribution, numbers of overnight stays are not spread homogeneously on the 70 000 km (inner) shoreline of the Baltic Sea, but focus on the southern BSR, especially on Denmark, Germany and southern Sweden, where climatic conditions are more suitable for beach and bathing tourism. However, in recent years there have been promising developments in Poland and the Baltic States.

The economic dependency on tourism is significant in many parts of the Baltic coast. Since coastal tourism's main activities – water sports, bathing and beach tourism – are closely linked to the coast and therefore to the weather, it has a very high spatial and seasonal impact. Benthien and Steingrube (2006) characterize it as follows: Coastal tourism in the Nordic countries comprises boat tourism, harbour attractions, accommodation with maritime flair and fishing including the important sector of second home tourism, whereas bathing tourism and increasingly boat tourism represent the main subsectors of the southern BSR. Business tourism plays a subordinate role and concentrates on coastal cities in the BSR (Benthien & Steingrube, 2006). However, this fact is not related to the characteristics of coastal cities in general, but to the fact that many of the BSR capital cities (Copenhagen, Stockholm, Helsinki, Tallinn and Riga) are directly located on the coast.

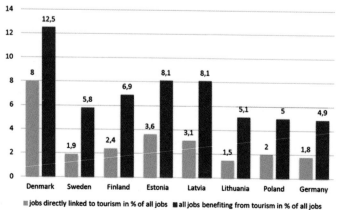

Figure 17.2 Relative effects of tourism towards job situation in BSR countries.
Source: compiled by the author, based on data of World Travel and Tourism Council 2012; www.wttc.org

17.3.2 History and key resources

The southern parts of the Baltic Sea Coast have been a touristic hotspot since the beginning of tourism in the early 1900s (Spode, 2009). Nevertheless, the southern BSR is not a homogeneous destination: The bordering countries with their individual cultures, different natural resources and landscapes are turning Baltic Sea coastal tourism into a diverse segment. The Baltic Sea borders on nine different countries (Denmark, Sweden, Finland, Russia, Estonia, Lithuania, Latvia, Poland and Germany), from which most have undergone dramatic changes within the past 20 years and whose economic and social framework conditions vary to a significant level.

The BSR was divided by the Iron Curtain until the early 1990s, splitting Europe into two different cultural, economic and social spheres, which from a touristic point of view was still visible until recent years. The planned tourism economy has left challenges, which still influence the tourism development in the former Soviet states. Economic problems, ecological threats, marketing problems – many people still project criminality on Eastern Europe – and in some places poor quality standards of traffic and hospitality infrastructure are challenging tourism development (Paesler, 2007). Still, economic growth and joining the European Union in 2004 have led to a promising development of tourism in Poland and the Baltic states with increasing tourism numbers (Jédrzejczyk, 2007; Standl, 2007).

A very similar development took place a couple of years earlier on the German Baltic Coast which hosted Eastern and Western tourism markets: In the federal state of Schleswig-Holstein tourism was booming until the early 1990s, with the beaches of the Baltic Sea being a desirable destination for West German tourists. On the other side of the border, in the federal state of Mecklenburg-Vorpommern, the yearning for coastal tourism was equal, but was monitored and controlled by the socialist leadership of the GDR. The touristic demand and the touristic market therefore never really matched. This changed with the fall of the Wall, when the beaches in Eastern Germany became accessible and tourism had the chance to build up touristic infrastructure based upon recent touristic demands, leading to a boom in tourism that still lasts.

The BSR's most important unique selling prepositions (USPs) are based on its history and its nature. From a mere climatic point of view, for tourists who are seeking bathing and beach tourism there are more attractive places in Europe. But history provided the southern BSR with a rich architectural heritage, many historic city centres and the Hanse, an alliance of merchants trading along the Baltic Coast from the 12th until the 17th century. The cultural richness of this history has survived the struggles of the past and is giving the coastal area of the southern Baltic Sea an attractive background for cultural tourism.

Europe's history in the second half of the 20th century, the Cold War and the Iron Curtain made a big part of the southern BSR (the areas which are now Lithuania, Estonia and Latvia, the former GDR and Poland) inaccessible to many tourists. The collapse of the Soviet Union changed this and a relatively new, unexplored coastal tourism destination popped up on the touristic map of Europe. Since tourism facilities and structures were owned and controlled by public institutions in the Eastern States, the collapse of these institutions were a welcome chance for private businesses. Economic, social and ecological challenges, as mentioned above, affected the development but have become less significant in recent years. The establishment of up-to-date tourism infrastructure and the restoration of old towns and cultural heritage are improving the BSR's attractiveness permanently.

Beside its cultural heritage, the natural richness of the area is one of the key resources for southern Baltic Sea tourism. The glacial past has left not only sandy beaches but also a dramatic coastline which is characterized by glacial lakes, cliffs, fjords and various other landscapes. The BSR is therefore a desirable destination for nature-based tourism. This fact is strengthened by the fact that – especially in the former Soviet countries – tourists have the chance to explore vast, unexploited landscapes (Paesler, 2007). Even though climate is not a key resource for tourism unlike for other coastal destinations, with the focus on nature-based tourism climate is a major framework condition for various activities in the BSR (Standl, 2007).

17.3.3 Regional challenges

In addition to the global challenges that are summed up in Section 17.2, coastal tourism in the BSR will

Table 17.1 Summary of the projected climate change impacts for the late 21st century in the Baltic Sea region

Air temperature	• total region: warming of the mean annual temperature 3 to 5 °C namely 4 to 6 °C in winter and 3 to 5 °C in summer.
	• largest warming in the northern part during winter and in the southern part during summer
	• extension of growing season by 20 to 50 days for northern areas and 30 to 90 days for southern areas.
Water temperature	• total region: increase in mean annual sea surface temperature 2 to 4 °C
Precipitation	• northern part: increase in winter by 25 to 75% and in summer by −5 to 35%
	• southern part: increase in winter by 20 to 70% and decrease up to 45% in summer
Wind	• projections of wind changes differ widely
	• increase of about 8 to 12% is more likely than a decrease
Sea-level rise	• 20–30 cm in the southern parts
	• 90–200 cm in the northern parts
River flow	• northernmost catchments: increase in mean annual river flow
	• southernmost catchments: decrease up to 50%
Salinity	• total region: decrease up to 45%
Ice extent	• total region: decrease 50 to 80 %

Source: Meier, 2006; MfWAT MV, 2007; BACC, 2008.

have to deal with changes not seen in the rest of the world. Referring to the challenge of climate change, Table 17.1 gives a summary of the projected direct and indirect climate impacts, some for the total region, some assigned to northern or southern parts.

An analysis of these changes has to be objective and must therefore not only deal with climate risks, but also with the chances a changing climate may bring (Becken & Hay, 2007). Even though today climate itself is not a major tourism resource for the BSR's appeal, climate change may alter that as direct changes, such as warmer summers and less precipitation during the peak season, might see a potential growth in tourism numbers. The southern Baltic Sea is one of the few coastal areas in the world where a change of temperature and precipitation may have the potential to improve some framework conditions for tourism.

The high population density of Europe (Collet, 2010) in comparison to other coastal tourism destinations is another advantage for tourism development, even though it is decreasing; the financial crisis in the last few years has indicated that domestic tourism numbers are growing in times of fiscal uncertainties. It is therefore likely that even if climate change is having a major negative impact on economic and social

key conditions, domestic tourism in the southern BSR will be a winning segment compared to some overseas destinations. The touristic potential 'on the doorstep' and climate-induced shifts in tourism patterns to closer destinations, leading to less travel expenses, can initiate growth of tourism in the southern BSR.

With favouring climatic conditions for nature-, beach- and bathing holidays it seems that coastal tourism in the BSR could face relaxed climatic variations. So why worry?

Upon closer examination the risks of indirect and induced climate changes will probably be higher than the chances. Regional geomorphological characteristics, for example, might strengthen the effects of sea level rise in the southern BSR more than in the northern parts of the BSR. In the latter, isostatic crustal movements as a result of the last ice age which ended 12 000 BP lead to land uplift that to some extent counters eustatic sea level rise. The southern shorelines are mainly stable and partly experience land subsidence (Meier, Broman & Kjellström, 2004). This could mean a direct threat for the coastal zone with the potential to worsen the effects of eustatic sea level rise compared to other coastal areas around the globe. Another threat is the vulnerability of the ecosystems in the region. The Baltic Sea as a semi-enclosed sea with a lower salinity than most seas is a very unique and therefore vulnerable ecosystem with very distinctive and specialized species (BACC, 2008). Changes in salinity, ice coverage and flow patterns as a result of warmer temperatures in air and water as well as in changes in coastal hydrology will probably challenge the fragile ecosystem (BACC, 2008). After determining nature as one of the BSR's most important tourism resources, changes in ecosystems can imply a direct threat to the attractiveness of the Baltic Sea as a destination.

17.4 Adaptation strategies for coastal tourism

Developing climate adaptation strategies presumes an understanding of what exactly adaptation means. It is directly linked to vulnerability (see Section 17.2) and involves all measures to handle actual and future changes in climate and to allow active risk control (Füssel, 2007). Up until recently, the main attention in climate politics was mitigation, with the primary intention to reduce CO_2 emissions and prevent or at least minimize climate change. Since mitigation processes did not reach a breakthrough within the past years, but instead proved to be insufficient, the call for adaptation is getting louder. Recent research has shown that climate change is inevitable, since the climate system is inertial and will react to the emitted CO_2 anyhow within the next few decades, even if emission were cut to zero immediately (Latif, 2008).

Tourism adaptation affects a diverse economic segment with many different actors on the market. The tourism market chain includes basically three groups: the tourists, travel businesses and the destinations (Simpson et al., 2008). When it comes to adaptation these three groups have to handle very distinct adaptation capacities: tourists as individuals can adapt to changes by choosing an alternative time to travel or destination to go to. Tour operators and transport businesses have the chance to serve these changes in demand by reorganizing their range of products. Destinations and local tourism companies, such as hotels and leisure infrastructure, will have to find different options to handle climate change. They have the

lowest adaptive capacity and will have to deal with climate change by organizational and financial tools. Mobility is therefore a key criterion for adaptation, as Figure 17.3 shows.

For tourism destinations and local tourism companies, it is obvious that dealing with climate change therefore requires thinking in terms of long time scales. Contrary to that, tourism planning is based on a time scale of only a few years (Wall & Mathieson, 2006) – and climate change is a much slower progress of decades or even centuries. Projections for 2050 or 2100 seem to be far away and tourism development may be affected by things other than climate change which cannot be predicted today. These two arguments of uncertainty of developments and different time scales are often used as counter-arguments against developing adaptation strategies in the near future. As evident as that might seem, it leaves out important characteristics of climate adaptation. First of all, climate change, adaptation and tourism development are highly dynamic processes; understanding and – if necessary – interfering with these processes will take time. Scenarios for specific points in the future are merely to highlight the intensity of these changes and must not be interpreted as tipping points, but as a snapshot of a constant, ongoing development. Furthermore, the efforts to adapt to these dynamic changes will increase with the intensity of the changes – to implement a successful adaptation strategy it is advisable to follow a step-by-step strategy rather than to prepare for a major adaptation measure at one point in the future. This strategy would also meet the short-term nature of tourism planning horizons. Beside this synchronization of adaptation and tourism development time patterns, it is of importance to monitor framework conditions influencing tourism development, which is highly dependent on regional and urban planning processes. Some of these processes presume long investment cycles and affect the tourist destinations directly, for example, coastal protection programmes. A successful adaptation strategy, therefore, does not only have to consider direct tourism development but also the time scales indirect influencing parameters are working within.

Financial aspects also have to be considered when it comes to time frames of touristic adaptation to climate

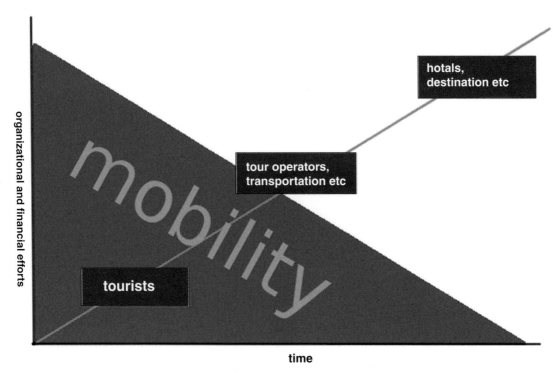

Figure 17.3 Adaptive capacity of different members of the touristic market.
Source: own illustration, modified after Simpson et al., 2008

change: some measures will only be of an organizational nature, but others will incur costs. Being able to handle these expenses will be a key condition in adaptation and will require a suitable financial leeway, since their dimensions are potentially large. Current extreme events, such as storms causing expenses into the millions to nourish eroded beaches, are an indicator of that.

Last but not least, climate change will challenge tourism destinations *in a new way*. The experience of it – like in any other sector – is limited. Building up capacities and creating competence is a matter of time and cannot be achieved within a few months. Even though tourism as a highly dynamic sector, which has had to adapt to changes in its framework conditions since its very beginning, the challenge of climate change differs because of its complex, multidimensional nature (Zebisch et al., 2005). Climate impacts have the potential to alter major framework conditions in a multidimensional way, and all at once. Nevertheless, the experience with adaptation in general can play an important role for tourism destinations

and could be a big advantage in coping with climate change and other future challenges.

A regional snapshot: expert interviews at the German Baltic Coast

Understanding the long-running nature of basic principles of adaptation – experience, networking and finance – raises the question of the current state in the Baltic Sea Region. The first minor signs of climate change are showing: sea level in the southern BSR is currently rising by two millimetres annually, the average temperature has risen about 0.85 °C within the last century and slight changes in long-term precipitation schemes are already discernible (BACC, 2008).

With the understanding of the complexity and the urgency to prepare and establish adaptation strategies, pan-Baltic projects like BaltCICA or national projects such as RADOST in Germany were initiated within the past few years. Nevertheless, it is important to underline that the aim of these projects is not the development of ready-to-implement adaption strategies, but to lay a foundation on which local stakeholders and regional decision makers can build and develop individually fitting strategies. The first approach is to establish networks, improve (or establish) communication structures between different network partners and to impart knowledge about climate change and adaptation.

To do this efficiently, it is essential to have an overview of the status quo of the tourism sector concerning perception and current reaction to climate change. Following this approach, nine expert interviews with different touristic decision-makers of federal, regional and local touristic institutions along the German Baltic Coast were held within a case study in the early spring of 2011. The aim of these interviews was to understand which role climate change is currently playing in the medium- and long-term planning of destinations, and their understanding of good communication and networking structures. The interviews were held individually at the office of the interviewee and were narrative in character. The questions asked were designed to initiate monologues and lasted between 20 and 90 minutes, with an average length of about one hour. The candidates were selected by the 'Gate-Keeper-Principle' based on candidates who were identified as 'motivated' in previous activities. This procedure is weakened by failing to address a representative range of candidates, since a selection always filters distinct interview partners. Nevertheless it became obvious during the interviews that the knowledge of climate change and the individual attitude towards touristic climate adaptation differed markedly. The objective of providing a snapshot of touristic decision makers was therefore accomplished.

The qualitative analysis showed different results: The importance of climate change destination planning is highly dependent on the question if the destinations have been affected by natural hazards, especially by storms, within the last years. The consequences of these disasters have been financially significant, since whole beaches, and with them the destinations' main attractions, had been washed away and required cost-intensive refilling-operations. Most touristic stakeholders who were affected by such extreme events showed a stronger awareness towards the development of the climate, even though the occurrence of storms cannot completely be linked to climate change. In fact storminess itself might decrease, whereas the intensity of storms might increase (Nikulin et al., 2010).

Still, even for those who have already been affected, adaptation is not a priority. The awareness may have risen, but conclusions have not been drawn so far – the strategy to handle climate change is based on direct reaction, not on proactive adaptation. According to the interviews, an explanation for that is lack of time, knowledge and, especially, financial resources. However, even though adaptation is not on the short-term agenda by now, the interviews clearly demonstrated concern about climatic developments over the next few decades, even though it was noticeable that only direct climate changes were attributed as being severe future problems. Indirect and induced climate changes were secondary, if mentioned at all. Social, cultural and economic changes as a result of climatic changes do not seem to play a part in the region's actual touristic strategies. This may be due to the adaptive character of tourism business as some indirect key conditions, such as political and economic, are subject to constant ongoing changes the sector has to adapt to anyhow.

The call for climate networks was another important result of the interviews: In the eyes of most destinations, the global challenge of climate change is a task which they delegate to federal or national authorities, since it needs cross-community action. Even though climate change does not seem to pose a big threat to the destinations at the moment, the readiness to become part of a network has been indicated by most interview partners. The dimension of this network was recommended as spatial and cross-institutional, and as broad as possible to bundle competences and knowledge. However, at this point specific knowledge of most of the touristic stakeholders in matters of climate change was stated to be limited, since adaptation is not rated as part of tourist-destination management core competences. The interviews also showed that the willingness to become part of a climate network is limited to a passive, expectant role – leadership is expected from governmental institutions, such as ministries or federal tourism associations. This result indicated a major discrepancy between the visions of corresponding projects to establish functional networks by the end of their life span and the interest of the potential network partners. Since networking is a vital part of the tourism sector's daily business, their know-how could favour the chances of establishing an efficient network – the main task is now to assign responsibility for adaptive networks to the stakeholders.

As for communication, the designated form within the network and between different institutions was clearly defined – scientific jargon and uncertain results were rejected with regard to limited time and organizational resources. Most touristic stakeholders cannot afford taking part in a detailed scientific discussion, but call for clear results and applicable suggestions for climate change adaptation.

Practical adaptation: Pioneers and possible action

Beside these 'preparing' steps to adaptation, some examples of practical adaptation strategies can already be found around the Baltic Sea.

With the implementation of a National Adaptation Strategy in 2005, Finland plays a pioneer role in European adaptation, with measures of for example, improving conservation procedures by directing tourists away from protected or potentially endangered areas, such as marine and coastal reserves (Ministry of Agriculture and Forestry of Finland, 2005). Germany and Denmark have followed Finland's example and have established own national adaptation programmes in recent years. Apart from these national attempts, Table 17.2 indicates examples of possible practical adaptation actions, applicable for tourism destinations.

The table makes no claim to be complete and only shows a small snapshot of the variety of possible adaptation strategies and measures. As stated earlier, adaptation requires a small-scaled approach to be efficient. The list demonstrates that climate change challenges are complex and handling them requires multidimensional actions to face the corresponding risks and to benefit from the chances.

17.5 Discussion

The analysis of the challenges for this area and the results of the interviews allow an estimation of the

Table 17.2 Climate change impacts and possible practical adaptation measures (compiled by authors)

Impact	Potential adaptation measure
increase in air temperature	– artificial shading – extension of natural shading – modification of business hours – constructional changes to cool buildings – extension of bathing season – development of new (northern) tourism destinations
changing bathing water quality	– information on algae, jellyfish and seaweed – creating bathing alternatives – intensification of mechanical beach cleaning – intensified monitoring of water quality with possible temporary and local bathing restrictions
changes in precipitation	– collection of rain water – information on sustainable use of drinking water – building ban on flood plains
increase of extreme weather events	– retraction from coastal flood plains – installation of early warning systems – reinforcement of coastal protection – protections of touristic infrastructure on the shorelines – new marketing offers, e.g. 'Experience the Wild Baltic Sea'

vulnerability of the German Baltic Coast. Further, the challenges can be associated with the potential of the tourism sector to change and adjust to the needs of altered conditions in the touristic market.

The interpretation of the results has to be done with regard to the time dimensions of adaptation, the specifics of Baltic Sea tourism and the complexity of the tourism sector, as stated in the previous sections of this chapter. Bearing these characteristics in mind, the results of the interviews portray a state of climate adaptation within the examined regional tourism sector which seems insufficient yet. Since climate adaptation is a relatively new field, this fact is neither surprising nor an exclusive touristic attribute. Aiming at developing independent and efficient adaptation networks is therefore adequate. For coastal tourism, this task seems to be challenging though; the willingness to play an active role has to be extended amongst tourism stakeholders and the knowledge about climate change and the specifics of Baltic Sea tourism challenges has to be made accessible to as many decision makers as possible. It is essential not just to focus on the direct risks, but also to underline the possibility of social and cultural changes, which are major framework conditions for tourism. The chances for southern Baltic Sea tourism have to be evaluated in the context of the risks. Regionally applied projects will be most successful by realistically visualizing the future of coastal tourism. This requires detailed research and communication of feasible strategies, including the fact that the actual tourism situation cannot be preserved but will undergo severe changes.

The fact that climate adaptation is hardly playing any role in current tourism planning indicates that the need for timely preparation has not been communicated in a way that appealed to tourism stakeholders. The time patterns of climate change adaptation strategies, which were introduced in Section 17.3 of this chapter, are different from those of tourism business plans. Nevertheless, gaining experience, organizational changes of working procedures, building up financial resources and establishing networks will take time – especially for touristic destinations, which are located on the end of the tourism market chain which needs most time to adapt. The first steps of the ongoing process of adaptation should therefore be quickly implemented into current destination business plans. It is not necessary to develop a sophisticated, func-tional strategy within a short time period, but it is crucial to start as soon as possible, since the first effects of climate change are already recognizable – time *is* a matter, even though most interview partners pointed to the long-term character of climate change.

Tourism as a complex and multidimensional sector has to develop a strategy, which needs to be developed with regard to regional and system-inherent specifics. Some propositions of a successful climate change adaptation for coastal tourism can be considered favourable: Adding to the inherent ability and experience of tourism to adjust to changes quickly, the willingness to work within networks and the existence of already established networks can be of advantage for future adaptation strategies. Initiating topical networks with respect to climate adaptation or presenting the topic to existing tourism networks can only be a first step though. However, the results of regional activities show promising developments which may form a successful basis for regional climate adaptation initiatives. Participation of tourism stakeholders in these initiatives and a growing interest can constitute a solid base on which climatic adaptation can build. It is obvious on the other hand that many stakeholders which are part of the tourism sector are currently lacking important competences such as sufficient knowledge of touristic climate adaptation, as well as adequate financial resources to work efficiently within these networks.

Furthermore, the results of the interviews did not show many new impulses from the tourism sector to solve the specific challenges for Baltic Sea tourism. The approach still seems to be rather reactive than active; there are no answers to increasing influences of sea level rise and the vulnerability of sensitive ecosystems. The questioned Baltic Sea tourism decision makers do not seem to have solutions for the challenges which their most important resource, the Baltic coastlines, will probably undergo. Assigning the responsibility to answer the most urgent questions of Baltic Sea tourism adaptation more or less to science or super-ordinate governmental institutions only, destinations deny their own responsibility as equal parts of the alliance to face climate change, although the individual knowledge and direct on-site observations could prove to be valuable tools to work out effective strategies.

Another important result of the interviews was the apparent attempt to preserve early 21st century tourism. When talking about climate change impacts, the discussion turned into minimizing those effects that change the current tourism situation; contrary to the development of tourism as a result of ongoing change within the past 100 years, touristic stakeholders seem to shy away from future changes, which means neglecting the risks, but also withdrawing potential chances of climate change. This desire for preservation is an unusual and remarkable development in the otherwise dynamic tourism sector.

Once again, it is important to point out that the results of the interviews portray the situation on examples of the German Baltic Coast. Some of the results are likely to be transferable to other riparian states, such as the placement of climate change amongst other challenges and the handling of uncertainties, but others have to be analysed individually. Even though the Baltic Sea is characterized as one destination in this chapter, the differences that were pointed out in Section 17.3 have to be considered; the Baltic Sea is not a homogenous tourism destination, but consists of very distinct national or even micro-regional tourism markets. The readiness to work in networks, existing network structures, as well as organizational and financial prepositions varies. The described experiences made in Germany are thus not necessarily applicable to all Baltic Sea destinations.

The results show that even in regions where tourism is well established, climate adaptation seems to be seen as a secondary problem. However, appraising it as a 'luxury problem' might be a misjudgement of the situation and provokes a wait-and-see attitude in other regions around the Baltic Sea. The pan-European attempt of projects such as BaltCICA and baltadapt to sensitize stakeholders in all riparian states is, therefore, a promising attempt to provide a network between different Baltic regions to exchange information and to communicate the urgency of concerted climate adaptation.

17.6 Conclusion

The results of the interviews in context with the analysed situation around the Baltic Sea indicate that climate change adaptation within the Baltic Sea tourism sector is a possible, but challenging attempt. The heterogenic tourism sector in the BSR has to be a major stakeholder in the attempt to face climate change – climate science and governmental institutions will not be the key to successful adaptation. Due to the fact that climate change effects will be comparable around the Baltic Sea, the chance to learn from experiences of other destinations can be advantageous – if communication and networks are set up properly. The impacts of climate change may vary, but if an overall adaptation strategy for the BSR is established as a mosaic of all the different experiences of the touristic regions, the know-how pool will be a valuable source to identify appropriate adaptation options. The knowledge of networks and the adaptive nature of this sector give tourism an enormous advantage towards other economic sectors. However, tourism as an interdisciplinary field can benefit from these advantages only for a limited time; once they are exceeded by the increasing complexity of tourism adaptation it will be getting more challenging to establish successful adaptation.

Acknowledgements

The work has been supported by the projects BaltCICA (Climate Change: Costs, Impacts and Adaptation in the Baltic Sea Region), part-funded by the EU within the Baltic Sea Region Program, the German project RADOST (Regional Adaptation Strategies for the German Baltic Coast), funded by the Federal Ministry of Education and Research (BMBF) within the activity 'KLIMZUG' (grant number 01LR0807K), and baltadapt (Baltic Sea Region Climate Change Adaptation Strategy), also funded by the EU within the Baltic Sea Region Programme.

References

Alcamo, J., Moreno, J.M., Nováky, B., Bindi, M., Corobov, R., Devoy, R.J.N., Giannakopoulos, C., Martin, E., Olesen, J.E. and Shvidenko, A., 2007: Europe. In: M.L. Parry, O.F. Canziani, J.P. Palutikof, P.J. van der Linden and C.E. Hanson, eds. 2007. *Climate*

Change 2007: Impacts, Adaptation and Vulnerability – Contribution of Working Group II to the Fourth Assessment Report of the Intergovernmental Panel on Climate Change. Cambridge, UK: Cambridge University Press, pp.541–580.

Bacc Author Team, 2008. *Assessment of Climate Change for the Baltic Sea Basin.* Berlin, Heidelberg: Springer.

Becken, S. and Hay, J.E., 2007. *Tourism and Climate Change – Risks and Opporunities.* Clevedon, Buffalo, Toronto: Channel View Publications.

Benthien, B. and Steingrube, W., 2006. Some footnotes on tourism in the Baltic region. In: Österreichische Gesellschaft für Wirtschaftsraumforschung, eds. 2006. *Unterwegs in touristischen Landschaften.* Festschrift für Univ.-Prof. Dkfm. Dr. Felix Jülg zum 70. Geburtstag, Wien: WUV Universitätsverlag.

Collet, I., 2010. *Portrait of EU Coastal regions.* [pdf] Available at: <http://epp.eurostat.ec.europa.eu/cache/ITY_OFFPUB/KS-SF-10-038/EN/KS-SF-10-038-EN.PDF> [Accessed 15 December 2011].

Dow, K., Downing, T. and Schellnhuber, H.J., 2007. *Weltatlas des Klimawandels – Karten und Fakten zur globalen Erwärmung.* Hamburg: Europäische Verlags-Anstalt.

Ehmer, P. and Heyman, E., 2008. Climate change and tourism: Where will the journey lead? In: Deutsche Bank Research, 2008. *Energy and Climate Change: Current Issues.* Frankfurt am Main: Deutsche Bank.

ETC (European Travel Commission) ed., 2006. *Tourismus Trends für Europa.* [pdf] Available at: <http://www.etc-corporate.org/resources/uploads/ETC_Tourismus_Trends_fuer_Europa_02-2007 GER.pdf> [Accessed 17 December 2010].

Füssel, H.-M., 2007. Adaptation planning for climate change: concepts, assessment approaches, and key lessons. *Sustainability Science,* 2 (2), pp.265–275.

Gee, K., Kannen, A. and Licht-Eggert, K., 2006. *Raumordnerische Bestandsaufnahme für die deutschen Küsten- und Meeresbereiche.* [pdf] Kiel: Forschungs- und Technologiezentrum Westküste der Universität Kiel. Available at: <http://iczm.ecology.uni-kiel.de/servlet/is/524/060821_BBR_Bestandsaufnahme.pdf?command=downloadContent&filename=060821_BBR_Bestandsaufnahme.pdf > [Accessed 27 November 2010].

Jedrzejczyk, I., 2007. Tourismus in Polen im Wandel der letzten 20 Jahre. In: Becker, C., Hopfinger H. and Steinecke, A., eds. 2007. *Geographie der Freizeit und des Tourismus.* 3rd ed. Munich: Oldenbourg, pp.568–581.

Latif, M., 2008. *Bringen wir das Klima aus dem Takt? Hintergründe und Prognosen.* 5th ed. Frankfurt am Main: Fischer-Taschenbuch-Verlag.

Matzarakis, A. and Tinz, B., 2008. Tourismus an der Küste sowie in Mittel- und Hochgebirgen – Gewinner und Verlierer. In: J.L. Lozán, H. Grassl, G. Jendritzky, L.

Karbe and K. Reise, eds. 2008. *Warnsignal Klima: Gesundheitsrisiken: Gefahren für Pflanzen, Tiere und Menschen.* Hamburg: Wissenschaftliche Auswertungen, pp. 254–259.

Meier, H.E.M., Broman, B. and Kjellström, E., 2004. Simulated sea level in past and future climates of the Baltic Sea. *Climate Research,* 27, p.59–75.

Meier, H.E.M., 2006. Baltic Sea climate in the late twenty-first century: a dynamical downscaling approach using two global models and two emission scenarios. *Climate Dynamics,* 27, pp.39–68.

Ministry of Agriculture and Forestry of Finland, 2005. *Finland's National Strategy for Adaptation to Climate Change.* [pdf] Available at: <http://www.mmm.fi/attachments/mmm/julkaisut/julkaisusarja/5g45OUXOp/MMMjulkaisu2005_1a.pdf> [Accessed 28 February 2012].

Nikulin, G., Kjellström, E., Hansson, U., Strandberg, G. and Ullerstig, A., 2010. Evaluation and future projections of temperature, precipitation and wind extremes over Europe in an ensemble of regional climate simulations. *Tellus,* 63A (1), pp.41–55.

O'Hare, G., Sweeney, J. and Wilby, R., 2005. *Weather, Climate and Climate Change – Human Perspectives.* Pearson: Harlow.

Paesler, R., 2007. Der Wandel des Tourismus in den Transformationsländern Ostmittel- und Osteuropas durch die politische Wende. In: Becker, C., Hopfinger, H. and Steinecke, A., eds. 2007. *Geographie der Freizeit und des Tourismus.* 3rd ed. Munich: Oldenbourg, pp.555–567.

Schumacher, S. and Stybel, N., 2009. Auswirkungen des Klimawandels auf den Ostseetourismus – Beispiele internationaler und nationaler Anpassungsstrategien. In: EUCC, ed. 2009. *International approaches of coastal research in theory and practice.* Warnemünde, Leiden: EUCC, Coastline Reports 13, pp.23–46.

Simpson, M.C., Gössling, S., Scott, D., Hall, C.M. and Gladin, E., 2008. Climate change adaptation and mitigation in the tourism sector: Frameworks, tools and practices. [pdf] Paris, France: UNEP, University of Oxford, UNWTO, WMO. Available at: <http://www.uneptie.org/shared/publications/pdf/DTIx1047xPA-ClimateChange.pdf> [Accessed 28 November 2010].

Standl, H., 2007. Die Integration der baltischen Staaten in den internationalen Tourismusmarkt: Potenziale – Strukturen – Perspektiven. In: Becker, C., Hopfinger H., Steinecke, A., eds. 2007. *Geographie der Freizeit und des Tourismus.* 3rd ed. Munich: Oldenbourg, pp.555–568.

Spode, H., 2009. Der Aufstieg des Massentourismus im 20. Jahrhundert. In: Haupt, H. and Topf, C., 2009. *Die Konsumgesellschaft in Deutschland 1890 – 1990. Ein Handbuch.* Frankfurt am Main: Campus Verlag.

Wall, G. and Mathieson, A., 2006. *Tourism – Change, Impacts and Opportunities*. Harlow, UK: Pearson.

Welzer, H., Soeffner, H. and Giesecke, D., 2010. *KlimaKulturen – Soziale Wirklichkeiten im Klimawandel*. Frankfurt am Main: Campus Verlag.

Werner, P.C. and Gerstengarbe, F.-W., 2007. Welche Klimaänderungen sind in Deutschland zu erwarten? In: W. Endlicher and F.-W. Gerstengarbe, eds. 2007. *Der Klimawandel – Einblicke, Rückblicke und Ausblicke*. Potsdam: Potsdam-Institut für Klimafolgenforschung e. V., pp.56–59.

Zebisch, M., Grothmann, T., Schröter, D., Hasse, C. Fritsch, U. and Cramer, W., 2005. *Klimawandel in Deutschland – Vulnerabilität und Anpassungsstrategien klimasensitiver Systeme*. [pdf] Available at: <http://www.umweltdaten.de/publikationen/fpdf-k/k2947.pdf> [Accessed 13 December 2010].

18 Tourists' Perception of Coastal Changes – A Contribution to the Assessment of Regional Adaptation Strategies?

Larissa Donges[1], Inga Haller[2] & Gerald Schernewski[1,3]

[1]Leibniz Institute for Baltic Sea Research Warnemünde, Rostock, Germany
[2]EUCC - The Coastal Union Germany, Rostock, Germany
[3]Coastal Research & Planning Institute, Klaipeda University, Klaipeda, Lithuania

18.1 Introduction

There is clear evidence of an on-going global warming trend but we are still far from being able to understand the complex interaction of all relevant parameters involved (IPCC, 2009; 2012). Climate change is a global issue with high geographical variability. Impacts of global change vary at national level but also on a regional scale, such as in the Baltic area, where we encounter heterogeneous natural and social conditions and, thus, varying climate impacts. Due to this variability as well as with regard to the individual sensitivity and vulnerability of a region, specific case studies are required. General information about climate change and its possible impacts are often of limited use in practice. Therefore we can only learn about ongoing changes and influences of climate change from specific analyses that take into account the regional or local conditions and characteristics.

In our study we focus on the German Baltic coast and take a closer look at coastal tourism in the federal state of Mecklenburg-Vorpommern. In addition to industry, craft and trade, tourism is the most important economic sector in this state with gross sales of more than 5.1 billion Euros. In 2010, tourism constituted about 10% of the primary income and employed about 173 000 people. The period 1998 to 2009 registered a continuous increase of overnight stays – except for the years 2004-2006. Approximately 6.92 million visitors and 28.4 million overnight stays were recorded in 2009. In particular, coastal tourism that represents 73.3% of the turnover plays an important role. However, in 2010 the number of visitors (6.67 million) and overnight stays (27.8 million) decreased compared to the previous year. Possible reasons might be the strong winter, unfavourable weather conditions during the summer season and increasing competition between destinations (Ministerium für Wirtschaft, Arbeit und Tourismus Mecklenburg-Vorpommern, 2010; Statistisches Amt Mecklenburg-Vorpommern, 2011). It is assumed that tourism will be strongly influenced by global warming (Zebisch et al., 2005). Currently, possible benefits and disadvantages are discussed in research and public. Furthermore, the discussion and research on climate change adaption strategies that will be essential in the future are increasing (Parry et al., 2007; Schuchardt et al., 2008). Different international, national and regional projects such as BaltCICA and RADOST deal with this topic in the region of the case study area. It is clear that climate change will pose a challenge to beach and tourism management as managing tourism resources notably requires meeting the demands of tourists.

Climate Change Adaptation in Practice: From Strategy Development to Implementation, First Edition.
Edited by Philipp Schmidt-Thomé and Johannes Klein.
© 2013 John Wiley & Sons, Ltd. Published 2013 by John Wiley & Sons, Ltd.

For the development of successful strategies and policies, perception analyses are more and more involved in decision-making processes. Taking into account the opinion, perception and preferences of citizens, residents or tourists can contribute to the development of successful strategies and policies and is therefore a popular method applied in sectors such as spatial planning or tourism management (Vogt & Andereck, 2003; Acar & Sakici, 2008; Frochot & Kreziak, 2008; Raymond & Brown, 2011). Analyses are mostly realized through surveys based on questionnaires or interviews. In Germany, for example, annual surveys are undertaken on a national and regional level by the Tourist Association and the German Association of Industry and Trade (DIHK) in order to improve the range of touristic offers and services. Recent studies for the German federal state of Mecklenburg-Vorpommern were published by the Mecklenburg-Vorpommern Tourist Board (2009). On the local level, communities as well as hotels and lodges pay particular attention to their guests' opinion and demands. Considering beach management, surveys can provide crucial hints for planning processes (Cervantes, Espejel, Arellano & Delhumeau, 2008; Marin et al., 2009; Roca, Villares & Ortego, 2009; Vaz, Williams, da Silva & Philipps, 2009).

Until now, analyses and surveys mainly contributed to the development of short-term strategies, for example, to the development of new tourist offers and improvements for the next season. In the context of climate change the question of interest is which role tourists' perception plays in long-term planning.

Our survey deals with the following aspects: Changes are taking place. But how do tourists perceive coastal environments along the Baltic Sea? Are they aware of changes and local climate impacts? Could tourists' perception and opinion be involved in future decision-making processes as well as in the development of future adaptation strategies? Which conclusions can be drawn with regard to medium or long-term adaptive management strategies?

18.2 Climate change and coastal tourism – the connecting parameters

Weather and climate play an important role for tourism as they significantly influence tourists' travel patterns and decision-making processes. While good weather conditions represent a vital resource to be exploited, bad conditions can constitute a risk for the tourism industry. It is assumed that climate change will have profound impacts on tourists' demands and, hence, pose a challenge to the tourism sector. For this reason, there is an increasing demand for detailed climate information. A large number of physical parameters will be directly influenced by global warming. Direct impacts and potential changes are derived from regional climate change models that depend on used forcing scenarios and downscaled global circulation models. However, it is substantially more complex to reliably predict indirect impacts such as changes in the coastal ecosystems. Furthermore, not all parameters have the same relevance for coastal tourism. Table 18.1 provides a schematic overview of the coastal physical parameters that are under climatic influence (primary impacts), the parameters that are influenced indirectly (secondary impacts), and the resulting effects on coastal tourism destinations (induced impacts). In the following section, we look at a selection of the parameters that on the one hand allow meaningful interpretations in the context of past and future climate induced changes and, on the other hand, play a major role for beach and nature related tourism.

18.2.1 Primary impacts

The following section on climate impacts primarily focuses on the study area at the German Baltic coast. For climate impacts on a regional Baltic Sea scale also see the introductory chapter.

Air surface temperature (AST)

Long-term observations of the Baltic Sea Basin mean AST from 1871 to 2004 indicate positive seasonal trends. For the southern area (latitude < 60° N) trends are significant in spring, autumn and winter. The largest seasonal trends (0.11 K/decade) in this area are observed for spring (HELCOM, 2007; Heino et al., 2008). Time series of summer AST from 1947 to 2005 for Warnemünde (Germany) also show a positive trend of 0.17 K/decade (Hagen & Feistel, 2008). Figure 18.1(a) presents modelled 2 m AST from Warnemünde for the period 1960-2010 (summer), based on the A1B greenhouse gas (GHG) emissions scenario proposed by the IPCC (Nakicenovic et al.,

Table 18.1 Climate impacts on coastal tourism

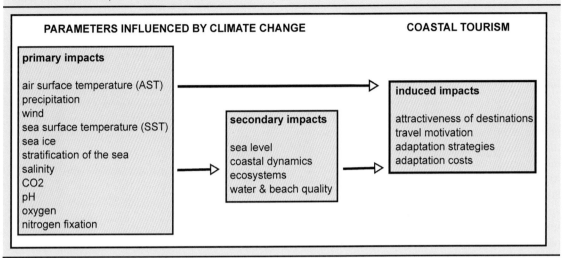

PARAMETERS INFLUENCED BY CLIMATE CHANGE   COASTAL TOURISM

primary impacts

air surface temperature (AST)
precipitation
wind
sea surface temperature (SST)
sea ice
stratification of the sea
salinity
CO_2
pH
oxygen
nitrogen fixation

secondary impacts

sea level
coastal dynamics
ecosystems
water & beach quality

induced impacts

attractiveness of destinations
travel motivation
adaptation strategies
adaptation costs

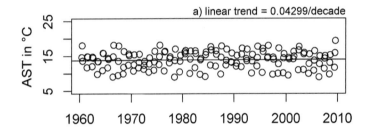

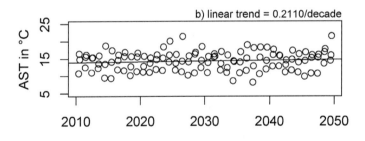

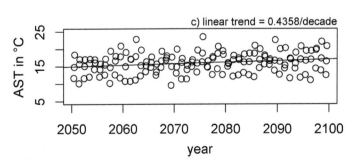

Figure 18.1 Simulated summer (JJA) AST of Warnemünde for the periods 1960-2010 (a), 2010-50 (b) and 2050-2100 (c); scenario A1B with linear trends.
Source: Data was provided by T. Neumann (personal communication, 10 October 2011).

2000). The simulated temperatures show a trend of 0.04 K/decade. The dynamic downscaling was carried out by the Max-Planck-Institute for Meteorology (2008) together with the Climate Limited-area Modelling Community (CLM Community, 2008). It has to be considered that data is biased due to the RCM forcing.

With respect to the future, Figures 18.1(b) and (c) show simulated summer AST of Warnemünde for the periods 2010-50 and 2050-2100 with distinct positive trends. Simulations with the dynamic regional climate models COSMO-CLM (Rockel, Will & Hense, 2008), REMO (Jacob et al., 2008) and RCAO (Döscher et al., 2002) also projected increasing temperatures until the end of the 21st century. Based on these results, summer AST in 2071-2100 could increase by 2.8 °C compared to the period 1961-90, which is consistent with the results presented in Figure 18.1. The number of warm days ($T_{max} > 25$ °C) during summer is expected to rise by 11.7 days during the same time interval (Helmholtz-Zentrum Geesthacht, 2011).

Sea surface temperature (SST)

For Warnemünde, summer SSTs were simulated by Neumann (2010) and are presented in Figure 18.2(a) for the period 1960-2010. In this time interval, temperatures show a slightly positive trend of 0.10 K/decade for the A1B scenario. It has to be considered that there is a bias in SSTs, introduced by the RCM forcing (Neumann & Friedland, 2011).

Current simulations of the SST in the Baltic Sea Basin do all project a positive warming trend for the next few decades. Future development scenarios by Gräwe and Burchard (2011) show an average SST warming of 0.9 K for the period 2020-50 and 2.5 K for the period 2070-2100 (both scenario A1B) in the western Baltic Sea. These results are consistent with those of Neumann and Friedland (2011) who assume an increase of 2 to 2.3 K at the end of the 21st century compared to the period 1960-99. The warming trend depends on the scenarios used as well as the seasons and regions investigated. Simulated summer SSTs for Warnemünde in the periods 2010-50 and 2050-2100 show a positive warming trend of 0.29 and 0.33 K/decade (Figures 18.2(b) and (c)). Simulations based on data from the Max-Planck-Institute for Meteorology were carried out by Neumann (2010).

For tourism, projected warming trends are relevant as they allow conclusions concerning the duration of the bathing season. It is assumed that the requirements for a bathing day are met when water temperatures exceed 15 °C. Based on this definition, the bathing season in the years from 1961 to 1990 in Travemünde (Germany) lasted about 100 days. Until 2050 the season could increase by 25 days, until 2100 by 60 days (Matzarakis & Tinz, 2008).

Precipitation

Compared to other parameters, precipitation varies greatly in time and space. Due to that and the poor data coverage, it is difficult to establish long-term trends for the Baltic Sea Basin. Long-term observations indicate patterns of seasonal changes and beyond, variability from region to region. For each season, both increasing and decreasing trends can be found for the period 1976-2000 compared to the period 1951-75. In western and southern parts of the Baltic Sea Basin, summer precipitation decreased slightly during this period (Heino et al., 2008).

Based on WETTREG-simulations, Spekat, Enke and Kreienkamp (2007) project a decreasing trend for the amount of summer rainfalls along the German Baltic coast until the end of the 21st century (2071-2100 compared to 1961-90). These findings are consistent with simulations of the dynamic RCMs COSMO-CLM, REMO and RCAO. Based on these projections, a decrease of mean summer precipitation of about 17.0% until the end of the 21st century (2071-2100 compared to 1961-90) is assumed for the study region. Correspondingly, rainy days are projected to decrease by 7.9 days in the same time interval (Helmholtz-Zentrum Geesthacht, 2011).

18.2.2 Indirect impacts

Sea level

For more than 100 years, the Baltic Sea Region has been subject to dynamic processes, affecting sea level. One of the most important factors is the isostatic effect, due to post-glacial rebound, which results in an uplift of the Scandinavian plat with simultaneous lowering of the southern Baltic coast. Consequently, sea level in the southern part is rising by about 1.7 mm/year. Secondarily, there is an eustatic sea level rise that can

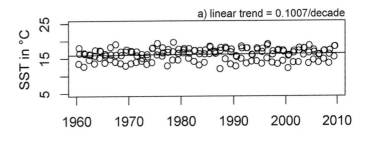

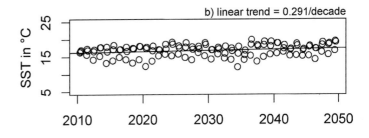

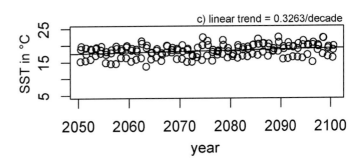

Figure 18.2 Simulated summer (JJA) SSTs of Warnemünde for the periods 1960-2010 (a), 2010-50 (b) and 2050-2100 (c); scenario A1B with linear trends.
Source: Data was simulated and provided by T. Neumann (unpublished data, for methodology see Neumann, 2010).

be explained by changes in either the volume of water in the world oceans or net changes in the volume of the ocean's basins. Both factors have to be considered, discussing changes in sea levels along the Baltic coast (Heino et al., 2008).

Jensen and Mudersbach (2004) analysed annual water level data of four Baltic gauging stations in Germany (Travemünde: 1826-2001, Warnemünde: 1883-2001, Wismar: 1910-2001 and Sassnitz: 1931-2001) and found that the mean sea level (arithmetic mean of the daily measurements at 12 o'clock) of the southern Baltic Sea is increasing. These findings are consistent with those of Hofstede (2007). His results show a positive linear trend in the annual mean sea level at five gauges of the German federal state of Schleswig Holstein since 1925. For the period 1990-2005, the positive trend can be confirmed. Along the German Baltic Coast of Schleswig-Holstein, a linear trend of 1.3 mm/year was calculated for this time interval. At the gauge of Warnemünde, an average sea level rise of 1.23 mm/year has been observed during more than one century (1856-2005) (Fröhle et al., 2011).

Regional simulations of future sea level rise (SLR) highly depend on the driving global models and the scenarios used. Meier, Broman and Kjellström (2004) used the Rossby Centre Atmosphere Ocean (RCAO) model to calculate sea level scenarios for the Baltic Sea. For a global SLR scenario of 0.88 m (higher case scenario), they assume a SLR (winter mean) of about 1 m for the eastern and south-eastern coast of the Baltic Proper and Gulf of Finland (for 2071-2100 relative to the annual mean sea level for 1961-90).

With regard to beach tourism, as well as coastal protection, future SLR plays an important role as it may, combined with storm surges, lead to flooding and increasing coastal retreat.

Water and beach quality

Bathers pay scrupulous attention to the water and beach quality at their holiday destination. Hence, the occurrence and future development of factors such as cyanobacteria, vibrio-type bacteria, macroalgae or jellyfish play an important role. However, due to the complexity of ecosystem interactions, it is very difficult to predict possible future changes.

Cyanobacteria have a long evolutionary history and show a high diversity of taxa. Some species may be harmful to humans and animals as they form massive blooms on water surfaces that produce toxins. Furthermore, they may cause oxygen depletion and alter food webs. Their development is directly or indirectly influenced by many parameters such as temperature, radiation, nutrients, CO_2, wind and salinity. Cyanobacteria are known as the oldest oxygen-evolving photosynthetic microorganisms that are able to fix atmospheric nitrogen. Their production is thus only dependent on phosphate, especially in late spring and early summer. Due to over-enrichment caused by urban, agricultural and industrial development, accelerated rates of primary production were observed in the few last years. There is no direct link to global climate change. However, it is assumed that global warming might favour cyanobacteria as their production is optimal when water temperatures exceed 25 °C (Paerl & Huisman, 2009). For the Baltic Sea, Neumann (2010) simulated that the cyanobacteria season could be prolonged under the scenario A1B by about one month. At the same time, however, he emphasizes that modified nutrient loads due to catchment changes could influence phytoplankton biomass much more than changes in climate. Another factor potentially favouring the growth of cyanobacteria is the salinity that is expected to decrease until 2100. Up to now, it is still very uncertain what future changes in algal blooms will look like.

Like harmful cyanobacteria, vibrio-type bacteria, also contribute to the health risks of bathers as some of them cause, for example, cholera (*vibrio cholerae*). Roijackers and Lürling (2007) assume the risk of infec-

tion to be slightly increased as a result of anticipated climate change.

Another crucial factor for the quality of beaches is the impact of climate change on macroalgae which are part of the so called 'beach wrack'. Occasionally, tourists feel disturbed by accumulations washed ashore especially by the foul smell the material produces when rotting on the beach. Communities along the Baltic Sea spend a lot of money on beach cleanings and need reliable information about future developments. Mossbauer, Haller, Dahlke and Schernewski (2012) analysed how climate change as well as environmental policy could influence macroalgae accumulations on beaches along the German Baltic coastline (see also Chapter 13). Global warming will lead to higher mean water temperatures and, thus, a decrease of sea ice cover, resulting in an extended growing season of macroalgae, though the impact will only be very small. Furthermore, it is expected that the projected sea level rise and extreme weather events will require more coastal protection measures to avoid coastal erosion. Protection measures such as breakwaters could increase the amount of artificial hard substratum in the nearshore zone. Even though these structures represent habitats for macroalgae communities, Mossbauer and colleagues found that their effect on macroalgae growth is marginal (see Chapter 13). In contrast, the authors assume the impact of the implementation of the Baltic Sea Action Plan (BSAP) to be much more serious as it will lead to a reduction of waterborne nitrogen and phosphorus. This reduction would result in a decrease of primary production, an increase in water transparency and, thus, favour the growth of macroalgae in deeper water. The authors consider a doubling of perennial macroalgae accumulation on German Baltic beaches possible.

Jellyfish are another criterion for tourists to judge when considering water and beach quality. Most of them dislike these animals and presume they are generally harmful or at least, a nuisance. This can be ascribed to a problem of perception and knowledge. The majority of European jellyfish are harmless, such as *aurelia aurita*, being the most common species in the Baltic Sea. Harmful *cyanea capillata* and *cyanea lamarkii* are, however, rare. In the context of climate change, tourist destinations fear that accumulations of jellyfish could become more frequent and decrease the number of visitors. Media also deal with this topic and regularly report about plagues of jellyfish that are correlated with climate change. It has to be noticed that such statements are highly speculative. Possibly, warmer water temperatures and reduced ice cover could favour the production of jellyfish as polyps attached to hard substratum are normally scraped off by moving ice. A reduction of ice would allow them to persist and to develop. However, reduced salinity could represent a disadvantage for their production (BACC, 2008; Baumann, 2010; Neumann, 2010). With regard to the complexity of interactions it is not possible so far to give reliable information about future changes of jellyfish occurrence.

18.3 Coastal tourism and climate change – a local case study

18.3.1 Methods and study area

To examine the perception of tourists of the German Baltic Coast, surveys were undertaken at the beaches of Nienhagen, Warnemünde and Markgrafenheide, in close proximity to the German coastal town of Rostock. The study area is located at the coast of the state of Mecklenburg-Vorpommern in the north of Germany (Figure 18.3). It is part of the so-called equilibrium coast and characterized by an alterna-

tion of beaches and cliffs. For many centuries the area has been subject to natural dynamics due to erosion and deposition of material. Tourism along the German Baltic Coast is one of the region's most important economic sectors. The three seaside resorts examined have a long tradition of bathing tourism and attract lots of visitors, especially during the summer season. However, their beaches differ in terms of width, infrastructure and the natural surroundings.

The surveys were carried out in June and July 2010, mainly between 10 am and 3 pm during the week as well as on weekends. The questionnaires were distributed to the beach visitors and collected after one hour. In total, 713 questionnaires were evaluated (Markgrafenheide: 232, Warnemünde: 235, Nienhagen: 246) and analysed with regard to the guests' profile, their perception and their awareness about local climate change at the German Baltic Coast.

18.3.2 The tourists' profile

The tourists questioned represent a heterogeneous group with regard to their age, education, origin and travel behaviour. The questionnaires were answered by 35.7% male and 64.3% female guests. Most of them were aged between 40 and 49 years, had an intermediate school leaving certificate and lived in Mecklenburg-Vorpommern. More than half of the

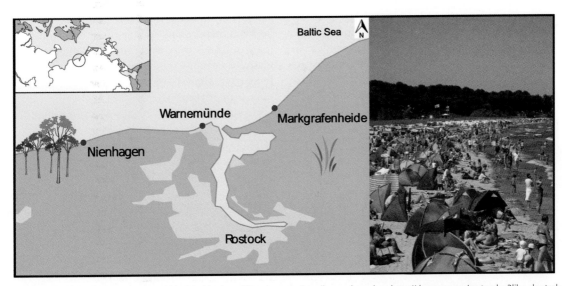

Figure 18.3 Map of the study area within Mecklenburg-Vorpommern (small map: based on http://d-maps.com/carte.php?lib=deutsch land_Lankarte&num_car=2014&lang=de, big map: based on www.planiglobe.com, Photograph by Stefanie Maack).

guests spent their annual leave ($\geq$ five days) at the Baltic Sea. About the same amount of tourists travelled with their family (35.4%) or with a partner (34.3%). Many are regulars and the majority has already visited the Baltic Coast during the last 10 years. Some of them own local holiday houses. Swimming and sunbathing were favourite activities.

18.3.3 Their perception and behaviour

With respect to tourists' perception, the questions were divided into two different parts. Firstly, guests were asked about phenomena perceived during their current holiday, secondly, about changes perceived within the last 10 years. For both parts, a list of possible answers was given. In their current holiday, tourists mainly perceived sun, heat and beach wrack (seaweed and macroalgae), followed by warm water, strong wind, cold water, clouds, eroded seacliffs, narrow beaches, jellyfish, rain, storm surges and the accumulation of sand / formation of dunes. Asked about perceived changes over the last 10 years, the guests named warmer summers (62.3%), warmer water (42.0%), and more beach wrack (32.8%) (Figure 18.4). However, it has to be noted, that 51.6% (368 people) of the tourists questioned did not answer the second part of the question. The majority of them clearly indicated that they are not able to answer it, because they did not visit the Baltic Sea Region often enough to appraise perceived changes.

In a second step, we asked the visitors to indicate whether the perceived phenomena or changes have consequences for their behaviour. Concerning phenomena perceived in the current holiday, the survey showed that only strong rainfalls and lots of jel-lyfish give rise to majority reaction, such as avoiding the beach or looking for another beach section. However, with respect to observed changes over the last 10 years, consequences for the behaviour of the majority of tourists could not be determined.

Additionally, tourists were asked about their perception of coastal protection and beach management measures. It became clear that visitors seldom perceived existing measures, even if they were present at all three beaches (such as groins).

18.3.4 Tourists' awareness and knowledge about local climate change

The survey revealed that the majority of the tourists (80.8%) were not informed about climate change at the Baltic Sea before leaving for the journey. Those who were informed mainly stated television (82.7%), newspaper (54.1%) and radio (29.3%) as information media. Subjects that dominated the media were changes in temperature (72.8%), coastal erosion (44.8%) and storm surges (30.4%). Tourists were also asked to assess the threat of climate change for the Baltic Sea. Most of the guests assumed the threat to be very high, high or moderate (overall 58.3%). About 20.6% thought that the threat is low, very low or non existent. The remaining persons were indecisive (Figure 18.5). Furthermore, the analysis showed that there is a significant correlation between these results and the pre-information of the tourists. Those who were informed about local climate change impacts before their holiday significantly more often assessed the vulnerability of the Baltic Sea to be high or very high.

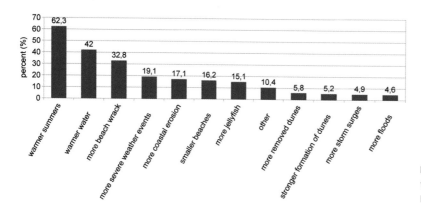

Figure 18.4 Perceived changes along the Baltic Sea within the last 10 years. Percentages refer to 345 cases.

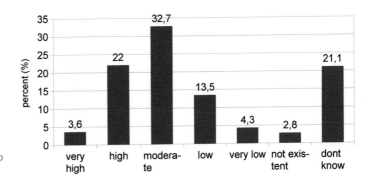

Figure 18.5 How strongly is the Baltic Sea threatened by climate change? Percentages refer to 673 cases (= 100%).

With regard to the observed changes, tourists believed that in general, changes in water and air temperature, as well as thunderstorms, are related to global change. However, again about half of the tourists questioned did not see themselves in a position to judge whether or not certain changes are linked with or caused by global warming. Only 14.4% (103 people) answered this question. Furthermore, climate change was often named as a reason for certain phenomena (like the massive occurrence of ladybugs or the accumulation of jellyfish) although the correlation is not supported by current science.

18.4 Discussion of findings

18.4.1 Perceiving coastal climate change as a tourist – the process and its limits

Human perception is a physiological and psychological process that is basically characterized by the stimulation of sensory receptors in the eyes, ears, nose, tongue or skin. Additionally, various factors like interests, age, feelings, expectations, experiences, morals and needs of a person influence their perception of the environment. Thus, every human being perceives a situation differently even if the objective conditions are the same. Everybody has his or her own subjective and filtered image of the reality that is constantly recreated and modified. Furthermore, perception is directly linked to the process of remembering. As we are able to retain information in the memory we can compare situations and experiences, though details are often forgotten or wrongly remembered. Our perception is also limited in time and space so that it is for example hardly possible to perceive very slow events. Yet, our behaviour and opinion is highly influenced

by what we perceive (Miller, 1998; Schermer, 2006; Myers, 2008; Werlen, 2008; Pritzel, Brand & Markowitsch, 2009).

Climate change is a slow process and changes can only be detected over a long time period. Long-term observations show that changes in physical parameters could be measured over the last few years, although they are still very small. Depending on the time interval analysed as well as the quality and quantity of data, changes are more or less pronounced.

In our perception analysis we found gaps and inconsistencies of which the majority can be ascribed to the complexity of the topic and to the process of perception itself. The survey showed that actual weather conditions are the major limiting factor and of most interest for the tourists questioned. Changes in the coastal climate within the last 10 years are, however, hardly perceived by the guests. Only 48.4% answered the corresponding question, with those being positive about it indicating that they mainly perceived warmer summers or warmer water. The majority of tourists responded that they did not see themselves in a position to answer the question. Next to changes in weather conditions, we found that also coastal protection and beach management measures were comparatively less frequently perceived even though obviously being present. Despite their uncertainties, almost all tourists questioned indicated an opinion concerning the general threat of climate change for the Baltic Sea.

The results make clear that tourists' perception and activities take place on other scales than climate change does. The latter is a slow, gradual process whereas the bathers' perception and behaviour very much depends on the situation and is limited in time and space. To detect changes in climate, averaged long-time data have to be analysed and evaluated.

However, tourists do not observe the coast on their holiday in detail, or for a long time, as they are only staying for some days or weeks. Even if guests return every year, it is hard for them to compare situations and to remember details. It is an immanent problem of perception and remembering that we forget details and create a filtered image of the reality. We are not like computers that memorize information in a database without loss. People especially remember events and situations that are extraordinary or have a high amplitude. Extreme events such as storm surges or floods are kept in mind, whereas small changes of the environment are not perceived or remembered.

Furthermore, perception is highly influenced by individual interests, sensitivity and knowledge. Former analyses show that bathers are very sensitive to issues such as beach wrack or jellyfish. They consider them as pollution of the beach and feel disturbed by them (Dolch, 2004). It is possible, that due to sensitivity, guests feel able to observe a change even though it does not exist. Scientifically, an increase in beach wrack cannot be proved. It can be assumed that single extreme events, such as algae blooms or the massive occurrence of jellyfish are remembered and generalized. The fact that coastal protection and beach management measures are also comparatively less perceived could be explained with the attention and interest of holiday guests. Having arrived at their destination, they want to relax, not pay immoderate attention to their surroundings and avoid thinking about complex issues such as climate change or coastal protection.

Other results can be interpreted in the context of information about climate change. As only a few guests were informed about the topic in advance, the majority of tourists were not able to answer the question whether or not certain parameters are influenced by climate change. With regard to the threat of climate change for the Baltic Sea, the opinion of the tourists was consistent with an objective evaluation of the regional planning association Vorpommern. Most of the tourists assumed the threat to be very high, high or moderate. The named association also indicates that the region is more vulnerable than others, especially with regard to projected sea level rise and the increase in temperatures (Regionaler Planungsverband Vorpommern, 2011). As indicated above, there is also a significant correlation between these results and the pre-information of the tourists. Those who were informed about local climate change impacts before their holiday, significantly more often assessed the vulnerability of the Baltic Sea to be high or very high.

This shows that the perception and opinion of tourists is not only influenced by the factors mentioned, but obviously also by the media that communicates scientific results to the public. The latter contributes to the construction of a certain image of climate change and its impacts. Still, often weather is confused with climate, so that single weather extremes are interpreted as climate change impacts in the media. Although the latest IPCC Special Report states that the 'Observed changes in climate extremes reflect the influence of anthropogenic climate change … ' (IPCC, 2012) there is no direct link between a single extreme event such as a heat wave or a storm surge and climate change. Furthermore, headlines in some media create the impression that there are doubts about an ongoing global warming – 'Proved: There is no climate crisis' (Science & Public Policy Institute, 2008), 'Climate chaos? Don't believe it' (Monckton, 2006). Such statements lead to uncertainties that were reflected in the results of the survey. Some guests wrote that they do not believe in climate change or that there is no climate change at all. Climate change proves to be a highly emotive subject that can provoke very subjective discussions or even defensiveness.

18.4.2 Adaptive recreational beach management – reflections and visions

With regard to climate change, coastal tourism in the South Baltic area might face opportunities due to more favourable climate conditions and an extension of the season, but also risks due to a rise of sea level and other secondary climate impacts. Coastal erosion, altered impressions of beaches due to intensified coastal protection measures, and increasing costs for maintenance of beaches might be the result. Recreational beach management will be in need of new and adapted strategies in times of climate change to keep the region's attractiveness of beach tourism destinations. Given that demand and satisfaction of guests are highly valuable for the tourism sector, visitors' perception of beaches and projected coastal climate changes could provide a possible instrument for the development of regional climate change adaptation strategies.

However, the findings show that the process of perception within holidays - as a period of emotion and expectations - holds clear weaknesses that might undermine its usefulness for rational and strategic adaptation concepts. Visitors to a destination can only be seen as temporary observers of site-specific conditions and changes as they only spend a limited time on-site. In the function of temporary observers, guests can for example reliably judge situational conditions of infrastructure and seasonal touristic offers at a given instant – their holiday period. Hence, visitor questionnaires are regularly and successfully used to monitor seasonal or new offers. In contrast, the reliability of the perception of changes over time is problematic as memories over the years naturally fade and are being mixed, even if visitors are regulars and willing to judge changes or processes at their destination objectively. Also, tourists cannot be seen as a chronologically stable population for perception surveys over time. Singular events (e.g. rainy holiday, hot period, large amount of jellyfish, etc.) meaning a change of the accustomed natural environment of a destination might result in a behavioural adaptation of guests that in turn constantly changes the survey population. Perception analysis on beach tourists only involves those guests who are satisfied with the given conditions at the relevant time. It excludes those who already changed their behaviour in favour for non-beach related activities such as cycling or walking or those who even switched to another destination due to unsatisfactory changes of the climate or the beach environment. Aiming at a stable population for perception analyses, it might be more meaningful to question coastal residents in place of coastal tourists as their perception is more constant over time and less influenced by emotional expectations. Residents are also less influenced by changing tourism trends, and could have more interest in the sustainable and adaptive development of the destination as it constitutes their home.

To comprehensively answer the question whether tourists' perceptions could be of general use and value for the development of adaptation strategies on a regional destination level, it is also necessary to take a look at tourism's strategic mechanisms. Medium or long-term strategy directions are mostly incorporated in tourism concepts with a larger spatial scale. Regionally or locally, tourism is highly capacity-oriented and subjected to economic and social parameters, and the interest of local politics. If existing infrastructure is used to capacity, medium or long-term strategic concepts comprising potential but arguable future conditions are often of secondary relevance. Touristic strategy concepts do regularly involve general future target groups but the level of information and awareness of future guests with respect to future challenges (such as climate change) is rather subordinate.

To get empirically sound results when addressing tourists as a target for surveys in the context of climate adaptation, it could be necessary to turn away from theory and develop comprehensive scenarios with pictures for future 'regional coastal tourism' (e.g. for 2050). They could also be used as tools to appraise the acceptance of prospective corresponding adaptation measures, for example, intensified coastal protection structures or smaller beach nourishments. Buzinde, Manuel-Navarette, Yoo and Morais (2010) analysed the tourists' perception of severely eroded coastal landscapes and of existing beach restoration measures (e.g. geo-tubes as breakwaters) and showed that tourists react in different ways to the measures. Surveys like this could also be used to visualize future adaptation measures.

18.5 Conclusions

The results of the conducted perception analysis show a heterogeneous picture with several difficulties, and do not allow clear conclusions. The statements of the visitors questioned reflect various uncertainties due to the limits and interactions discussed above (scale and amplitude, process of remembering, interest, sensitivity, attention and information of the guests, influence of media). The survey results indicate that looking at *present* visitors can give indicators for *future* adaptive management strategies only to a limited extent. We therefore conclude that the perception of tourists – even though being an important target variable for destinations – might only be of limited usefulness for long-term strategic adaptive concepts with regard to future climate adaptation. However, the discussion of results gives valuable clues for future analyses (e.g. residents' perception of changes). To appraise the tourists' acceptance of possible future adaptive beach management measures, the development of simple and comprehensive scenarios could be worth testing in future tourist surveys. This method could also be used to question future tourists in their home regions or countries or even at their present holiday destinations (e.g. the Mediterranean) to analyse the potential, interest and level of acceptance of future guests.

Finally, with regard to future adaptive strategies it will be essential for local tourism communities or regional tourism structures to develop their own strategic concepts and visions based on their given natural and economic resources. Regardless of present guests, the concepts might also have to include a more detailed picture of the desired future guests and define the willingness to share the developed visions for the destination/region with future guests. Sharing visions could be seen as a precautionary act for destinations and ensure and guarantee the support and acceptance of (beach) management measures by an important target variable - tourists. If adaptation strategies and measures have a sound explanation and justification from the tourist's perspective, changes and adaptive measures have a high potential of being accepted and supported.

With respect to the geographical relevance of the survey, future recreational beach management is likely to be defined by localized conditions as well as by the characteristics of regional climate impacts. This survey was based on the projected regional climate impacts on the South Baltic region. Yet, the results and recommendations are likely to be transferable to other Baltic and European regions.

Acknowledgements

The survey was supported by the projects RADOST (Regional adaptation strategies for the German Baltic Coast), funded by the German Federal Ministry of Education and Research (grant number 01LR0807K), BaltCICA (Climate Change: Impacts, Costs and Adaptation in the Baltic Sea Region) and Baltadapt (Baltic Sea Region Climate Change Adaptation Strategy), funded by the EU within the Baltic Sea Region Programme. Furthermore, the authors would like to thank Thomas Neumann for data provision and processing.

References

ACAR, C. and SAKICI, C., 2008. Assessing landscape perception of urban rocky habitats. *Building and Environment*, 43 (6), pp.1153–1170.

BACC Author TEAM, 2008. *Assessment of climate change for the Baltic Sea Basin*. Berlin, Heidelberg: Springer.

Baumann, S., 2010. *Quallen an der deutschen Ostseeküste – Auftreten, Wahrnehmung, Konsequenzen*. IKZM-Oder Berichte 59, Leiden: EUCC.

Buzinde, C.N., Manuel-Navarette, D., YOO, E.E. and Morais, D., 2010. Tourists' perception in a climate of change: Eroding destinations. *Annals of Tourism Research*, 37 (2), pp.333–354.

Cervantes, O., Espejel, I., Arellano, E. and Delhumeau, S., 2008. Users' perception as a tool to improve urban beach planning and management. *Environmental Management*, 42 (2), pp.249–264.

CLM Community, 2008. *Climate Limited-area Modelling Community*. [online] Available at: <http://www.clm-community.eu> [Accessed 08 November 2011].

Döscher, R., Willén, U., Jones, C., Rutgersson, A., Meier, H.E.M., Hansson, U. and Graham, L.P., 2002. The development of the coupled regional ocean-atmosphere model RCAO. *Boreal Environment Research*, 7, pp.183–192.

Dolch, T., 2004. Die Auswirkungen der Wasserqualität auf den Tourismus – Eine Studie am Beispiel des Oderästuars. In: Schernewski, G. and Dolch, T., eds. 2004. *The Oder Estuary – against the background of the European Water Framework Directive*. Marine Science Reports no. 57. Warnemünde: InstitutfürOstseeforschung.

Frochot, I. and Kreziak, D., 2008. Customers' perception of ski resorts' image: implications for resorts' positioning strategies. *Tourism and Hospitality Research*, 8 (4), pp.298–308.

Fröhle, P., Schlamkow, C., Dreier, N. and Sommermeier, K., 2011. Climate change and coastal protection: Adaptation strategies for the German Baltic Sea Coast. In: Schernewski, G., Hofstede, J., Neumann, T., eds. 2011. *Global Change and Baltic Coastal Zones*. Coastal Research Library. Berlin, Heidelberg: Springer, pp.103–116.

Gräwe, U. and Burchard, H., 2011. Regionalisation of climate scenarios for the western Baltic sea. In: Schernewski, G., Hofstede, J., Neumann, T., eds. 2011. *Global Change and Baltic Coastal Zones*. Coastal Research Library. Berlin, Heidelberg: Springer, pp.3–22.

Hagen, E. and Feistel, R., 2008. Baltic climate change. In: Feistel, R., Nausch, G. and Wasmund, N., eds. 2008. *State and Evolution of the Baltic Sea, 1952-2005: A Detailed 50-Year Survey of Meteorology and Climate, Physics, Chemistry,*

Biology, and Marine Environment. Hoboken: Wiley, pp.93–120.

Heino, R., Tuomenvirta, H., Vuglinsky, V.S. and Gustafsson, B.G., 2008. Past and current climate change. In: BACC Author Team, eds. 2008. *Assessment of the climate change for the Baltic Sea Basin*. Berlin, Heidelberg: Springer, pp.35–131.

Helmholtz-Zentrum Geesthacht, 2011. Norddeutscher Klimaatlas. [online] Availableat: <http://www.norddeutscher-klimaatlas.de> [Accessed 07 November 2011].

HELCOM (Helsinki Commission), 2007. *Climate Change in the Baltic Sea Area- HELCOM Thematic Assessment in 2007*. Baltic Sea Environment ProceedingsNo. 111.

Hofstede, J., 2007. Entwicklung des Meeresspiegels und der Sturmfluten: Ist der anthropogene Klimawandel bereits sichtbar? In: Gönnert, G., Pflüger, B. and Bremer, J.-A., eds. 2007. *Von der Geoarchäologie über die Küstendynamik zum Küstenzonenmanagement*.Coastline Reports 9. Leiden: EUCC, pp.139–148.

IPCC, 2012: Summary for Policymakers. In: *Managing the Risks of Extreme Events and Disasters to Advance Climate Change Adaptation*[Field, C.B., V. Barros, T.F. Stocker, D. Qin, D.J. Dokken, K.L. Ebi, M.D. Mastrandrea, K.J. Mach, G.-K. Plattner, S.K. Allen, M. Tignor, and P.M. Midgley (eds.)]. A Special Report of Working Groups I and II of the Intergovernmental Panel on Climate Change. Cambridge University Press, Cambridge, UK, andNew York, NY, USA, pp. 3–21.

Jacob, D., Göttel, H., Kotlarski, S., Lorenz, P. and Sieck, K., 2008. *Klimaauswirkungen und Anpassung in Deutschland – Phase 1: Erstellung regionaler Klimaszenarien für Deutschland*. Berlin: Umweltbundesamt.

Jensen, J. and Mudersbach, C., 2004. Analyses of variations in water level time-series at the southern Baltic Sea coastline. In: Schernewski, G. and Löser, N., eds. 2004. *BaltCoast 2004 – Managing the Baltic Sea*. Coastline Reports 2. Warnemünde: EUCC, pp.175–184.

Nakicenovic, N., Alcamo, J., Davis, G., deVries, B., Fenhahn, J., Gaffin, S., Gregory, K., Grübler, A., Jung, T.Y., Kram, T. et al., 2000. *IPCC Special Report on Emissions Scenarios (SRES)*. Cambridge University Press.

Marin, V., Palmisani, F., Ivaldi, R., Dursi, R. and Fabiano, M., 2009. Users' perception analysis for sustainable beach management in Italy. *Ocean & Coastal Management*, 52 (5), pp.268–277.

Matzarakis, A. and Tinz, B., 2008. Tourismus an der Küste sowie in Mittel- und Hochgebirge: Gewinner und Verlierer. In: Lozán, J.L., Grassl, H., Jendritzky, G., Karbe, L., Reise, K., eds. 2008. *Warnsignal Klima: Gesundheitsrisiken: Gefahren für Pflanzen, Tiere und Menschen*. Hamburg: Wissenschaftliche Auswertungen, pp.11–18.

Max-Planck-Institute for Meteorology, 2008. *Model and Data: CLM*. [online] Available at: <http://www.mad.zmaw.de/projects-at-md/sg-adaptation/clm/> [Accessed 07 November 2011].

Mecklenburg-Vorpommern Tourist Board – Tourismusverband Mecklenburg-Vorpommern, 2009. *Schriftenreihe des Tourismusverbandes MV e.V.: Tourismusstandort Mecklenburg-Vorpommern – Trends, Untersuchungen, Fakten*. Rostock: Tourismusverband Mecklenburg-Vorpommern.

Meier, H.E.M., Broman, B., and Kjellström, E., 2004. Simulated sea level in past and future climates of the Baltic Sea. *Climate Research*, 27, pp.59–75.

Miller, R., 1998. *Umweltspychologie*. Stuttgart: Kohlhammer.

Ministerium Für Wirtschaft, Arbeit und Tourismus Mecklenburg-Vorpommern, ed. 2010. *Fortschreibung der Landestourismuskonzeption Mecklenburg-Vorpommern 2010*. [online] Availableat: <http://service.mvnet.de/_php/download.php?datei_id=27585> [Accessed 14 November 2011].

Mossbauer, M., Haller, I., Dahlke, S. and Schernewski, G., 2012. Management of stranded eelgrass and macroalgae along the German Baltic coastline. *Ocean & Coastal Management*, 57, pp.1–9.

Myers, D.G., 2008. *Psychologie*. Berlin: Springer.

Neumann, T., 2010. Climate-change effects on the Baltic Sea ecosystem: A model study. *Journal of Marine Systems*, 81 (3), pp.213–224.

Neumann, T. and Friedland, R., 2011. Climate change impacts on the Baltic Sea. In: Schernewski, G., Hofstede, J., Neumann, T., eds. 2011. *Global Change and Baltic Coastal Zones*. Coastal Research Library. Berlin, Heidelberg: Springer, pp. 23–32.

Parry, M.L., Canziani, O.F., Palutikof, J.P., Van Der Linden, P.J., and Hanson, C.E., eds. 2007. *Climate Change 2007: Impacts, Adaptation and Contribution of Working Group II to the Fourth Assessment Report of the Panel on Climate Change (IPCC)*. Cambridge, UK and New York, USA: Cambridge University Press.

Paerl, H.W. and Huisman, J., 2009. Climate change: a catalyst for global expansion of harmful cyanobacterial blooms. *Environmental Microbiology Reports*, 1 (1), pp.27–37.

Pritzel, M., Brand, M. and Markowitsch, H.J., 2009. *Gehirn und Verhalten – Ein Grundkurs der physiologischen Psychologie*. Heidelberg:Spektrum.

Raymond, C.M. and Brown, G., 2011. Assessing spatial associations between perceptions of landscape value and climate change risk for use in climate change planning. *Climate Change*, 104 (3-4), pp.653–678.

Regionaler Planungsverband Vorpommern, 2011. *Raumentwicklungsstrategie – Anpassung an den Klimawandel und*

Klimaschutz in der Planungsregion Vorpommern. Greifswald: Regionaler Planungsverband Vorpommern.

Roca, E., Villares, M. and Ortego, M.I., 2009. Assessing public perceptions on beach quality according to beach users' profile: A case study in Costa Brava (Spain). *Tourism Management*, 30 (4), pp.598–607.

Rockel, B., Will, A. and Hense, A., 2008. The regional climate model COSMO-CLM. *Meteorologische Zeitschrift*, 17 (4), pp.347–348.

Roijackers, R.M.M. and Lürling, M., 2007. *Climate Change and Bathing Water Quality*. Wageningen: Environmental Science group, Aquatic Ecology and water quality chair.

Schermer, F.J., 2006. *Lernen und Gedächtnis*. Stuttgart: Kohlhammer.

Schuchardt, B., Wittig, S., Mahrenholz, P., Kartschall, K., Mäder, C., Hasse, C. and Daschkeit, A., 2008. *Deutschland im Klimawandel – Anpassung ist notwendig*. Berlin: Umweltbundesamt.

SPPI (Science & Public Policy Institute), 2008. *Proved: There is No Climate Crisis*. [online] Available at: <http://scienceandpublicpolicy.org/press/proved_no_climate_crisis.html> [Accessed 30 November 2011].

Siegel, H., Gerth, M. and Tschersich, G., 2008. Satellite-derived sea surface temperature for the periods 1990-2205. In: Fcistel, R., Nausch, G. and Wasmund, N., eds. 2008. *State and Evolution of the Baltic Sea, 1952-2005: A Detailed 50-Year Survey of Meteorology and Climate, Physics, Chemistry, Biology, and Marine Environment*. Hoboken: Wiley, pp.241–264.

Spekat, A., Enke, W. and Kreienkamp, F., 2007. *Neuentwicklung von regional hoch aufgelösten Wetterlagen für Deutschland und Bereitstellung regionaler Klimaszenarios auf der Basis von globalen Klimasimulationen mit dem Regionalisierungsmodell WETTREG auf der Basis von globalen Klimasimulationen mit ECHAM5/MPI-OM T63L31 2010 bis 2100 für die SRES-Szenarios B1, A1B und A2*. Berlin: Umweltbundesamt.

Statistisches AMT Mecklenburg-Vorpommern, 2011. *Statistische Berichte: Handel, Tourismus, Gastgewerbe*. Schwerin: StatistischesAmt Mecklenburg-Vorpommern.

Monckton, C., 2006. Climate chaos? Don't believe it. *The Telegraph*, [online] 5 November. Available at: <http://www.telegraph.co.uk/news/uknews/1533290/Climate-chaos-Dont-believe-it.html> [Accessed 30 November 2011].

Vaz, B., Williams, A.T., Da Silva, C.P. and Phillips, M., 2009. The importance of user's perception for beach management. *Journal of Coastal Research*, 56, pp.1164–1168.

Vogt, C.A. and Andereck, K.L., 2003. Destination perception across a vacation. *Journal of Travel Research*, 41 (4), pp.348–354.

Werlen, B., 2008. *Sozialgeographie: Eine Einführung*. Stuttgart: UTB.

Zebisch, M., Grohtmann, T., Schröter, D., Hasse, C., Fritsch, U. and Cramer, W., 2005. *Klimawandel in Deutschland – Vulnerabilität und Anpassungsstrategien klimasensitiver Systeme*. Berlin: Umweltbundesamt.

19 Experiences in Adapting to Climate Change and Climate Risks in Spain

Jorge Olcina Cantos

Alicante University, Alicante, Spain

19.1 Spain – a country at risk. Increasing vulnerability and exposure

Spain is a country with a high risk from natural hazards. Some of its territories are ranked at the top of the European classification of geographical areas at risk and have been included in the report on natural and technological hazards in Europe (Schmidt-Thomé, 2005). This is owing to the country's complex, difficult terrain and a dynamic, growing population that accumulates to a great extent along the coasts.

In fact, Spain is one of the territories of Europe that has been most affected by natural hazards, due to its geographical position as a peninsula surrounded by seas, its topography and historical settlement structure. When calculating the economic losses recorded annually in Spain from 1987 to 2001, earthquakes and flooding alone have cost the country €760 million, 98% of which has been attributed to flooding, the principal natural hazard in the country.

By regions, the Valencian Community, Catalonia, the Balearic Islands, the Canary Islands, Andalusia and Murcia have been the sorce of the majority of all economic natural losses (again, mainly flooding), between 1987 and 2001. In addition, it is estimated that in the years to come, losses caused by flooding will continue to be high in Spain, especially in the autonomous communities mentioned above. The Spanish Association of Professional Geologists has just updated the 'Economic and Social Impact of Geologi-

cal Risks in Spain', a report drafted in 1987 by the Geological Survey of Spain (IGME). According to the new report, the damage caused by flooding from 1986–2016 is estimated to reach € 55 billion. The cost of natural hazards in Spain represents 2% of the country's annual budget.

It is important to note that the risk caused by natural hazards increases when human exposure to such hazards rises. Thus, in addition to floods, droughts and windstorms, heat waves and avalanches have become two new risks that often claim a high number of victims. Tornados have also become much more frequent since 1995, causing severe property damage in the affected areas.

There were a total of 980 victims of natural hazards in Spain from 1995 to 2010. Of this total, 280 (28%) were victims of flooding, the natural hazard with the greatest social and economic impact in the country and the one which has led to the highest quantity of adaptation and mitigation projects (Table 19.1).

Since the mid-twentieth century, Spain has experienced an interesting phenomenon in terms of where the territories at risk are located: risk has become increasingly concentrated along the coastal areas. In other words, the risk has been 'coastalized'. The socioeconomic reduction of agriculture – the activity most exposed to climate hazards – has shifted the vulnerability and exposure to natural dangers from the countryside to the city, and within the city, the rise in recreational activities and tourism along the shores and islands have made these areas highly exposed

Climate Change Adaptation in Practice: From Strategy Development to Implementation, First Edition.
Edited by Philipp Schmidt-Thomé and Johannes Klein.
© 2013 John Wiley & Sons, Ltd. Published 2013 by John Wiley & Sons, Ltd.

Table 19.1 Number of fatalities in Spain caused by natural hazards (1995–2010)

	1995	1996	1997	1998	1999	2000	2001	2002	2003	2004	2005	2006	2007	2008	2009	2010	TOTAL
Flooding	22	110	40	0	5	14	9	13	9	7	8	9	11	6	5	11	280
Storms	19	13	14	2	20	28	17	12	8	6	8	9	4	3	11	6	180
Forest fires	8	1	4	4	8	6	1	6	11	4	19	8	1	1	11	9	102
Landslides	7	8	2	0	0	0	1	1	2	0	0	5	2	1	2	2	33
Heat waves	0	0	0	0	1	0	0	0	60	23	4	14	0	0	0	2	104
Avalanches	7	1	0	0	0	4	2	4	4	5	1	0	0	4	3	11	66
Cold spells and snow storms	0	2	5	1	0	2	4	0	0	3	3	0	0	0	1	1	22
Ocean storms	19	13	13	36	17	37	27	15	5	20	NDA	NDA	NDA	4	2	5	213
TOTAL	82	148	78	43	51	91	61	51	99	68	43	45	18	19	35	48	980

Source: Ministerio de Agricultura, Alimentación y Medio Ambiente, 2010. Available at: http://www.marm.es/es/calidad-y-evaluacion-ambiental/temas/informacion-ambiental-indicadores-ambientales/indicadores-ambientales-perfil-ambiental-de-espana/

and vulnerable. It is important to note that the risk of natural hazards, especially atmospheric ones, was greater in 2010 than it had been 30 years previously. This is not down to an increase in natural hazards – that is, a greater frequency of extreme events – but to the increase of the population and their exposure to natural hazards in urban areas along the Spanish coast.

In fact, since the 1980s, risks to humans have become the most significant ones when assessing climate hazards. An important factor in the rising exposure and vulnerability to extreme atmospheric events is the increase in the number of homes in certain Spanish regions since the 1980s, especially between 1995 and 2007. The so-called 'real estate boom' of the last decade and a half has mainly been focused on the Spanish Mediterranean coast. Outside the capital city of Madrid, the epicentre of residential construction in Spain has been along the Mediterranean (Figure 19.1).

The enormous development of residential constructions in Spain is one of the causes for the country's rising vulnerability and exposure to the climate hazards referred to above. In other words, in the past two decades, constructions have surpassed what is rationally sustainable in Spain, and some of these constructions have been built in areas at risk of natural hazards. Most of these areas are prone to flooding, but construction has also taken place in areas at risk of droughts, storm surges and landslides.

Many areas of the Spanish Mediterranean shoreline and the Canary Islands, especially Tenerife and Gran Canaria, have watched as their river margins, floodable areas and areas with poor drainage have been occupied by infrastructure, recreational areas (campsites) and residences over the past few decades. This has occurred in spite of the introduction of the 1985 Water Act – along with its Hydraulic Public Domain Regulation – and the 1998 Soil Act, not to mention laws passed in certain autonomous communities in the past 20 years related to soil, land planning and environmental impacts, all of which made such planning and development projects clearly illegal. There is yet another worrisome fact. Many of the recorded victims of floods in Spain have been foreign residents who were spending time in the country working or on holiday. This shows that these social groups are unaware of the risk and points to the lack of effective social communication to warn these individuals of the hazards arising from extreme atmospheric conditions in Spain.

19.2 Climate Change in Spain – an increase in extreme weather conditions

The expected effects of climate change in the Mediterranean basin will not help to reduce risk levels, quite

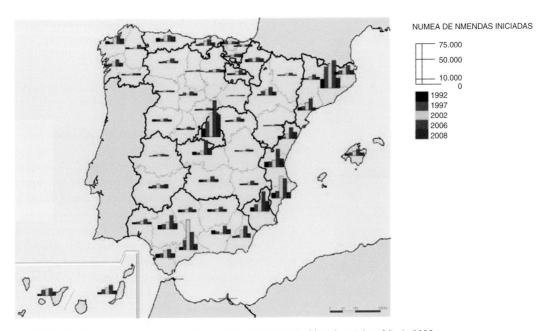

Figure 19.1 Residential constructions begun in Spain, 1992–2008. Sustainable Urban Atlas of Spain 2009.
Source: OSE, 2009. Available at: http://www.sostenibilidad-es.org/sites/default/files/_Informes/anuales/2009/sostenibilidad_2009-esp.pdf

the contrary. The Intergovernmental Panel on Climate Change (IPCC) (2007) indicates an increased frequency of developing extreme weather events, which will contribute to increased risks. The reduced availability of water for a growing population, rising temperatures and humidity in the air, that reduce climate comfort in this region, are other consequences of climate change that will affect people, settlements and economic activities in Spain.

We are witnessing a turning point in the recent history of the Mediterranean basin, because the consequences of climate change will create a scenario of greater risk in the coming decades (see Figure 19.2). So, any delay to programmes for adaptation and mitigation, climate change and associated risks, poses further risks to the socio-economic dynamics of the Mediterranean countries.

All these features make Spanish territory – in the Mediterranean region – an ideal laboratory for the analysis of climate change and its associated effects in the regional context (Olcina Cantos, 2009a). Climate risks already currently affect the functioning of societies and could become even more severe in the decades to come (see Figure 19.3).

Temperature change and the alteration of other climate factors that can be noted on the land surface have been registered over the past three decades in Spain, as part of global warming. Since the 1980s, temperatures and water levels have risen and ice coverage

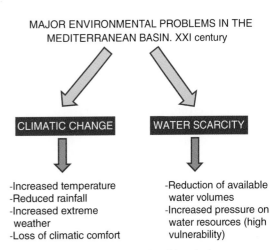

Figure 19.2 Major environmental problems in the Mediterranean basin in the 21st century.
Source: Compiled by the author

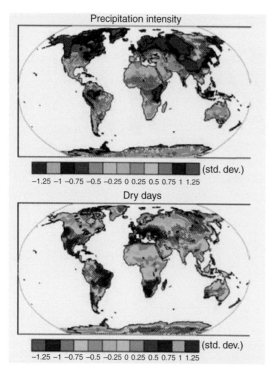

Figure 19.3 Changes in the quantity and quality of global precipitations. Note how rainfall is expected to become more torrential and the number of dry days per year is expected to increase along the Mediterranean.
Source: IPCC, 2007; http://www.ipcc.ch/publications_and_data/publications_ipcc_fourth_assessment_report_synthesis_report.htm.

has decreased, albeit slightly. These are the findings in the Ministry of Environment's official report, a preliminary assessment of the impact of climate change in Spain (MMA, 2005). The report was updated in 2007 to incorporate the regionalized climate scenarios to Horizon 2100 (AEMET, 2009a; b).

According to the climate models applied to the peninsula, these conditions will continue to worsen as the climate becomes increasingly irregular, a feature that is already common in climates with subtropical roots such as those in most of the Iberian Peninsula and the islands. The most outstanding features of the future climate evolution in Spain can be summarized as follows (INM & OECC, 2007):

• Progressive rise of average temperatures.
• More pronounced warming in summer than in winter.
• Inland's summer temperatures higher than along coasts and islands.

• Greater frequency of weather anomalies, especially when summer temperatures are at their highest.
• Diminished precipitation.
• Greater reduction of rainfall during spring; likely increase of winter rain in the west and autumn rain in the northeast.
• Likely **increase** of **climatic hazards** (torrential rains, heat waves, droughts, etc.).

If faced with a temperature rise of 3–4 °C and a 20–40% drop in precipitation compared to current values, human beings can adapt, though adapting will mean taking measures to mitigate these effects. However, for the country, its inhabitants and its economic activities, the worst effect would undoubtedly be the 'extreme' nature of weather conditions. The possibility of Spain frequently being subjected to torrential rainfall and flooding, severe droughts, intense summer heat waves, windstorms caused by extreme low pressure or occurrences on the fringe of anticyclones in the Mediterranean area creates a scenario of extreme socio-economic vulnerability. If the forecasts of the climate modelling prove correct, the increase in terms of annual economic damage, and possibly in the number of victims, could be striking. The area that could be most affected by the increase in climatic hazards is the Spanish Mediterranean shore.

According to the IPCC (2007) and to the Spanish Meteorological Office, climate change is likely to produce a negative effect on the existing water resources of the Spanish Mediterranean Coast. Models indicate that the increase in mean temperature (and therefore of evapotranspiration) and especially the reduction in precipitation (up to 20% towards the end of the 21st century) may add important stresses to the population and economic activities, especially agriculture and tourism. The decline in precipitation would be more intense in the southern parts of the study area. Presently (IPCC, 2007), Spain enjoys a resource availability of water per person per year estimated at 3000 m^3/person/year, for a demand of 2000 m^3/person/year. At the end of the 21st century, the IPCC estimates that water availability would be reduced to just 450 m^3/person/year, for a demand moderately higher than today.

Variations in climate can be translated into variations in total annual runoff. For example, an increase in the frequency of droughts will not only be caused by less rain, but also by increases in evapotranspiration and therefore in runoff reductions. The aim of this section is therefore to provide some figures regarding the future availability of water resources for the tourist sector in our study area. These figures give an indication of likely physical scarcity conditions which will have to be integrated with sensitivity and adaptive capacity to produce a vulnerability assessment of the Spanish Mediterranean coast towards water scarcity situations.

Our indicators in this case will be those climatic stimuli that are relevant for the estimation of the amount of surface water and groundwater potentially available for supplying the tourist demand of our study area. The climatic stimuli selected are precipitation, temperature and evaporation.

The Spanish Meteorological Agency (AEMET) published in 2009 a report, 'Generation of regionalised climate change scenarios for Spain,' including data on maximum and minimum temperatures and on precipitation patterns, using 10 regional models developed through the downscaling of five global models, and using two emission scenarios, A2 and B2. For the 2011–2040 period, emission scenarios show little influence on the increasing temperature, while in the latter period (2071–2100) the differences are substantial, with increases more marked for A2 than for B2. On the other hand, there is also a significant pattern over a year, with increases much more pronounced in summer than in winter and early spring. Temperature increases at the Spanish Mediterranean

coast for 2071–2100, compared with 1961–90, are approximately 1.5 to 3 °C in March for the Catalan, Valencia and Cadiz (Andalusia) areas, and more than 7 °C in August in the Balearic Islands and the Catalan coast. All these projections appear to have a high degree of confidence and spatial and temporal coherence. According to these figures, a significant increase in evapotranspiration, mostly in summer, can be expected.

As far as temperature is concerned, 90% of the projections show an increase between 3 °C and 6 °C for the period 2071–2100 compared to the period 1961–90. This increase is slightly smaller in the Balearic Islands due to maritime influences. On the other hand, precipitation patterns present more mixed results. During the first half of the 21st century there is no clear warning signal. However, in the second half of the century average precipitation in Peninsular Spain may decrease between 15 and 30% compared to the values of 1961–90.

In addition to increased temperatures and reduced rainfall, two climatic elements that will undergo significant changes are extreme temperatures and heavy rains within a year. They indicate, as noted above, an intensification of the extreme weather conditions in the Spanish Mediterranean coast and a loss of comfort climate with enormous importance for the development of tourism (JRC, 2009). The main changes in climatic elements for the Spanish Mediterranean regions on the horizon 2100 – A1B emissions scenario, are shown in the table below (see Table 19.2).

Table 19.2 Main changes in climatic elements of the Spanish Mediterranean regions, 2100

Region	Maximum Temperature (°C)	Minimum Temperature (°C)	Change in Warm Days Per Year (%)	Variation in Total Annual Precipitation (%)	Variation in Heavy Rainfall (%)
Catalonia	+4°	+3°–4°	+25–45	−10–25	+5
Balearic Islands	+3°	+3°–4	+−35–60	−10–30	+5
Valencian Community	+4°	+3°–4°	+30–50	−20–35	+5
Murcia	+4°–5°	+3°–4°	+35–50	−20–40	+10
Andalusia	+4°–6°	+4°	+35–50	−15–40	+10

Source: AEMET, 2009a; b.

In the Spanish context, several assessments have attempted to predict reductions of water flows as a consequence of climate change. The White Paper on Water in Spain (MMA, 2001) calculated a reduction of water for irrigation in the river basins of the Spanish Mediterranean coast. By 2030, that decrease would be between 3 and 31% in the Catalan Inner Basins, 9 and 22% in the Júcar basin, 11 and 28% in the Segura Basin, plus 3 and 31% in the Andalusian Basins. These scenarios agree with the projections of the mid to late 1990s when the river basin plans were being developed. Some authors estimated reductions in water flows in the study area of more than 30% compared to historic averages (Ayala-Carcedo & Iglesias, 2000).

More recent studies somewhat soften these grim predictions. Thus, the Spanish Plan for Adaptation to Climate Change of 2006 (MMA, 2006) calculated a reduction between 5 and 11% for different river basins in the Spanish Mediterranean coast towards the middle of the century. These figures were included in the review of river basin management plans undertaken by the Ministry of the Environment. The results of the application of hydrological models, such as SIMPA, by the Centre for Hydrographic Studies (CEDEX) to climate data indicate overall reductions in water availability. More specifically, the average runoff in Spain will decrease from 75 mm to 25 mm at the end of this century under a standard scenario.

As shown in Table 19.3 Water Planning Instruction (MMA, 2008) adapted to the development of new Demarcation Hydrological Plans in Spain, adapted to the Water Framework Directive (2000/60), assigns the percentages of decline in natural input of natural water resources for Mediterranean basins up until 2027.

Some regional organizations have in turn generated, climate scenarios and hydrological forecasts. For example, in Catalonia, a report by the Catalan Water Agency foresees a reduction in water resources from 495 hm^3 to 420 hm^3 (15%), throughout this century with a more noticeable decrease in the second half (Calbo et al., 2009). Recently, Mas-Pla (2010), based on the analysis of climatic series (temperature and precipitation) from the territory of Catalonia in the framework of the PRUDENCE project, has indicated reduction in river runoff between 3.5 and 51.7%, in relation to current hydrological conditions.

For the central sector of the Spanish Mediterranean coast, occupied by the Jucar and Segura Basins, Quereda, Monton and Escrig (2009) have developed two possible scenarios of water resources behaviour for the year 2030 that consider changes in runoff. The first scenario envisages an increase by 1 °C in mean annual temperature and no change in precipitation patterns. Under these climatic conditions, the Jucar and Segura Basins would experience a reduction of 12% (550 hm^3/year) of their current water resources (4596 hm^3/year). The second scenario implies a temperature rise of 1 °C and a reduction of 5% in average rainfall. This reduction would be translated into a runoff of only 57.6 mm, that is 23% lower than the actual runoff of 74.3 mm. In turn, water availability would fall 22% or over a 1000 hm^3/year. Under both scenarios, but especially the second, the availability of water resources would cover at best 60% of the estimated water demand in 2020, put in some 7000 hm^3/year according to the Jucar and Segura river basin plans.

For Andalusia, an 8% reduction in available water resources would represent, according to Corominas (2008), a decrease of 12% in irrigation supplied with regulated water and 16% with unregulated water. These values are duplicated in 2050. For its part, the increase in years of severe drought in this area is estimated at 7% compared to the current situation for 2027, and up to 40% in 2050.

Until more detailed figures on water availability reduction may be generated by modelling, it appears reasonable to assume a gradual trend in which water in rivers and aquifers in our study area may decrease by 5% in the period 2011–20; 10% for the two

Table 19.3 Reduction in natural water resources of the Mediterranean Hydrological Demarcations. Horizon 2027

	Decrease percentage (%)
Ebro	5
Júcar	9
Segura	11
River Basins of Andalusia	8

Source: MMA, 2008.

following decades; 15% for the decades 2041–50 and 2051–60, and at least 20% for the remaining decades of the 21st century. Generally, reductions in runoff will have a clear latitudinal component. Reductions in summer may double while reductions in winter and autumn may be minimal, especially in the Northern basins. For the purposes of this study and taking into account all the estimations presented above we assume the latitudinal decrease in the availability of water resources appearing in all sources and we quantify this decrease to be 5% in the tourist areas of Catalonia and the Balearic Islands (low decrease); 10% in the tourist areas of Valencia, except for the Costa Blanca (medium decrease), and 15% in the tourist areas of Costa Blanca, Costa Cálida and Andalusia (high decrease).

19.3 Adapting to climate hazards and climate change in Spain – some experiences

In shortfall or excess, water is the element that constitutes the main natural hazard in Spain. Flooding and droughts are the risks that have merited the greatest number of adaptation and mitigation projects. The possible increase in climate hazards predicted for the Mediterranean in climate change models will only serve to exacerbate this risk in the future. The problem, however, has already manifested itself in Spain and for this reason measures based on different strategies have been implemented for some time to help mitigate the risk.

In Spain, adaptation and mitigation projects have mainly taken two forms: structural measures and preventative measures (land-use planning). The structural measures (works of infrastructure) have been those most used when natural hazards have occurred. Preventative measures (land-use planning, risk mapping, education and communication about the risk) were not taken until the mid-1990s, when the Biescas campsite disaster (August 1996) helped to change the perspective on land-use planning in the country as a way to mitigate risk.

One item that is worth noting is that to date, Spain has not put into action any structural changes for the adaptation and mitigation of the expected effects of climate change. The one thing the country has done is approve plans for climate change adaptation at the national and regional level. These plans are mainly based on adopting energy strategies in the middle-term, that is, by encouraging projects aimed at reducing greenhouse gas emissions in the industrial and transport sectors. Alternative energy projects (wind and sun) have been encouraged. In addition, funding has been provided for research on climate change – both for research focused solely on atmospheric aspects as well as studies on how climate change affects economic sectors and activities – through national plans for R + D, regional plans and sector-specific plans. However, for the time being, strategies for adapting to climate change have been limited to structural or non-structural (land-use planning) measures. Similarly, at the local (municipal) level, climate change has not been incorporated in land-use planning policies quickly enough, nor have natural hazards been incorporated as an essential component.

Over the past two decades, there have been important shifts in the way natural hazards have been approached in Spain. The situation has shifted from overlooking risks in the processes of spatial planning to approving regulations which establish a requirement for risk analyses in the documentation required for developments. Flooding and drought events have received particular attention in the risk mitigation policies implemented in the territories of Spain and Europe. The approval of EU Floods Directive 2007/60 on the management of flood risks, on the one hand, and the new National Soil Act (Royal Legislative Decree 2/2008) on the other hand, will bring about a radical change in the administrative processing of planning and development projects in Spain in the years to come, given that a risk table will have to be drafted and consulted as an essential requirement. Other natural risks such as windstorms, tornados and forest fires have not been addressed as thoroughly as flooding or drought. However, in the current context of climate change caused by the greenhouse effect, they will have to be incorporated in future land-use planning, given that the extreme climates of Southern Europe are expected to become even more extreme.

Thus, in just a few years' time, there has been a significant shift in terms of how risk mitigation policies are viewed. There has been a shift from using infrastructure works as the main tool for mitigating natural hazards towards proposing measures related to land organization and management.

As mentioned above, the Biescas camping catastrophe (in August 1996) marks a before and after in the way risk is considered in land-use planning processes. Since the mid-19th century, only structural measures were taken to reduce natural hazards in Spain. Channelization, rerouting and changes to river course in urban areas, dams, water diversions and the implementation of desalination plants have been some of the State's (i.e. national) measures taken in areas at risk. Table 19.4 below shows the most important structural works done in Spain in the second half of the 20th century in response to catastrophic droughts or flooding events. It is important to note that the structural works were always begun after a major natural hazard that led to severe economic losses or to the loss of human lives. One important factor to keep in mind in terms of the effects of structural works in response to flooding or drought is that they create a false sense of security among the population. This is because they are designed for events that comply with statistical parameters related to their frequency (return periods) that generally do not reflect the disorderly, extreme behaviour of the Mediterranean or subtropical climates present in certain Spanish regions (the Mediterranean shore, the Atlantic shore of Andalusia and the Canary Islands). In these areas, climatic factors often exceed the median upon which the structural works are based (Olcina Cantos, 2007).

Since the end of the 1980s, some of the Autonomous Communities have approved land-use laws and plans to develop their competences in land-use planning and included the requirements for considering natural risk (essentially floods) when approving new planning and development projects in their regions. Thus, new non-structural methods have begun to be implemented for the adaptation or mitigation of natural risk. This has been the case in the Basque Country, Navarra, Andalusia, the Valencian Community, the Balearic Islands and Catalonia.

At the national level, the modification of the 1992 Soil Act and the approval of the 1998 Act were two steps towards genuinely incorporating risk analysis in land-use planning. However, the 1998 Soil Act which was later adapted by the Autonomous Communities turned out to be no more than a statement of intent, because the obligation of classifying certain lands that had a 'verified' natural risk (Art. 9) as

Table 19.4 Major structural works to respond to flooding and drought risks in Spain – 1950 to the present

Natural Hazard	Structural Work	Catastrophic Event which causes their construction
FLOOD	'Sur' Plan on the Turia River (Valencia)	Flood in October 1957
	Channelization of the Llobregat	Flood in September 1962
	New dam in Tous	Flood in October 1982
	Structural works along riverbanks in the Basque Country	Flood in August 1983
	Flooding defence plan for the Segura river basin	Floods in October 1973 and November 1987
	Structural works along riverbanks in Almería and Málaga	Floods in September and November 1989
	Structural works along riverbanks in Santa Cruz de Tenerife	Floods in March 2002
	Plan for fluvial sediments in the city of Barcelona	Autumn storms in different years of the 1990s
	Anti-flooding plan in the city of Alicante	Flood in September 1997
DROUGHT	Water diversion Tajo-Segura	Droughts in the southeast of the peninsula in the 1960s
	Water diversion of the Zadorra	Drought in the Basque Country 1988–90
	Desalination plants, Metadrought Plan	Drought 1990–95
	Ship operation to Palma de Mallorca	Drought 1990–95
	Desalination plants, AGUA Program	Dry years at the start of the 2000s

Source: Compiled by the author.

Table 19.5 Non-structural measures for natural risk adaptation or mitigation in Spain

Non-Structural Measure	Natural Risk Addressed	Work Scope
-Amendments to the Soil Act (1998 and 2008)	All natural hazards	Nationwide with effect on municipal spheres (local)
-EU Directive 60/2007	Floods	Nationwide. River basin district sphere
-Civil Protection emergency plans (since 1995)	Flooding, earthquakes, volcanic eruptions	Nationwide. Application in regional and local spheres
-Weather warning systems (since 1989)	All atmospheric hazards (torrential rain, cold and heat waves, storms)	Nationwide and regional scale
- Regional land-use planning	All natural hazards	Regional
-Regional plans for mitigating the risk of flooding through land-use planning	Floods	Regional
-Drought management plans (since the passage of the National Hydrological Plan Act in 2001)	Drought	Nationwide. Hydrographic Demarcation Sphere
-News forest laws (national and regional). Since 2003	Forest fire	Nationwide and regional scale

Source: Compiled by the author.

'non-urbanizable' involved the need for risk tables for verification in each case. In those cases – throughout most of Spain – where the risk table was not available, the act became null and void, which is what in fact occurred. In this regard, the recently approved Soil Act (Royal Legislative Decree 2/2008) represents a true revolution, as Art. 15 of the act requires that new urban planning processes include a map of 'existing risks'.

In this context, it is necessary to move *from the analysis of natural hazards towards studies on the vulnerability and exposure* to these natural hazards. The study of the social and economic impact associated with natural hazards has become one of the main research guidelines within risk analysis in the past few years. In fact, awareness of hazards has risen significantly in the past few years across the world, though research in the field of vulnerability has not kept pace. In Spain, there are very comprehensive studies on climatic hazards but few approaches to vulnerability associated with extraordinary atmospheric events.

As outlined above, natural hazard mitigation in Spain has mainly been focused on the effects of droughts and inundations. Nevertheless, some actions

have been taken to reduce the potential impacts of other natural hazards. For earthquakes and volcanism national civil protection directives have led to the establishment of norms and regulations on a regional level. These include detailed hazard and vulnerability maps, as well as emergency plans. The national and the regional meteorological services (Cataluña, Basque Country and Galicia) have developed early warning systems to detect meteorological extreme events such as cold and heat waves, heavy precipitation, storms and storm surges. Also the communication structure between the meteorological and civil protection services, as well as public information services for potentially affected areas has been improved (see Table 19.5).

The particularly important forest fire hazard is addressed by early warning systems on atmospheric conditions that strongly support this hazard and civil protection plans on a regional and local level. But certainly the most innovative aspect is being developed to reduce the exposure and the vulnerability towards forest and wildfires by developing legislations on the management of areas in the urban-forestal interface, including infrastructure, agricultural and urban areas.

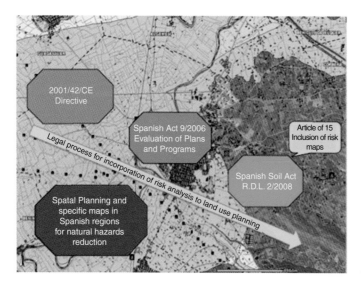

Figure 19.4 Incorporation of risk analysis within land-use and urban planning.
Source: Compiled by the author

The Ley de Montes (Law of the Mountains) from 2003 obliges a specific focus on the land use in areas of high forest fire risk and several regions are preparing legislations on the management of urban-forestal interface areas.

In addition to the regulations for reducing natural hazards at the level of the national government, it is important to note that the environmental and territorial legislation of the autonomous communities, which are highly empowered in these areas and which have, in some cases, approved regulations as well as plans for reducing risks through land-use organization. It is also important to mention the documents (the European Territorial Development Strategy, 1999, and the Territorial Agenda of the European Union, 2007) along with regulations approved in Europe (the EU Water Directive, 2000, and the EU Flood Directive, 2007) which, when adapted to national legislation in varying degrees, must also be observed in the processes of land-use planning. Currently, any land action plan or programme that is approved in Spain must include a risk analysis along with a risk map. Thus, if there is no applicable regulation at the level of the autonomous community, the stipulations of Act 9/2006 must be complied with for the environmental assessment of plans and programmes and those of Royal Legislative Decree 2/2008 on soil, mainly what is stipulated in Art. 12 and 15. However, cooperation is required among competent authorities in terms of land-use organization and the

environment, both horizontally (at the same administrative level) and vertically (among the different levels of the national government) (see Figure 19.4).

In compliance with Directive 60/2007, a large mapping databank of flood risks is currently being developed at the national level. The databank is called the National Flood Zone Mapping System of Spain (SNCZI), and it is being drafted by the Ministry of Environment. The various river basin districts in

Figure 19.5 Fluvial systems (on the peninsula of Spain and the Balearic Islands) with flood studies incorporated to the SNCZI.
Source: Ministerio de Agricultura, Alimentación y Medio Ambiente, 2012. Available at: http://sig.magrama.es/snczi/visor.html?herramienta=DPHZI

Figure 19.6 Sample site form showing areas with a significant potential of flood risk.
Source: URA, 2011. Available at: http://www.uragentzia.euskadi.net/u81-0003/eu/contenidos/informacion/2011_epri/es_doc/index.html

Spain draft flood risk maps to a scale of 1:10 000, following the stipulations of the aforementioned directive (see Figure 19.5). Before completing the databank with the mapping data, the river basin districts were required to draft the 'Preliminary Risk Assessments' for their respective territories. These reports, which are required according to EU Directive 60/2007 and were adopted by the Spanish judiciary in Royal Decree 903/2010 (on 9 July) were completed at the end of 2011. The map boundaries of the 'Areas with Significant Potential Risk' were assigned to the SNCZI. In 2013, the official map of flood risk must be completed, within the parameters established in said directive.

At the end of 2001, the river basin districts drafted the Preliminary Flood Risk Assessments in their respective territories, a requirement established in EU Directive 60/2007 that was adopted by the Spanish judiciary in Royal Decree 903/2010 (on 9 July). The map boundaries of the 'Areas with Significant Potential Risk' were assigned to the SNCZI. In 2013, the official map of flood risk must be completed, with the parameters established in the EU Directive (see Figure 19.6).

In the past few years, some of the autonomous communities (regional level) have developed land-use organization plans and laws to reduce risks, that is, mainly floods. Thus, the drafting of risk maps and the application of specific measures for risk mitigation have become common procedures in land-use and urban planning. The most outstanding cases are those of the Basque Country, Catalonia, the Valen-

cian Community, Andalusia and the Balearic Islands, which have approved respective land-use plans and laws for the mitigation of risks. First and foremost, this has involved risk mapping (hazards and vulnerabilty + exposure) and the adoption of measures that forbid urban developments and infrastructure in areas with a high risk of flooding.

To reduce the risk of drought, the solutions traditionally adopted in Spain have been mainly structural, as noted above, and usually aimed at increasing the availability of existing resources in the affected areas. The use of subterranean waters, the construction of dams and conduits for the diversion of water and, more recently, the start up of desalination plants have been the structural measures adopted from the end of the nineteenth century to date.

When a drought begins to develop in Spain, a series of measures are taken by the government. Special decrees provide economic support for the sectors most affected by the lack of water (mainly, agricultural producers) and Drought Committees are created in order to track the situation over time. In addition to these measures, the policies have mainly been aimed at increasing existing resources in the areas with a structural deficit of resources (the southeast of the peninsula) or those with deficits arising at specific moments in time (the Júcar River, the Segura River, Southern Spain, the Catalonia River Basins). The construction of dams and the interconnection of basins have been the main focus of the water resource policy in order to help extend the irrigation surface and ensure the supply of water to cities. The National Hydrological Plan of Spain (2001) reflects this philosophy of water resource planning and its most important achievement to date has been the diversion of waters from the Ebro to the Mediterranean shoreline regions (Catalonia, Valencia, Murcia and Almería).

This plan has involved major works that have been the subject of controversy. In fact, the diversion of the Ebro as laid out in the National Hydrological Plan of Spain in 2001 was ultimately repealed in June of 2004.

The approach has changed in Spain over the course of the past decade, since the enactment of the EU Water Framework Directive 60/2000, the revision to the National Hydrological Plan Act (Law 10/2001) and the implementation of the 2004 'Water' Program. The

goal of droughts measures is to optimize the availability of water, that is, to rationally use the existing resources of a region or river basin through environmental education, rational planning of agriculture, policies to reduce the demand for water (in agriculture and cities), and the reuse of wastewater.

The 'Water' Program aims for a rational use of existing resources in the different river basins (cleaning and reuse, improved irrigation) and in the regions with a lack of natural resources – generally, the Mediterranean shoreline regions – a decision has been made to set up desalination plants for urban and agricultural use. The Figure 19.7 summarizes the set of measures adopted in the 'Water' Program. This new philosophy in resource planning incorporates the principles of the

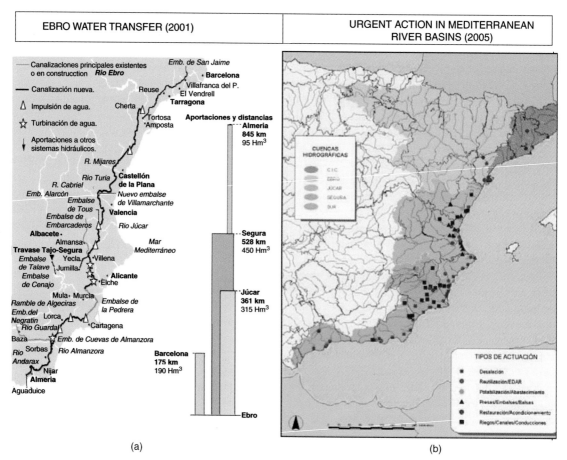

(a)　　　　　　　　　　　　　　　　(b)

Figure 19.7 Comparison of water resource planning alternatives along the Spanish Mediterranean shoreline. (a) Diversion of the Ebro as contemplated in the 2001 National Hydrological Plan, later repealed. (b) Structural works contemplated in the 'Water' program (2005). *Source*: MMA, 2005.

EU Water Framework Directive and attempts to avoid territorial conflicts that often lead to the diversion of water resources among river basins.

Another fundamental part of efficient drought management involves educating the population about this natural event, this is a major focus of water resource policies, but advances have been slow in this regard. In Spain, certain distributors of drinking water in major cities (Madrid, Seville, Barcelona, Valencia, Alicante, Murcia) carry out awareness campaigns and encourage residents to reduce their use of water at home during droughts. During the 1990–1995 drought, the Ministry of Environment joined forces with the river basin authorities in the south and centre of Spain to launch a campaign for citizen information and awareness about how to save water at home. The same campaign was carried out again during the last drought in 2005. In some cities, these measures have led to important reductions of 20% in water consumption.

The terms of the National Water Resource Plan Act (Act 10/2001; MMA, 2001), Art. 27.2 states that: within a maximum of two years following the present Act coming into force, the River Basin Entities shall draft specific action plans for situations in which there is a risk of drought and in the case of an eventual drought, including regulations on the use of the systems and the measures to be applied to the public water domain, as part of the Water Resource Plans of the corresponding basin. The aforementioned plans, after being presented to the Water Council of each basin, will be sent for approval to the Ministry of the Environment.

In compliance with this rule, the different basin entities have drafted Drought Management Plans. In these documents, the most significant events that have occurred in the past decades are analysed in the corresponding area of the river basin districts. All existing resources are studied along with possible water volumes in the case of droughts. Drought scenarios are described with regard to precipitation and the possible water volumes at each moment and the different drought levels are described. Finally, the measures to be taken are defined along with mechanisms for follow-up and control of the plan and the set of the basin's environmental requirements, which are included in the emergency planning and management. This is a very important step towards more efficient management of droughts and has served as the

basis for the drafting of the new Water Basin Resource Plans (river basin districts) that are being revised or will be revised in the coming years in compliance with EU Water Framework Directive 2000/60. Finally, these will be important tools in future water planning projects in Spain. In this regard, certain Spanish municipalities, entities working in collaboration or at the regional level, and their water distribution organizations (companies, associations, etc.) have drafted emergency plans for droughts in the past years. This is a very positive initiative that should be a legal requirement in all municipalities with more than 20 000 inhabitants. The stipulations laid out drought plans drafted by the hydrographic confederations will have to be adapted to the local scale, with specific scenarios designed for each sphere of action.

Finally, in addition to non-structural measures to respond to climate hazards, communication and education about risks are essential. However, these initiatives have not found widespread support in Europe or in Spain, in spite of the fact that they are the most economical and have the most visible effects to mitigate risks in society.

19.4 Conclusions

Spain is a risk country in the context of Europe. Some of its territories have a high level of risk due to the frequent appearance of diverse natural hazards and the intense level of occupation, the economic activities or infrastructure in affected areas. In the last 30 years risk towards natural hazards has strongly increased (Olcina Cantos, 2009b). This trend is not based on changes in natural hazard patterns but due to anthropogenic activities and settlements in hazard-prone areas. This process has been particularly intensive in areas with strong urban development, such as Madrid, the Mediterranean coast as well as the Balearic and Canaries Archipelagos. The extension of urban areas and settlements into hazard-prone areas was endorsed by the absence of any legislation that would prohibit such activities and has lead to a strong increase of exposure and vulnerabilities towards natural hazards. An urban normative that demands the elaboration of local risk maps prohibits the occupation of hazard-prone areas was implemented as recently as 2008.

Floods and droughts are the natural hazards with the greatest economic and social impact in Spain, and they have also been the focus of the majority of planning and development projects based on adaptation and mitigation strategies. Other geo- and hydrometeorological hazards have so far been principally addressed by actions on risk reduction and emergency plans.

In Spain, the principal measures for risk mitigation have targeted infrastructure (structural measures). Preventative measures based on incorporating the risk in the process of land-use planning have only been implemented since the mid-1990s. Risk mapping and its inclusion in land-use planning represents the hope for the future in Spain in the years to come. A national databank with flood risk maps is currently being constructed in compliance with EU Directive 60/2007.

Climate modelling has shown that climate hazards are expected to increase across the Mediterranean in the decades to come. Extreme atmospheric events could become more frequent and this will increase the risk unless the necessary measures are taken to adapt to these changes. To date, no land-use organization measures have been taken to adapt to climate change. National and regional plans to combat climate change have been almost exclusively aimed at reducing greenhouse gas emissions and encouraging the use of clean energies.

The structural measures carried out in Spain to reduce the risk of flooding and drought over the past 50 years have proven efficient, though they have come at a high economic and environmental cost in the areas where they have been implemented. On the other hand, it is too early to examine the success of the planning and development projects aimed at adapting to risk in land-use organization, as they have only been in place for the past five years.

The next two decades will be decisive in terms of verifying current climate models. We must be aware that natural hazards will become increasingly important in the immediate future for all societies that inhabit the earth. Unfortunately, this is a trend that will continue. It is essential for us to work to counter this trend, because we are capable of rationally mitigating this risk. The Mediterranean regions which are already at risk must work to implement measures aimed at adapting and miti-

gating risk as much as possible, given that climate hazards are expected to become even more extreme in the future.

The current steps to mitigate hazards and risks in Spain are pre-active, in comparison to earlier ones that were re-active. The new regulations thus have a high potential to be further applied to climate change adaptation purposes, probably not in the coming years but in the near future when changing patterns in climatic conditions might begin to be statistically relevant. The regulations could then be amended to changing demands in order to safeguard settlements, industries, service sectors and infrastructures and to reduce respective risks.

References

AA.VV., 2006. *Cambios de ocupación del suelo en España – Implicaciones para la sostenibilidad*. Alcalá de Henares: Observatorio de la Sostenibilidad e España, Ministerio de Medio Ambiente.

AEMET (Agencia Española de Meteorologia), 2009a. *Proyecciones regionalizadas de cambio climático generadas por el proyecto ENSEMBLES para un escenario de emisiones medio (A1B)*. [html] Madrid : Ministerio de Medio Ambiente. Available at : <http://www.aemet.es/es/elclima/cambio_climat/proyecciones> [Accessed November 2011]

Aemet, 2009b. *Generación de escenarios climáticos regionalizados para España*. [online] Available at: <http://www.aemet.es/documentos/es/elclima/cambio_climat/escenarios/Informe_Escenarios.pdf> [Accessed November 2011]

Ayala Carcedo, F.J. and Iglesias, A., 2000. *Impactos del posible cambio climático sobre los recursos hídricos, el diseño y la planificación hidrológica en la España Penínsular*. Madrid: BBVA, Servicio de estudios,.

Ayala-Carcedo, F.J., 2000. La ordenación del territorio en la prevención de catástrofes naturales y tecnológicas – Bases para un procedimiento técnico-administrativo de evaluación de riesgos para la población. In : Asociación de Geógrafos Españoles, ed. 2000. *Boletín de la Asociación de Geógrafos Españoles, 30*, Madrid, pp.37–49.

Ayala-Carcedo, F.J., 2002. El sofisma de la imprevisibilidad de las inundaciones y la responsabilidad social de los expertos – Un análisis del caso español y sus alternativas. In : Asociación de Geógrafos Españoles, ed. 2002. *Boletín de la Asociación de Geógrafos Españoles, 33*. Madrid, pp.79–92.

Ayala-Carcedo, F.J. and Olcina Cantos, J. eds., 2002. *Riesgos Naturales*. Barcelona: Editorial Ariel. Col. Ciencia.

Beck, U., 2002. *La sociedad del riego global*. Madrid: Edit. Siglo XXI.

BOE (Boletín oficial del Estado), 1998. *Ley 6/1998, sobre el régimen del suelo y valoraciones*. [pdf] Available at: <http://www.boe.es/boe/dias/1998/04/14/pdfs/A12296-12304.pdf> [Accessed may 2012]

Boe, 2008. Real Decreto Legislativo *2/2008, de 20 de junio, por el que se aprueba el texto refundido de la ley de suelo*. [pdf] Available at: <http://www.boe.es/boe/dias/2008/06/26/pdfs/A28482-28504.pdf> [Accessed may 2012]

Burton, I., Kates, R. and White, G., 1993. *The Environment as Hazard*. New York: Oxford University Press.

Calbo, J., Cunillera, J., Llasat, C., Llebot, J.E. and Martin-Vide, J., 2009. Proyecciones climáticas para Cataluña. In: *Agua y cambio climático. Diagnosis de los impactos previstos en Cataluña*. Barcelona: Generalitat de Catalunya, Agencia Catalana del Agua, pp.91–100.

Calvo García-Tornel, F., 2001. *Sociedades y Territorios en riesgo*. Barcelona: Ediciones del Serbal.

Corominas, J., 2008. *Los nuevos Planes Hidrológicos de las Cuencas Andaluzas*. [online] 1° Seminario Nacional sobre 'Los nuevos planes de Cuenca según la Directiva Marco del Agua'. Madrid: Fundación Botín. Available at: <http://www.fundacionmbotin.org/seminarios-nacionales-ponencias_observatorio-del-agua_publicaciones.htm> [Accessed November 2011]

Dauphiné, A., 2003. *Risques et catastrophes – Observer, spatialiser, comprendre, gérer*. Paris: Armand Colin.

Diez Herrero, A., Laín-Huerta, L. and Llorente-Isidro, M., 2009. *A Handbook on Flood Hazard Mapping Methodologies*. Madrid: IGME.

Greiving, S., Fleischhauer, M. and Wanczura, S., 2007. Planificación territorial para la gestión de riesgos en Europa. In: Asociación de Geógrafos Españoles, ed. 2007. *Boletín de la Asociación de Geógrafos Españoles*, 45. Madrid, pp.49–78.

Herrero, G., 2011. Las interfaces urbano-forestales como territorios de riesgo frente a incendios forestles – Análisis y caracterización regional en España. [online] Ph. D. (unedited). Universidad Autónoma de Madrid. Available at: <http://www.interfazurbanoforestal.com/index.php/tesis-doctoral-gema-herrero/resumen> [Accessed December 2011]

IPCC, 2007. *Climate Change 2007 – Syntesis Report*. [online] Available at : <http://www.ipcc.ch/publications_and_data/publications_ipcc_fourth_assessment_report_synthesis_report.htm> [Accessed May 2012]

JRC (Joint Research Centre), 2009. *The PESETA project. Impact on climate change in Europe*. [online] European Commission. Available at: <http://peseta.jrc.ec.europa.eu/> [Accessed May 2012]

Lamarre, D. ed., 2002. *Les risques climátiques*. Paris: Ed. Belin.

Martín-Vide, J. ed., 2007. Aspectos económicos del cambio climático. In: *Estudios Caixa de Cataluña*. Barcelona: Caixa Cataluña, Ch.4.

Mas-PLA, J., 2010. Vulnerabilitat territorial dels recursos hidrològics al canvi climàtic. In: *Segon Informe sobre el canvi climatic a Catalunya*. Barcelona: Generalitat de Catalunya, Institut d'Estudis Catalans, pp.309–342.

Mata Olmo, R. and Olcina Cantos, J., 2010. El sistema de espacios libres. In: L. Galiano and J. Vinuesa, eds. 2010. *Teoría y Práctica para una ordenación racional del territorio*. Madrid : Ed. Síntesis, pp.87–127.

Ministerio de Agricultura, Alimentacion Y Medio Ambiente, 2010. *Perfil medioambiental de España, 2010*. [online] Available at: <http://www.marm.es/es/calidad-y-evaluacion-ambiental/temas/informacion-ambiental-indicadores-ambientales/indicadores-ambientales-perfil-ambiental-de-espana/> [Accessed May 2012]

Ministerio de Agricultura, Alimentacion Y Medio Ambiente, 2012. *SNCZI/Inventario de Presas y Embalses*. [online] Available at: <http://sig.magrama.es/snczi/visor.html?herramienta=DPHZI> [Accessed May 2012]

MMA (Ministerio de Medio Ambiente), 2001a. *Libro Blanco del Agua en España*. [online] Madrid. Available at: <http://hercules.cedex.es/Informes/Planificacion/2000-Libro_Blanco_del_Agua_en_Espana/> [Accessed May 2012]

MMA, 2001b. *Ley 10/2001, del Plan Hidrológico Nacional*. [pdf] Available at : <http://www.boe.es/boe/dias/2001/07/06/pdfs/A24228-24250.pdf> [Accessed May 2012]

MMA, 2005. *Principales. conclusiones de la evaluación preliminar de los impactos en España por efecto del cambio climático*. Madrid: Secretaría General para la Prevención de la Contaminación y del Cambio Climático, Oficina Española de Cambio Climático.

MMA, 2006. *Plan de Adaptación al Cambio Climático en España*. [pdf] Madrid. Available at: <http://ec.europa.eu/clima/policies/adaptation/docs/pna_v3_en.pdf> [Accessed may 2012]

MMA, 2008. *Instrucción de Planificación Hidrológica*. [pdf] Madrid. Available at: <http://www.boe.es/boe/dias/2008/09/22/pdfs/A38472-38582.pdf> [Accessed May 2012]

Olcina Cantos, J., 2007. *Riesgo de inundaciones y ordenación del territorio en España*. Murcia : Instituto Euromediterráneo del Agua.

Olcina Cantos, J., 2009a. Cambio climático y riesgos climáticos en España. In: Instituto Universitario de Geografía, ed. 2009. *Investigaciones Geográficas*, 49. Universidad de Alicante, pp.197–220.

Olcina Cantos, J., 2009b. Hacia una ordenación sostenible de los territorios de riesgo en Europa. In: J. Farinós, J. Romero and J. Salom, eds. 2009. *Cohesión e inteligencia territorial. Dinámicas y procesos para una mejor planificación y toma de decisiones.* Valencia : Publicaciones de la Universitat de Valencia, pp.153–182.

ONU (Organización de las Naciones Unidas), 2004. *Living with Risk: A Global Review of Disaster Reduction Initiatives.* Nairobi: ISDR.

OSE (Observatorio Español de la Sostenibilidad), 2009. *Atlas de la Sostenibilidad en España.* [pdf] Available at: <http://www.sostenibilidad-es.org/sites/default/files/_Informes/anuales/2009/sostenibilidad_2009-esp.pdf> [Accessed May 2012]

Quereda, J., Monton, E. and Escrig, J., 2009. *Evaluación del cambio climático y de su impacto sobre los recursos hídricos en la cuenca del Júcar.* Valencia: Generalitat Valenciana, Fundación Agua y Progreso.

Schmidt-Thomé, P. ed., 2005. *The spatial effects and management of natural and technological hazards in Europe.* [online] Luxemburg: ESPON, (thematic project 1.3.1.). Available at: <www.espon.eu> [Accessed November 2011]

URA (ur agentzia), 2011. *Evaluación preliminar del riesgo de inundación (E.P.R.I) en la demarcación hidrográfica del cantábrico oriental.* [online] Available at: <http://www.uragentzia.euskadi.net/u81-0003/eu/contenidos/informacion/2011_epri/es_doc/index.html> [Accessed May 2012]

Villevieille, A., ed., 1997. *Les risques naturels en Méditerranée – Situation et perspectives.* Paris: Les Fascicules du Plan Bleu.

20 Developing Adaptation Policies in the Agriculture Sector: Indonesia's Experience

Daisuke Sano[1], S.V.R.K. Prabhakar[1], Kiki Kartikasari[2] & Doddy Juli Irawan[2]

[1]Institute for Global Environmental Strategies (IGES), Japan
[2]Center for Climate Risk and Opportunity Management in Southeast Asia and Pacific, Bogor Agriculture University, Indonesia

20.1 Introduction

The agricultural sector contributes to a significant part of Indonesia's economy with a share of 12.9% in 2006 and 43.3% in 2004 in the total GDP and employment generated, respectively (ADB, 2009a). The agriculture production in Indonesia has been steadily increasing mostly due to increase in rice production and expansion in the irrigated area in the past few years (FAO, 2003; Fuglie, 2010). The country was able to divert significant investments in rural infrastructure, including developing irrigation systems, and attained self-sufficiency in rice and maize in 2009 due to a record level harvest (GIEWS, 2010). Despite these gains, there have been reports of a decreasing trend of land fertility, water availability and a steep increase in food prices in the past three years, severely impacting food security (Hadar, 2009; Prabhakar, Sano & Srivastava, 2010). The country is still a net importer of other major food crops such as sugar, soybean and wheat.

Historically, Indonesian agriculture has not been free from weather-related abnormalities. The El Niño/La Niña-Southern Oscillation (ENSO) events have significantly affected river flows and water reservoirs, particularly during the dry season (Las, Boer & Syahbudin, 1999). As a result, crop failures have often been reported during extreme drought events (GIEWS, 1998). Climate change could bring an additional dimension to the agriculture sector. While there will be variation from region to region and uncertainties in climate projection, the southern region of Indonesia is expected to experience a drier climate (USAID, 2010). The climate change could decrease rice production in Java between 2025 and 2050 by about 1.8 and 3.6 million ton from the current production level respectively (GOI, 2009a) and there is a possibility that major crop pest/ diseases will shift and/or intensify (FAO, 2010a).

Keeping the above context in view, the chapter reviews the current status of climate change adaptation in the agriculture sector; examines challenges and opportunities in adaptation actions; and identifies lessons for promoting climate resilience of the farming sector in Indonesia. This chapter is based on a review of information from various national plans and associated documents of the GOI in response to the climate change and interviews with government officials. Government officials interviewed include the National Development Planning Agency (Bappenas), Ministry of Public Works (PU), State Ministry of Environment (KLH), Agency of Meteorology, Climatology and Geophysics (BMKG), Ministry of Agriculture (MOA), National Council on

Climate Change Adaptation in Practice: From Strategy Development to Implementation, First Edition.
Edited by Philipp Schmidt-Thomé and Johannes Klein.
© 2013 John Wiley & Sons, Ltd. Published 2013 by John Wiley & Sons, Ltd.

Table 20.1 Key policy documents related to climate change in Indonesia

Document	Publisher	Year	Content
National Action Plan Addressing Climate Change (NAP-CC)	State Ministry of Environment	2007	Provides principles covering immediate (2007–2009), short-term (2009–2012), medium-term (2012–2015) and long-term (2025–2050) time frames for both mitigation and adaptation.
National Development Planning: of Indonesia's Responses to Climate Change	Bappenas	2008	Serves as a bridge document between the National Mid-term Development Plan (RPJM 2004–2009) and the next RPJM (2010–2014).
Second National Communication to UNFCCC (SNC)	State Ministry of Environment (with support from UNDP)	2010	States the latest national circumstances, GHG inventory, needs and policies both for Mitigation and Adaptation policies till CY2020.
Indonesia Climate Change Sectoral Road map (ICCSR)	Bappenas (with support from GTZ)	2010	Sets priority issues and key policy actions in four 5-year periods till CY2030.

Source: GG21 and IGES, 2010; KLH, 2010.

Climate Change (DNPI), institutions or experts on the agriculture sector and agricultural offices at the provincial/district levels. Interviews were conducted between 2008 and 2010 as a part of monitoring activity of Indonesia Climate Change Program Loan (ICCPL) by the Government of Japan.[1] The authors were engaged in the programme loan as members of the advisory and monitoring team for the loan activity for the agriculture sector in Indonesia.

20.2 Recent development in climate change adaptation in Indonesia

20.2.1 Initiatives by the Government of Indonesia (GOI)

The GOI has instituted several initiatives to address climate change during recent years (Table 20.1) with the establishment of key institutes such as the National

Council on Climate Change (NCCC) in 2008 and the Indonesian Climate Change Trust Fund (ICCTF) in 2009 (GG21 & IGES, 2010). In addition to the GOI's coordinating ministry such as Bappenas, these institutes are expected to facilitate mainstreaming responses to climate change and communicating with foreign agencies on climate talk. The NCCC manages the funds of approximately US$ 213 million to improve capacity of handling climate change mitigation and adaptation in 2009 (NCCC, 2009). This figure is equivalent to about 29% of the budget allocated for environmental development by the GOI (US$ 745.5 million) and 0.28% of the GOI's total budget (US$ 76.3 billion) in 2009.[2]

The ICCTF was established to achieve Indonesia's goals of low carbon economy and greater resilience to climate change and to improve the management in addressing climate change issues (ICCTF, 2010). It currently handles the fund of approximately US$ 8.5 million, 54% of which is allocated to priority issues such as energy conservation, sustainable peat land management and public awareness on climate change (ICCTF, 2010). Although it has only been established recently, the ICCTF fund is limited in scale compared to the national budget.

---

[1] ICCPL is a programme loan to the Government of Indonesia from the Government of Japan designed to provide assistance on the basis of bilateral policy consultations to developing countries that aim to achieve emission reductions and economic growth and to contribute to climate stability. It was the first large-scale programme loan (three tranches over three years) under the 'Financial Mechanism for Cool Earth Partnership' launched in 2008. USD 300 million was disbursed to the GOI for 2008.

[2] The same exchange rate used in NCCC report (NCCC, 2009) was applied.

Setting adaptation policy priorities

In 2007 the Ministry of Environment has published the National Action Plan Addressing Climate Change (NAP). The plan emphasized six key areas for adaptation in the agriculture sector: (1) data and information management, (2) improvement of farming activities, (3) improvement of irrigation management, (4) institutional/capacity development, (5) research, and (6) socialization and advocacy (KLH, 2007). Across these key areas, water-saving farming measures were emphasized. The recommended measures include specific activities such as the development of a System of Rice Intensification (SRI) and the rehabilitation and improvement of the irrigation network. The empowerment of farmers' groups was also mentioned in the effort of improving water use efficiency. The need for research, dissemination/management of information and technologies, and associated capacity-building are also mentioned as a part of scaling up of national efforts.

In 2007, the Ministry of Agriculture (MOA) also published the 'Strategy of Adaptation and Mitigation to deal with Climate Change and Strategy and Technology Innovation to Cope with Global Climate Change.' Since the MOA's then five-year plan (Indonesian Agricultural Development Plan 2005–2009, published in 2006) did not list measures addressing climate change, this strategy is one of the first official climate policies developed by the MOA. The strategy emphasizes the improvement of the resilience of farm production and reduction of drought risk. It pays special attention to cropping patterns and the irrigation management system/technique (JBIC, 2008). Adaptation strategies that have been and that continue to be developed include: (1) a technology and information system to predict climate change, including an agricultural early warning system, (2) Climate Field School, (3) infrastructure, particularly rural irrigation and the associated realignment of farming areas and water–use-saving techniques, (4) cropping calendar, based on climate change and adaptation farming practices, (5) good agricultural practices (GAP), (6) development of new varieties for withstanding warming climate, and (7) farming intensification for staple crops.

In 2008, the Bappenas published the National Development Planning Response to Climate Change (called the Yellow Book) in which the agriculture sector has been identified as a priority area in adaptation. Priority activities that have been identified under adaptation are: (1) Implementation of good agricultural practices (GAP), (2) introduction and integration of a food and nutrition security system (SKPG) to prevent and minimize food crises that result from climate change, (3) increased production and consumption of local-specific food to reduce rice dependency, and (4) expansion and strengthening of Climate Field School (Bappenas, 2008).

In 2009, the draft Second National Communication recommends that farmers should consider altering their cropping pattern from rice–rice to rice-non-rice and notes the need for improvement in water storage and irrigation (GOI, 2009a). In the same year, the Indonesian Climate Change Sectoral Roadmap (ICCSR) Synthesis Report indicated possible impacts on estate crops such as coffee, cacao, rubber and palm oil, due to the export importance these crops have. The ICCSR identified two primary targets for adaptation: (i) reducing the uncertainties from climate change as well as increasing awareness, (ii) reducing risks and impacts of climate change on infrastructure, agricultural production system and the socio-economic aspect of the sector. The document also presents two corresponding goals: (i) detailed analysis of the climate change impacts through improved human resources, and (ii) preservation of fertile/productive land and application of adaptation technologies (GOI, 2009b). It is noteworthy that the ICCSR proposes several other activities beyond conventional adaptation measures found in the aforementioned documents such as production of non-rice crops (integrated crop management on rice as well as maize, soybean and peanut, estate crops), improvement of agricultural goods distribution (such as storage system and cold-chain), and food security (such as food independent village programme and diversification of food consumption). As shown above, climate change adaptation policies have been developed and streamlined with a few concrete measures.

Climate vulnerability assessments

There is a basic institutional capacity to collect and process climate and weather data at the national level; however, downscaling and translating the information into adaptation measures in the agriculture sector, including early warning, are still in a nascent stage. The Agency of Meteorology, Climatology and Geophysics (BMKG) is in charge of collecting climatologic data (exposure) and assessing climate sensitivity, adaptive capacity and vulnerability. The agency has already created a Map of Maximum Precipitation Frequency Forecast; Flood and Drought vulnerability maps as well as an Atlas of Agroclimate Suitability. For the agriculture sector, the BMKG plans to make a map of the agriculture sector using data such as farmer's capacity and available water resources. The Indonesian Agroclimate and Hydrology Research Institute of the MOA has created rice cropping calendar maps to provide alternative planting times in response to climate change for all provinces. The maps were created in steps including (i) collection of meteorological data, (ii) simulations for three weather patterns of rain (El Niño, La Niña and average) using a meteorology model, and (iii) printing the results in 1/25 000 maps. These maps are considered helpful in designing cropping strategy to improve farmers' income during ENSO years.

Development of irrigation infrastructure and community-level organizations

In the wake of projected water scarcity under climate change, good attention has been given to developing irrigation with aid from multiple donor agencies. The investment will be paid off when institutional capacity for both government officials and water users are developed enough to be able to duplicate necessary practices. Irrigation asset management was selected as one of the GOI's 100-day programmes after President Yudhoyono took office in 2009. Irrigation has also been identified as one of the priority technologies for adaptation by the Asian Development Bank (ADB, 2009b).

Merging of the water users' association and the farmers' group to establish a new farmers' economic entity is being attempted. This initiative has largely been inspired by the advances made in China and elsewhere (Ministry of Water Resources, China, 2009). Formation of water groups is based on the idea that giving more responsibility and ownership to farmers on the way the water resources are managed has resulted in efficient use of water.

Transfer of knowledge and technologies

The implementation of System of Rice Intensification (SRI) and Climate Field School Programs has been considered as an effective vehicle to provide farmers with opportunities to gain knowledge and specific farming skills for adaptation (Stoop, Adam & Kassam, 2009; Mishra, 2009; FAO, 2010b; IIED, 2009). Effectiveness would increase if the operation takes place on a national scale through expanded training of trainers and provision of comprehensive, but tailored curriculum matching to the local needs.

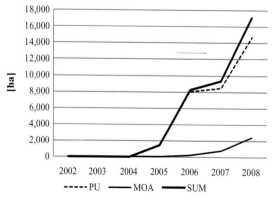
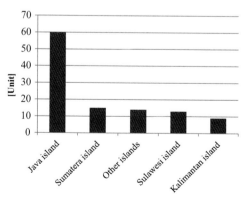

Figure 20.1 Increasing SRI area* (left, in hectares) and locations** (right) in Indonesia.
Source: GG21 and IGES, 2010. *Data source*: unpublished data from the Ministry of Agriculture as of February 2010.
* SRI areas implemented by projects by donor agencies are included in the PU's figure. ** By MOA in 2009.

Table 20.2 Estimated nationwide impacts of the SRI programmes in Indonesia*

Estimated impact parameter	Range		
	Min	Max	Average
Gain in paddy production in programme areas (million tonnes)	0.37	0.76	0.56
Additional income generated (million US$)**	107	216	162
Irrigation water saved (million m^3)	665	1328	996
Potential additional irrigated paddy area that could be created with the above water savings (ha)	26 569	53 140	39 854

Source: GG21 and IGES, 2010.
*Calculated by the authors based on the total areas of 177 133 ha under SRI, using estimates available.
**Paddy prices were the government purchasing price for wet paddy from farmers in 2009.

Ever since the SRI was first conducted in Indonesia before the turn of the 21st century as an on-farm trial (Sato & Uphoff, 2007), the technology has expanded rapidly with aid from donor agencies such as JBIC and ADB since 2002 (Figure 20.1).

Overall, SRI is considered to contribute to improving the resilience of farm production/income and drought risk reduction. Although the extent of the effects still requires robust scientific analysis,[3] overall favourable impacts on water saving and yield[4] have been observed. The latter is important to encourage farmers to accept SRI. Estimated impacts of SRI on the national scale are presented in Table 20.2.

In addition to adaptation benefits, SRI can also contribute to greenhouse gas (GHG) emissions reduction (Table 20.2). Intermittent irrigation technique is reported to be able to reduce the amount of methane gas emissions by 35 to 90% (Sass et al., 1992) or 83% (Setyanto, Surmaini & Boer, 2009) compared with continuous flooding. The GHG mitigation potential from SRI alone could surpass all other technologies such as zero tillage, leaf colour charts, and composting together.

Climate field schools have been chosen as a vehicle to train farmers on climate-resilient farming practices. The current model for the Climate Field School

follows as that of the MOA's Farmers Field School on plant protection implemented since 1998. The Climate Field Schools started in 2002 with the purpose of disseminating knowledge and skills to farmers for adapting to climate change. Typically, training for trainers is provided by central government officials to provincial officials; the provincial officials (extension agents) train district officials; and district officials train farmer groups. In 2009, a total 177 units of Climate Field Schools were conducted in which 20–25 farmers participated (on 20–25 ha of farming areas) per unit on average.

Policy coordination in mitigation and sustainable agriculture

Although mitigation potential from the energy and forestry sectors in Indonesia is significant, the agriculture sector could also contribute to GHG emission reduction. There are four main sources of GHG emissions from Indonesia's agricultural sector: (i) rice cultivation (CH_4), (ii) domestic livestock (CH_4), (iii) agricultural soils (N_2O), and (iv) field burning of agricultural residues (CO_2)[5] (ADB, 2009b). The Indonesia Second National Communication highlights CH_4 emissions from the agriculture sector, which accounted for about 65% of the total emissions of the sector mainly from rice paddy fields (69%) and livestock (28%), followed by 31% from N_2O and 4%

[3]Technical challenges include lack of water gauges at the tertiary canals (terminal canals to farmland).
[4]Varying 1.1% decrease to 35% increase when compared to conventional farming (GG21and IGES, 2010).

[5]However, CO_2 emissions from biogenic sources are not counted in the GHG inventory.

from CO_2 (GOI, 2009a; KLH, 2010). The fertilizers use increased nine-fold between 1975 and 2002 with heavy dependency on urea in rice production (FAO, 2005; Rochayati & Husnain, 2010) and the overall increase in GHG emissions in the sector was recorded by 6.3% between 2000 and 2005 (GOI, 2009a).

There are already various recommendations made to reduce GHG emissions: (i) introduction of permanent agriculture system such as palm, rubber, cacao and fruits to reduce pressure on forest – for example, fruit trees could sequester carbon in the range of 53–254 tC/ha (GOI, 2009a), (ii) palm oil plantation or other sources of biofuel conducted in collaboration with other sectors for a sink enhancement (Bappenas and Gesellschaft für Technische Zusammenarbeit (GTZ), 2009), (iii) recycling livestock wastes in the form of energy and/or fertilizer – bioenergy from manure could reduce methane emissions up to 80% (GOI, 2009a), (iv) appropriate fertilizing – for instance N_2O emissions could be curbed by slowing nitrogen release by using tablet urea and polymer-coated urea (CRM), or nutralene (slow release nitrogen fertilizer) – and no tillage for rice field and horticulture (ADB, 2009b). According to the projections for mitigation potential of CH_4 emissions from rice paddies provided in the Second National Communication, the most effective mitigation scenario would be the adoption of less methane-emission varieties when compared with other options such as intermittent irrigation and fertilizer supplement, it also indicates that the mitigation potential of CH_4 from livestock seemed limited (KLH, 2010).

Since Indonesia's domestic fertilizer production, except for urea, does not meet domestic demand and its import jumped by 20% or more than two million tons in 2007 (US Embassy Jakarta, 2008), supplying organic fertilizers to supplement the synthetic ones is beneficial not only from the point of reducing the dependency on imported fertilizers but also promoting more sustainable farming methods. The potential of organic agriculture in developing countries has been well documented (Sano & Prabhakar, 2010; Prabhakar, Sano & Srivastava, 2010) and Indonesia is gearing up its promotion. The MOA plans to increase organic fertilizer supply as a part of fertilizer subsidy reform: subsidy switch from synthetic to organic fertilizers as well as from fertilizer manufacturers to farm-

ers groups.[6] A part of the savings from fertilizer subsidies will be used for subsidizing organic fertilizers production.[7] In addition, meat consumption is on the rise in Indonesia as income increases and dietary patterns become more westernized. The GOI set a target of beef self-sufficiency by 2010[8] in response to increased domestic demand.

20.2.2 Assistance from donor agencies

Prior to the beginning of ICCPL, there was little technical assistance by donor agencies that was specifically designed for climate change. The average annual Official Development Assistance (ODA) that the GOI received during the last 10 years was estimated at about US$ 2 036 million (NCCC, 2009) and several projects and programmes are implemented for general improvement in governance and rural development. Though these are not directly aimed at climate change adaptation, these programmes do enhance the adaptive capacity of the agricultural sector in a broader sense.

GTZ set the precedence among donor agencies in conducting a Climate Change Adaptation Strategy and Action Plan for the Water Sector in Indonesia from June 2007 to June 2008 that aimed at mainstreaming climate change issues into development planning by first focusing on the water sector and then developing proposals for strengthening institutional cooperation on adaptation (GTZ, 2007).[9] GTZ has published the sectoral roadmap in 2010. UNDP has been providing technical assistance in developing National Communication to UNFCCC (KLH, 2010).

---

[6]The proposed ministerial decree is to introduce new regulations for fertilizers that aim to subsidize fertilizers to poor farmers and to stabilize their market prices. Under the current regulations, subsidies are paid to fertilizer companies, resulting in imbalanced fertilizer supply. Under the proposed regulation, farmers' groups will receive subsides based on their reporting on the fertilizer needs, aiming at reducing excess fertilizer use and government expenditure on subsidies. This change will give more responsibility to farmers.

[7]Subsidies will be paid to increase the capacity of organic fertilizer manufacturing (compost) through the compost house, livestock (cows), machinery (grass chopper) and training.

[8]One of the GOI's 100-days programmes after the President Yudhoyono took office in 2009.

[9]The project was integrated into the then existing programme initiated in 1999 and conducted in coordination with UNDP and the government of The Netherlands.

Other major contributors include organizations like World Bank, ADB, Japan International Cooperation Agency (JICA), Australian Government Overseas Aid Program (AusAID), Canadian International Development Agency (CIDA), and International Fund for Agricultural Development (IFAD). World Bank's notional annual investment in Indonesia is about US$ two billion of which 4% goes to the agriculture sector and that from the ADB in 2009–2011 is US$ 1 237 million of which 5% goes to the agriculture and natural resources sectors (NCCC, 2009). Compared to the GOI's national budget for the environmental measures, the foreign aid still plays a significant role in the agriculture sector development. Assistance from both organizations in agriculture focuses on rural community development and empowerment, water resources and irrigation management, and land management. Their specific projects that have adaptation implications in the agriculture sector include the Water Resources and Irrigation Sector Management Project (WISMP, World Bank) and Participatory Irrigation Sector Project (PISP, ADB).

JICA is the largest ODA donor to the GOI estimated at US$ 1 129 million per annum (NCCC, 2009). In addition, JICA disbursed annual 300 million to the GOI as a loan called the Indonesia Climate Change Program Loan (ICCPL) in 2007–09, which was the first large-scale programme loan under the government of Japan (GG21 and IGES, 2010).[10] The loan aims at strengthening institutional and regulatory framework in five key sectors, that is, water resources, forestry, agriculture, waste and sanitation, and disaster risk reduction.[11]

To respond to adaptation needs in the agriculture sector in Indonesia, the government agencies, including the Ministry of Planning (Bappenas) and Ministry of Agriculture (MOA), have identified key issues with the help of ICCPL to support climate policy development. The loan's focus in the agriculture sector was set to improve resilience in farm production and to reduce drought risk and five specific actions were selected (i)

the development of an irrigation asset management information system, (ii) issuing a ministerial decree on merging the farmers' association and the water users' association for better water resources and farming management, (iii) carrying out System of Rice Intensification (SRI) practice, (iv) carrying out the Climate Change Field School Programs, and (v) making a 'Dynamic Cropping Calendar Map' with a long-term meteorological forecast (GG21 & IGES, 2009).

The priority issues in agriculture indicated by the GOI and corresponding assistance from donor agencies are summarized in four categories, that is, (i) development of scientific basis, (ii) development of infrastructure and organization, (iii) transfer of knowledge and technologies, and (iv) other issues to identity gaps between needs and assistance (Table 20.3). It may appear that all priority areas have been addressed by the donor agencies, including ICCPL. However, it is not possible to conclude whether or not the assistance provided within each issue area is sufficient in terms of geographical and agricultural population coverage.

20.3 Challenges

Challenges of adaptation actions in Indonesia are examined based on the observations and interviews made during the monitoring activity of the ICCPL between 2008 and 2010. The analysis was conducted in five categories: (1) policy priority setting, (2) development of scientific basis for adaptation, (3) development of infrastructure and community-level organizations, (4) transfer of knowledge and technologies, and (5) policy coordination.

20.3.1 Policy priority setting

Analysis of all the above documents indicates that the country has been able to take forward the climate change adaptation from research to the policy and further identifies the gaps in achieving a full climate-proof agriculture sector. While a comprehensive list of issues was developed and main areas of focus were narrowed down in the process of documentation, the identified actions somewhat gave the impression of being sporadic although these existing actions can be worth continuing and considered as low-hanging fruits for achieving the goals with the available fund sources. Either possible sources of funds for initiating

---

[10]By definition, the assistance was made in the form of a loan to the national general account of the government of Indonesia and the allocation of the fund to these activities were determined by the Government of Indonesia.

[11]Monitoring and advisory services were provided by the Institute for Global Environmental Strategies, Japan.

Table 20.3 Priority issues in agriculture indicated by the GOI and corresponding assistance from donor agencies

Category of issue	Priority issues indicated in existing GOI policies	Assistance from major donor agencies
(1) Development of scientific basis for adaptation such as climate forecasting and down-scaling, impact/ vulnerability assessment, agricultural R&D	Climate change information collection and utilization (including early warning system, etc.) Research on farming technologies	Adaptation strategy and action plan for the water sector (GTZ) Climate forecasting and dynamic crop calendar map making for rice production (ICCPL) n/a
(2) Development of infrastructure and community-level organizations	Development of irrigation and its management, water harvesting, etc.	The Participatory Irrigation Sector Project (PISP, ADB) and Water Resources and Irrigation Sector Management Project (WISMP, World Bank), Irrigation Asset Management Project* (SIIAM, JICA), and ICCPL Merging water users' association and farmers' groups (ICCPL)
(3) Transfer of knowledge and technologies	Improvement of farming techniques and good practices such as System for Rice Intensification (SRI) Institutional/capacity development (including Field School, etc.)	SRI by the Decentralized Irrigation Sector Improvement Management Project (DISMP, JBIC), the Participatory Irrigation Sector Project (PISP, ADB), and ICCPL Climate Field School (ICCPL)
(4) Policy coordination and other	Food and nutrition security system (SKPG), promotion of locally-grown products, advocacy	Climate Change and Environment component under Human Security and Stability (AusAID)

Source: Official documents published by the GOI and project database of donor agencies.
* JICA started the Supporting Implementation of Irrigation Asset Management Project (SIIAM) in July 2009.

some of the additional activities or plans for assessment of current institutional mechanism that will enable the GOI to carry out the proposed activities was not yet clearly identified. In addition, how to improve the scientific basis for the climate change projections, which is fundamental to developing adaptation policies, has not been sufficiently discussed especially at the regional/sub-regional scale due to Indonesia being a large state in part. Furthermore, climate-proof measures and general agricultural actions were not clearly differentiated although there is intrinsically an overlap between these two.

Whether or not the GOI is securing enough funds to address climate change adaptation is a difficult question to answer. However, the sum of the currently available funds both from the GOI and foreign donors hardly matches to the estimated cost of sea level rise itself in Indonesia (about US$ 25.5 billion by the year

2100, EEPSEA, 2008), let alone the overall climate change adaptation measures.

20.3.2 Development of scientific basis for adaptation

There is an extent of repetition of efforts among various agencies involved in generating climate information products. For example, both the meteorological agency (BMKG) and the MOA intend to create useful but similar drought maps and cropping maps that require information from both agencies. This was largely due to limitations in sharing the information and partially due to inter-agency prejudices/sectionalism which is often found elsewhere as well. Interviews conducted by the authors have indicated a lack of resources to install equipment for collecting and disseminating weather information at the district/village level and to support training of trainers for the government officials and experts both at the provincial and district levels to use the information generated by the BMKG. The drought impacts data and other farming records accumulated at the local agricultural offices (province and district levels) are still not being fully utilized by either of the agencies above. By improving the inter-agency coordination, these records could become available and help in refining vulnerability and cropping maps at a finer scale. This could also help in developing downscaled climate change impact projections.

Currently the MOA is mainly focused on adaptation measures for rice production. Although rice is the staple food in many parts of Indonesia and the improvement of rice varieties such as high-yielding, drought/flood-tolerant, and early mature crops is needed, there are other important staple food and cash crops that needs greater attention especially in the wake of the country increasingly importing maize, sugar and soybeans, undermining national food security. More opportunities will be opened up after long-term efforts of crop diversification, varietal development and adoption in these crops.

20.3.3 Development of infrastructure and community-level organizations

The water scarcity scenario under climate change, the development and upgrading of irrigation infrastructure and its user associations are critical for water resources management in the agriculture sector, the largest water user in Indonesia. Main challenges are two-fold: (i) financial and human resources constraint and (ii) insufficient institutional capacity for efficient water resource management. There are 241 irrigation schemes under the responsibility of the central government and only a few of them have completed the inventory making and planning of the irrigation asset management, as of February 2010. The remaining 32 000 small-scale irrigation schemes under provincial and district governments would require proper management for which capacity building of local officers is necessary.

Insufficient institutional capacity for water resources management within the government is partly due to a combination of lack of skilled human resources and too much dependency on experts from the donor-supported projects. The project implementation relies heavily on temporary experts hired by the donor agencies and no sufficient efforts were made to transfer the skill and knowledge pertaining to these projects to the government officials.

Another reason for insufficient institutional capacity, especially on the part of the MOA officials and farmers, comes from the fact that the authority on agricultural water (tertiary canals) had just been transferred from the PU to the MOA in 2007 and hence the capacity of water resource management has not fully developed within the MOA and shift of authority and responsibilities are still in transition. Pricing irrigation water remains a sensitive issue in many developing countries (WRI, 2001). A case study conducted in Indonesia found that farmers in the Brantas River Basin were not charged for use of irrigated water, in fact water tariffs would impose a substantial burden on their welfare and water saving would remain modest (Rodgers & Hellegers, 2005). Yet, appropriate water resources management, including water pricing in close coordination with an agricultural development plan would become possible when water users' capacity in resource management is improved.

20.3.4 Transfer of knowledge and technologies

One of the major challenges faced with SRI and Climate Field School programmes has been their limited expansion. So far, the area under SRI expanded to 170 000 ha compared with the total area of 12 million hectares under paddy (FAOSTAT, 2008).[12] Similarly, the total number of farmers trained only reached approximately 3500 – 4400 farmers out of a total agricultural population exceeding 88 million (FAOSTAT, 2010).[13] Therefore scaling-up of these programmes is needed to increase the effectiveness.[14]

Another challenge, in the case of SRI, is that the farming method is considered to be more labour-intensive and knowledge-intensive than conventional systems because of the need for precise levelling, regular spacing for tender seedling transplantation, frequent weeding and precise water management (GG21 & IGES, 2009). Thus, there is a considerable learning curve required to master the techniques. Obtaining

---

[12] Paddy area harvested in 2008.
[13] Data of 2007.
[14] A total of 265 units of CFS were planned in 2010.

the full potential of the technology also requires greater support from the irrigation infrastructure to regulate precise quantities of irrigation water. Therefore, along with the improvement in irrigation infrastructure mentioned above, increasing the number of trainers essential to scaling up the implementation of these technologies is essential.

In the long-run, more opportunities to gain knowledge and technologies for adaptation will be provided when the Climate Field School becomes able to offer a more comprehensive adaptation programme from which each province or district can choose appropriate modules of courses to tailor to their specific needs. A comprehensive programme can include not only rice cropping calendars developed by the MOA and associated pest management and farming methods but also those for other staple and cash crops and other useful resource conserving farming technologies. As Indonesia's demand for high-value agricultural products such as meat and organic/processed products grow, the Climate Field School should be able to accommodate these emerging dietary trends and advocate best practices to farmers.

20.3.5 Policy coordination

Coordination of policies in mainstreaming collective adaptation efforts across sectors is needed to increase synergy and avoid zero-sum game among ministries. The Second National Communication reflected that 'the involvement of sectors during the development of the first national action plan was very low and the sector failed to mainstream climate change concerns' (GOI, 2009a), and a certain degree of sectionalism still exists, leading to fragmentation of actions and inefficient deployment of resources. Some adaptation measures in the agriculture sector are inseparable from other related measures and they should be rationalized in the nation's broad and long-term development plans. Bioenergy from energy crops such as palm or recycled bio wastes has a large potential to partially meet the rapidly increasing demand for energy in the nation, replacing a part of fossil fuels' use or CH_4 emissions at the same time. Improved farming methods can also contribute to GHG emissions reduction by themselves, including SRI mentioned above. This requires policy coordination among the MOA, Department of Energy and Mineral Resources, and Ministry of Research and Technologies. Introduction of peren-

nial estate crops into unused land, marginal forests, or reclaimed peat land[15] would create opportunities to diversify the economic activities of farmers and those who cannot sustain their livelihood by the conventional methods; however implementing this idea may not be so simple.

For example, the complex issue of peat land conservation, a hot spot of GHG emissions in Indonesia, is handled by multiple ministries based on its functions. Peat land is primarily under the authority of the Ministry of Forestry, but it comes under the Ministry of Public Works or Agriculture if it is a water resource (swamp) or used as farmland.[16] In addition, peat land can be subject to the environmental management law of the Ministry of Environment. Although it is not uncommon that management of a land falls under multiple regulations, a lack of single consistent regulation or coordination makes it difficult to implement monitoring and management of a land – peat land in this case. To make matters worse, changing the nature of peat land status also makes it difficult to keep track of the status of the land combined with ambiguous land ownership often found in Indonesia. This is one of the reasons why biofuel is considered a double-edged sword and some palm plantation is under criticism. A net GHG emissions reduction would become negative if biofuel (palm) is produced from converted peat land. Although the GOI puts a special emphasis on mitigation efforts and identified its potential across sectors, the expected benefits may not realize fully or might possibly be reduced if such measures were not coordinated among ministries and rationalized as a collective effort in the nation's long-term development plans.

Challenges also include insufficient policy coordination within the ministries and with regional governments under the on-going process of decentralization. One example is found in the effort of promoting domestic organic fertilizer use. It can have multiple benefits such as potential GHG emissions reduction from reduced chemical fertilizer production and use as well as reduced dependency on

---

[15]The latest regulation tightens the requirement for peat land utilization of oil palm plantation (GOI, 2009b).

[16]Peat land with peat thickness over 3 m is for conservation. Those with less than 3 m becomes either for protected or production area.

imported fertilizers and improved soil conditions that can enhance the productivity and resiliency to natural environmental changes. In addition to a financial challenge to make a significant investment on digester technology deployment already identified by the GOI (GOI, 2009a),[17] scaling up of effective compost making would be impossible if it were not for either appropriate manure management/treatment facilities for livestock or a mechanism of quality control for the produced compost. This would require better coordination within the MOA, otherwise the effect would remain marginal or unsafe compost produced may cause soil pollution. Furthermore, developing the capacity of farmers' group as well as individuals would need coordinated efforts with regional governments. Local needs for improved input management (water and agricultural chemicals use), support agro-business to take advantage of greater opportunities, or a safety net or financial mechanism such as weather derivatives, micro-finance, and other forms of credit schemes should be assessed in the local context and linked to national policy making.

20.4 Conclusions

The GOI has been developing adaptation policies for the agriculture sector in recent years. In the process of preparation, many issues are identified, discussed and addressed by related projects with the aid of donor agencies. It goes without saying that the GOI needs to secure appropriate fund sources to scale up the existing programmes and/or implement additional activities, prioritizing actions and developing associate institutional capacity/mechanism would be essential. The NCCC and ICCFT are expected to play an important role in coordinating with ministries and foreign donors.

Collecting and analysing climatic data is crucial in the sector. There is a basic institutional capacity to collect and process climate and weather data at the national level; however, downscaling and translating them into adaptation measures in the agriculture sector, including early warning, are still in a nascent

stage. This is a challenging task if the size of nation is considered. The sector would also need increased efforts and capacity to develop varieties such as high-yielding, drought/flood-tolerant, and/or early mature crops not only for rice but also for other crops to minimize the climate change risks.

Attention has been given to agricultural irrigation with aid from multiple donor agencies. The investment will be paid off when institutional capacity for both government officials and water users are also developed enough to be able to duplicate necessary practices.

System of Rice Intensification (SRI) and Climate Change Field School Programs are considered to be a well-structured and effective vehicle to provide farmers with opportunities to gain knowledge and specific farming skills for adaptation. The impact of these practices will become more visible when they are scaled-up through expanded trainings for trainers and provision of comprehensive but tailored curriculum matching local needs. These programmes should try to reach the poorest sections of society since they are likely to be affected most by the climate change.

Some adaptation measures in the agriculture sector are inseparable from other sectors and should be rationalized in the nation's broad and long-term development plans. Shared roles by the local governments and private sector should also be encouraged. It is important for the country to identify and promote the mitigation co-benefits of adaptation actions considering the potential of GHG emission reduction from the agricultural sector. Some mitigation measures could generate additional benefits to the sector or rural communities such as provision of organic fertilizers and renewable energy. To do so, more coordinated policies across sectors are needed for collective mitigation efforts and to avoid zero-sum games.

More importantly, institutional capacity development is essential for at both the national and local levels for sustainability of the adaptation efforts. In Indonesia there are already good mechanisms of technology transfer to farmers and there exist opportunities to learn from experts and consultants from donor agencies. This type of investment may seem invisible but will pay off in the long run. Strengthened capacity of farmers' groups will become the basis of development for agro-business in Indonesia for years to come.

[17]Some private sector firms (such as PT. Ajiubaya, a plywood manufacturer in Sumatra) are already using small (4–6 MW) biomass energy plants (ADB, 2009a).

Acknowledgements

Authors gratefully acknowledge the inputs from numerous government officials interviewed during monitoring and advisory activities under ICCPL. Authors also acknowledge valuable support provided by, Dr Jun Ichihara, Mr Yoshitaro Fuwa and Ms Maiko Yoshizawa in coordinating and collecting policy documents and monitoring data from various government agencies in Indonesia during the implementation of ICCPL.

References

ADB (Asian Development Bank), 2009a. *The Economics of Climate Change in Southeast Asia: A Regional Review.* Manila: Asian Development Bank.

ADB, 2009b. *RETA 6427 A Regional Review of the Economics of Climate Change in Southeast Asia (RRECCS) Country Report: Indonesia.* Manila: Asian Development Bank.

BAPPENAS (Badan Perencanaan Pembangunan Nasional), 2008. *The National Development Planning Response to Climate Change.* Jakarta: Badan Perencanaan Pembangunan Nasional.

BAPPENAS and GTZ (Gesellschaft für Technische Zusammenarbeit), 2009. *Roadmap mainstreaming of climate issue in national development planning responding to climate change in forestry sector* (draft), November 2009. Jakarta: Bappenas.

EEPSEA (Economy and Environment Program for Southeast Asia), 2008. *Proceedings of EEPSEA Conference on Climate Change: Impacts, Adaptation, and Policy in Southeast Asia.* Bali, Indonesia: Economy and Environment Program for Southeast Asia, 13th–15th of February 2008.

FAO (Food and Agriculture Organization), 2003. *Agricultural reform in Asia.* In: FAO, 2003. *Trade Reforms and Food Security: Conceptualizing the Linkages.* Rome: Food and Agriculture Organization. Ch.13.

FAO, 2005. *Fertilizer Use by Crop in Indonesia.* Rome: Food and Agriculture Organization.

FAO, 2010a. *'Climate-Smart' Agriculture: Policies, Practices and Financing for Food Security, Adaptation and Mitigation.* Rome: Food and Agriculture Organization.

FAO, 2010b. *Advancing Adaptation through Communication for Development. Proceedings of the technical session on Communication in the Third International Workshop on Community-Based Adaptation to Climate Change.* Dhaka, Bangladesh: February 2009.

FAOSTAT, 2008. *Paddy area harvested in Indonesia.* [online] Available at: <http://faostat.fao.org/default.aspx> [Accessed 18 May 2008]

FAOSTAT, 2010. *Total agricultural population in Indonesia.* [online] Available at: <http://faostat.fao.org/default.aspx> [Accessed 26 February 2010]

Fuglie, K., 2010. Indonesia: From food security to market-led agricultural growth. In: Alston, J.M., Babcock, B.A. and Pardey, P.G. eds., 2010. *Shifting Patterns of Agricultural Production and Productivity Worldwide.* Ames, Iowa: Midwest Agribusiness Trade Research and Information Center, Iowa State University. Ch.12.

GG21 (Global Group21) and IGES (Institute for Global Environmental Strategies), 2009. *Draft Final Report on the Advisory and Monitoring Activity for the Climate Change Programme Loan to the republic of Indonesia.* Tokyo: Global Group21 Japan Co., Ltd (GG21); Hayama: Institute for Global Environmental Strategies (IGES).

GG21 and IGES, 2010. *Draft Programme Evaluation Report on Indonesia Climate Change Program Loan 2007–2009.* Tokyo: Global Group21 Japan Co., Ltd (GG21); Hayama: Institute for Global Environmental Strategies (IGES)

GIEWS (Global information and early warning system on food and agriculture), 1998. *Special Report: FAO/WFP Crop and Food Supply Assessment Mission to Indonesia.* [html] Rome: Food and Agriculture Organization (FAO). Available at: <http://www.fao.org/docrep/004/w8458e/w8458e00.htm> [Accessed 6 December 2010].

GIEWS, 2010. *GIEWS Country Brief: Indonesia.* [pdf] Rome: Food and Agriculture Organization (FAO). Available at: <http://www.fao.org/giews/countrybrief/country/IDN/pdf/IDN.pdf> [Accessed 6 December 2010].

GOI (Government of Indonesia), 2009a. *Summary for Policy Makers: Indonesia Second National Communication Under The United Nations Framework Convention on Climate Change (UNFCCC).* Jakarta: Government of Indonesia, November 2009.

GOI, 2009b. *Indonesia Climate Change Sectoral Roadmap (ICCSR) Synthesis Report.* Jakarta: Government of Indonesia (GOI), December 2009.

GTZ, 2007. *Adapting to Climate Change in Indonesia.* Fact sheet, Jakarta: GTZ Office Jakarta.

Hadar, I.A., 2009. Food security in RI: Time for policy change. [online] *The Jakarta Post*, 28. April 2009. Available at: <http://www.thejakartapost.com/news/2009/04/28/food-security-ri-time-policy-change.html> [Accessed 21 May 2012]

ICCTF, 2010. Indonesia Climate Change Trust Fund (ICCTF). [online] Available at: <http://www.icctf.org/site/> [Accessed 7 December 2010]

IIED (International Institute for Environment and development), 2009. *Participatory Learning and Action 60.*

London: International Institute for Environment and development.

JBIC (Japan Bank for International Cooperation), 2008. *Background and policy note on climate change program loan (Cool Earth Program Loan) to the Republic of Indonesia.* Tokyo: Japan Bank for International Cooperation.

KLH (Kementerian Lingkungan Hidup), 2007. *The National Action Plan Addressing Climate Change.* Jakarta: Kementerian Lingkungan Hidup.

KLH, 2010. *Second National Communication to UNFCCC (SNC),* Jakarta: Kementerian Lingkungan Hidup.

Las, I., Boer, R., Syahbudin, H., 1999. *Analisis peluang penyimpangan iklimdan ketersediaan air pada wilayah pengembangan IP Padi 300 (Probability analysis of climate extreme in major rice growing areas).* Bogor: Pusat Penelitian Tanah dan Agroklimat, Laporan Proyek ARMP-II, Badan Penelitian dan Pengembangan Pertanian.

Ministry of Water Resources, China, 2009. *Farmer Water Users' Association in China – Making a Difference.* [pdf] Available at: <http://files.inpim.org/CBP/SS%20Learning%20WUA-Flyer.pdf> [Accessed 25 December 2009].

Mishra, A., 2009. System of rice intensification (SRI): a quest for interactive science to mitigate the climate change vulnerability. IOP Conf. Series: *Earth and Environmental Science* 6 (2009) 242028.

NCCC (National Council on Climate Change), 2009. *National Economic, Environmental and Development Study (NEEDS) for Climate Change Final Report.* Jakarta: National Council on Climate Change, Republic of Indonesia.

Prabhakar, S.V.R.K., Sano, D. and Srivastava, N., 2010. *Food Safety in the Asia-Pacific Region: Current Status, Policy Perspectives and a Way Forward.* In: *Sustainable Consumption and Production in the Asia-Pacific Region: Effective Responses in a Resource Constrained World.* Institute for Global Environmental Strategies, White Paper III, pp 215–238. Hayama, Japan: Institute for Global Environmental Strategies.

Rochayati, S. and Husnain, 2010. *Fertilizer Management for Improving Lowland Sawah Productivity in Indonesia: Integrated Plant Nutrient Management System.* International Conference of Balanced Nutrient Management For Tropical Agriculture, 12–16 April 2010. Kuantan Pahang, Malaysia.

Rodgers, C. and Hellegers, P., 2005. *Water pricing and valuation in Indonesia: Case study of the Brantas River Basin.* EPT Discussion Paper 141. Washington D.C.: International Food Policy Research Institute (IFPRI).

Sano, D. and Prabhakar, S.V.R.K., 2010. Some policy suggestions for promoting organic agriculture in Asia. *Journal of Sustainable Agriculture,* 34(1), pp.80–98.

Sass, R.L., Fischer, F.M., Wang, Y.B., Turner, F.T. and Jund, M.F., 1992. Methane emission from rice fields: the effect of floodwater management. *Global Biogeochem. Cycles,* 6, pp.249–262.

Sato, S. and Uphoff, N., 2007. A review of on-farm evaluation of system of rice intensification (SRI) methods in eastern, Indonesia. CAB Reviews: Perspectives in Agriculture, Veterinary Science, *Nutrition and Natural Resources,* 2(54), pp.1–12. Commonwealth Agricultural Bureau International, Wallingford, UK.

Setyanto, P., Surmaini, E. and Boer, R., 2009. *Mitigation of Methane Emission from Rice Field.* In: MoE, 2009. *Projection of GHG Emission and Mitigation Analysis for Indonesia's Second National Communication.* Jakarta: Ministry of Environment and United Nations Development Programme.

Stoop, W.A., Adam, A. and kassam, A., 2009. Comparing rice production systems: A challenge for agronomic research and for the dissemination of knowledge-intensive farming practices. *Agricultural Water Management,* 96(11), pp.1491–1501.

USAID (United States Agency for International Development), 2010. *Asia-Pacific Regional Climate Change Adaptation Assessment Final report: Findings and Recommendations.* Washington D.C.: United States Agency for International Development.

U.S. EMBASSY JAKARTA, 2008. *Petroleum Report Indonesia 2007–2008.* [online] Available at: <http://www.usembassyjakarta.org/econ/(PR_7_2008)_Petrochemicals_and_Fertilizer.pdf> [Accessed 22 December 2009].

WRI (World Resources Institute), 2001. *Earth Trends: Featured Topic – Will There Be Enough Water?* Washington DC: World Resources Institute.

21 'Climate Refugee' Is Not a Hoax. But We can Avoid it. Empirical Evidence from the Bangladesh Coast

M. Mustafa Saroar[1,2] & Jayant K. Routray[1]

[1]School of Environment, Resources and Development (SERD), Asian Institute of Technology (AIT), Bangkok, Thailand
[2]Urban and Rural Planning, School of Science, Engineering and Technology (SET), Khulna University, Bangladesh

21.1 Climate change and climate refugees – the research agenda

There is widespread speculation that climate change induced extreme events would trigger huge emigration from the fragile densely populated coasts even by the end of the first quarter of this century (Paul, 2010). This chapter aims to present a case from Bangladesh which is one of the deltaic countries most susceptible to climate change induced extreme events and sea level rise (SLR). First, it presents the likely scenario of climate change induced migration although it is hard to distinguish the effect of climate change on migration from the effects of other underlying drivers. Second, it identifies the key concerns over livelihood security of the coastal population which might trigger mass emigration from the vast coastal tract. Finally, it explores the specific measures that have the potential to slow down the pace of emigration and hold back prospective emigrants who otherwise would migrate for specific reasons that relate to climate change vulnerability in their natural resources-based livelihood security. It is expected that the findings will help policy makers and planners to effectively deal with the 'climate refugee' issues through anticipatory adaptation planning.

For some parts of the globe, accelerated SLR and frequent extreme events such as floods, droughts, cyclones and storm surges induced by climate change is no longer a contested issue (IPCC WG I, 2007), although there is considerable uncertainty about the timing and extent of the events. The Intergovernmental Panel on Climate Change (IPCC) for instance, presents a SLR between 92 cm (Leggett, Pepper & Swart, 1992), 88 cm (IPCC WG I, 2001) and 59 cm (IPCC WG I, 2007) by the end of this century in its second, third and fourth assessment report, respectively. Various other studies have shown an SLR of a few centimetres to even a few metres by the same date. This wide uncertainty is attributed to the use of various plausible scenarios of the future world in their respective models. Although it is expected that scientific advances would reduce the uncertainty, complete elimination of uncertainty is never possible in any way. The bottom line is humanity has to deal with the issue of climate change induced SLR assuming certain levels of uncertainty.

Most projections of SLR are for 20 to 30, 50 to 80, and almost 100 years from now. The local people, however, prefer to deal with the immediate problem than a problem which would be felt decades from now. The rational choice theory supports this position. Contradiction arises for the accelerated SLR, however.

Climate Change Adaptation in Practice: From Strategy Development to Implementation, First Edition.
Edited by Philipp Schmidt-Thomé and Johannes Klein.
© 2013 John Wiley & Sons, Ltd. Published 2013 by John Wiley & Sons, Ltd.

Although the SLR itself is a very slow process, many of its associated climatic events, for example, cyclonic storms, tidal surges and salinity intrusion might be amplified both in terms of frequency and magnitude/intensity over short time spans. This inherent linkage of SLR with other climatic events makes it very distinct from other slow processes such as desertification and other forms of land degradation. This peculiar hybrid nature of SLR leaves no room for nations which have vast coastal areas to wait for a clearly visible effect of SLR and respond afterwards. Time is very critical for nations with great coastal areas as Middleton (1999) warned more than a decade ago when he said that 170 million people would be affected from 22 low-lying countries. Over 150 000 km^2 of land would be lost, including 62 000 km^2 of coastal wetland. The most severely affected countries would be small island countries and countries having vast coastal areas (Nicholls, 1995; Mimura, 1999; IPCC WG II, 2007). Bangladesh is one of the countries that would be worst affected by SLR (Nicholls, 1995; UNDP, 2004; IPCC WG II, 2007; UNISDR 2009).

Drawing from the study of Ahmed and Alam (1998), the World Bank (2000) has shown a projected SLR (high-end scenario) of 30 cm for 2030 and 50 cm for the year 2050 along the coast of Bangladesh. An average rise of one centimetre per year is estimated. However, projected SLR of 14 cm, 32 cm and 88 cm along the Bangladesh coast for the years 2030, 2050 and 2100, respectively, are adopted in National Adaptation Programmes of Action (NAPA) for Bangladesh to prepare a national strategy for climate change adaptation (GOB, 2005; 2006; 2008). Such an SLR along the coast of Bangladesh is likely to amplify the devastating power of currently occurring disasters, such as coastal floods, salinity intrusion, tidal surges and cyclonic storms, that ravage the country recurrently (Castro-Ortiz, 1994; Nicholls, Leatherman, Dennis & Volonte, 1995; Huq, Karim, Asaduzzaman & Mahtab, 1998; Ali, 1999; 2003; Ali Khan, Singh & Rahman, 2000; World Bank, 2000; Singh, Ali Khan, Murty & Rahman, 2001; Cannon, 2002; GOB, 2006, 2008). Each of these have various effects on the coastal topography or morphology causing a series of reactions on various aspects of coastal livelihood, including agriculture, fisheries, forestry, food security, human health, transport and infrastructure as well as settlement and housing (CARE, 2003; Saroar & Routray,

2012). Therefore, climate change and SLR are considered to be the most pressing problem against which the population of Bangladesh in general and coastal population in particular have to adapt in this century and beyond (GOB, 2008). Failure to adapt will surely result in a huge number of climate refugees, most of whom would originate from vast coastal areas. Therefore, the issue of climate refugees is a grave concern not only for the long-term future, but also for the present (Ingham, Ma & Ulph, 2005; Stern, 2006).

Adaptation involves processes and actions in order to better cope with, manage or adjust to changing conditions, stress, hazard, risk or opportunity (Smithers & Smit, 1997; Fankhauser, Smith & Tol, 1999; Smit, Burton, Klein & Wandel, 2000; Smit & Skinner, 2002; Brooks, Adger & Kelly, 2005; Smit & Wandel, 2006). For adaptive response, there are only three options, viz. protection, accommodation and retreat for the population who have been living in coastal settings (Klein et al., 2001; Sterr, 2008; Tol, Klein & Nicholls 2008). 'Protection' through the construction of some form of coastal defence to minimize exposure often works well but beyond the private adaptive response (World Bank, 2000). 'Accommodation' by reducing sensitivity and enhancing adaptive capacity can be an adaptation option where there is enough room for both private and public to play a role. The last option, 'retreat' or pulling back (or simply emigration) to a safer place is basically evacuation which might involve the relocation of 25 millions of people from the coastal plains of Bangladesh to the inlands which are already congested (GOB, 2008). Experts already warn that for a country short of land like Bangladesh and which may even lose one-fifth of its landmass due to SLR, retreat may not be a sustainable solution (Paul, 2010). Rather, adaptation *in situ* which is just the opposite of 'retreat' is considered to be the best option as it does not require evacuation and relocation (Castro-Ortiz, 1994; Saroar & Routray, 2010a).

While it is well understood that anticipatory adaptation *in situ* against the impacts of climate change induced extreme events including SLR would be a plausible solution for a country short of land like Bangladesh, such response might not take place automatically, especially when there is uncertainty. Unlike reactive or concurrent responses, anticipatory adaptation is not just something to be undertaken in the moment, when and if the event occurs, these are instead prevention measures to harness the benefit for the future (Kelly & Adger, 2000; Grothmann & Patt, 2005). As the uncertainty and length of time to accrue the benefit from anticipatory adaptation against SLR is high it is less likely for the people to go for anticipatory adaptation when they are pressed overwhelmingly with immediate needs (Smith, 1997; Tol, Klein & Nicholls, 2008).

However, there are empirical evidences that private anticipatory adaptive responses work well if publicly planned adaptations are undertaken beforehand. Unfortunately, most research in this regard is done in the context of developed countries or where the SLR has no impact. In the United Kingdom (UK) and the Netherlands, for example, the national government constructed the hard coastal defence to encourage soft measures from private individuals (Tol, Klein & Nicholls, 2008). In Germany, coastal defence against flood is done by the state government and the local communities (Sterr, 2008), while in Ireland the government takes measures that encourage private adaptive responses (Burbridge & Humphrey, 2003). Although none of these qualitative studies unveil exactly what kind of public initiative (i.e. exposure and sensitivity minimization, and adaptive capacity maximization measures) influences the private adaptive responses to what extent, yet these studies have encouraged us to hypothesize that visible public initiatives in terms of exposure minimization, sensitivity minimization and adaptive capacity maximization would encourage private anticipatory *in situ* adaptation in coastal Bangladesh. Eventually with this the likely trend of climate stress induced emigration could be avoided to a greater extent if complete avoidance is not possible.

In the case of Bangladesh, a knowledge gap in the area of anticipatory *in situ* adaptive research is notably apparent. Only scanty research is observed that identifies the changes in the hydro-meteorological processes (Islam, Huq & Ali, 1998; Ali Khan et al., 1999; Singh, 2001; Singh, Ali Khan, Murty & Rahman, 2001), the locations particularly at risk (Castro-Ortiz, 1994; Huq, Karim, Asaduzzaman & Mahtab, 1998; Ali, 1999; GOB, 2006; 2008) and that characterizes the impacts (Nicholls, Leatherman, Dennis & Volonte, 1995; World Bank, 2000; Cannon, 2002; Mirza, 2002; Agrawala et al., 2003; Choudhury et al., 2005; Faruque & Ali, 2005; Islam, Ahmad, Huq &

Osman, 2006; Patt, Daze & Suarez, 2009; Saroar & Routray, 2010b; 2012). Apart from the mentioned issues, Khondker (1996), Leaf (1997), Rashid and Michaud (2000), Schmuck (2000), Matin and Taher (2001), Cannon (2002), Paul (2005; 2010), Khan (2008), Assan, Caminade and Obeng (2009) and Azad, Jensen and Lin (2009) have investigated the various aspects of current coping strategies with current natural calamities and hazards. None of them, however, addresses the issue of private anticipatory *in situ* adaptation in the context of future SLR and its associated events as a potential strategy to avoid mass emigration from the coastal areas of Bangladesh. This is one of the grey areas of an existing body of anticipatory adaptation research in Bangladesh which this research has already addressed. Specifically, this chapter has first identified the nature of the likely impacts of SLR and its associated events that might force the coastal population to emigrate. Secondly, it has identified the exposure and sensitivity minimization, as well as the adaptive capacity enhancement measures that have the potential to restrain people from migrating for specific reasons linked to climate change induced livelihood insecurity. This analysis is expanded with the aid of statistical modelling, which significantly predicts the influence of various vulnerability reduction measures (i.e. measures related to exposure, sensitivity and adaptive capacity) in altering the affected-person's intentions from emigration to anticipatory adaptation. The findings imply that it might be a good idea to devise public adaptive responses to encourage anticipatory adaptation *in situ* among the population who otherwise might emigrate due to SLR induced events.

21.2 Study area and the methods

21.2.1 Selection of study area, respondents and survey procedures

Earlier research has confirmed that among the coastal areas of Bangladesh, Patuakhali District is most affected by storm surges, salinity intrusion, tidal floods and may continue to experience these hazards given the scenarios of future SLR (Castro-Ortiz, 1994; Ali & Chowdhury, 1997; Huq, Karim, Asaduzzaman & Mahtab, 1998; Ali Khan, Singh & Rahman, 2000; World Bank, 2000; Singh, Ali Khan, Murty & Rahman, 2001). Three 'Union Parishads' (lowest tier

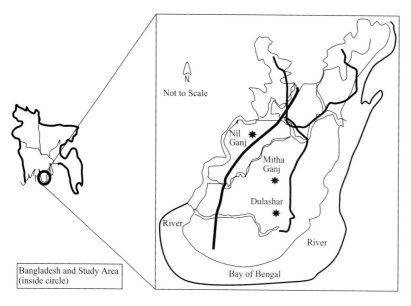

Figure 21.1 Study sites in Kalapara Upazila (Sub-district) in relation to map of Bangladesh.

of local government) from 'Kalapara Upazila' (sub-district) of Patuakhali District which is flanked by the Bay of Bengal and from 0 to 30 cm (one foot contour) above mean sea level were selected for this study (see Figure 21.1). The whole study area is under tidal influence due to the presence of hundreds of natural channels and creeks connected to three river systems and the Bay of Bengal. People who have been living there, are historically prone to various natural calamities. Altogether 285 randomly selected households were surveyed during January and April 2009. To get a comprehensive and representative picture of the study area, population belonging to various occupation categories, age groups and gender compositions (175 male and 110 female) were included in the survey process. Interviews were usually conducted with the head of the household through administering semi-structured questionnaires. The Bengali version of the questionnaire was administered to facilitate the survey process.

21.2.2 Framework of research inquiry

The objective in this section is firstly, to identify the coastal people's causes of concerns about their livelihood security that is related to SLR induced events which have immense potential to force them to emigrate. Secondly, the respondents had to identify measures they believe to persuade people from emigrating.

Accordingly the following working procedures were pursued:

(i) First, the respondents were introduced to the likely scenarios of SLR for the year 2020–30, 2050–80 and 2100 and were asked what they or their decedents would probably do when faced with the future SLR and its associated impacts on their livelihood security. Numerous responses inspired us to identify the likely trend of future emigration from the coastal areas. The whole process is discussed in more detail in the respective section.

(ii) Secondly, the respondents were asked to nominate a maximum of five key sources of concern that relate to their livelihood security, which they believe are the triggering factors for their or their decedents' (e.g. children or grandchildren) likely emigration from the coastal areas in the event of future SLR and its associated events. Therefore, it was assumed that if these sources of concern could be adequately addressed through various vulnerability reduction measures, the respondents would change their mind, that is, follow adaptation *in situ* instead of emigration from the coast.

(iii) Thirdly, following Mimura (1999), Klein et al. (2001) and the place-based vulnerability analysis framework of Cutter (2003), the respondents were provided with a list of 18 measures that have the potential to reduce livelihood vulnerability. Among these measures, six relate to exposure minimization, five relate to sensitivity minimization and the remaining seven relate to adaptive capacity maximization. Broadly, these measures include technological, managerial, policy and other supportive measures that have the potential to ameliorate the negative impacts of SLR and its associated events on coastal livelihood.

The respondents were asked to identify appropriate measures that in their estimation would stop possible emigration (due to causes identified earlier) by minimizing livelihood vulnerability. Considering the level of the rural, mostly illiterate, population's understanding, the questions were asked in a lay public language/word. In many cases they were given a clue or hint to facilitate their responses. The use of such techniques (i.e. providing a list of the technical solutions in a common/lay public language) at group level were observed in many instances (Nicholls, Leathermann, Dennis & Volonte, 1995; Mimura, 1999; Nicholls & Lowe, 2004).

(iv) Fourthly if a particular measure is believed to stop the future trend of emigration (would have been triggered by any of the five causes identified earlier) it is rated by one and by zero if the measure fails. Therefore the responses are binary coded. Effectiveness of all the 18 measures in terms of stopping emigration was tested against each of the five causes of livelihood vulnerability. With the help of this we got the respondent's attitude towards a decisional change in the case of emigration if appropriate (vulnerability reduction) measures are in place. Preference for these 18 measures are used as independent variables (Table 21.3) and effectiveness to stop emigration (most likely of the five causes) of respondents are used as dependent variables in the Binomial Logistic Regression (BLR) model.

Before using all the 18 independent variables in each BLR model, variables that are constant (i.e. 100% of responses are the same) and variables that have colinearity or multi-colinearity (i.e. correlation coefficient with one or more variables in excess of 0.70) (Bryman & Cramer, 2001) have been identified and dropped. Therefore, independent variables that were constant and dropped are 'construction and maintenance of cyclone/flood shelter', 'awareness campaign on CC-SLR', and 'cash incentive for post disaster rehabilitation'. Similarly the variable 'enhanced/quicker

and better access to cyclone/flood shelter' is excluded because of its higher colinearity ($r = 0.83$, $p < 0.01$) with the variable 'emergency rescue and recovery unit at community level'. Finally 14 variables were used in each BLR model as independent variables (see all the 18 measures in Table 21.3).

(v) Therefore finally, four separate binomial logistic regressions (BLR) models were developed to predict the likely influence of each of the three kinds of measures, viz. exposure minimization, sensitivity minimization and adaptive capacity maximization, on restraining people from emigrating for a particular reason (source of concern). We developed four sets of BLR models for the four sources of concern. These are discussed in detail in the next sections.

21.3 Survey findings

21.3.1 Socio-demographic profiles of the respondents

About 61% of the respondents are male and 39% are female. The average age of the respondents is 49 years and their average duration of stay in the same locality is 44.45 years, meaning that a respondent's spatial mobility in terms of permanent migration is not very common. The latter figure does not consider the permanent shift of female residents from their parental houses to their husbands' houses after marriage due to cultural practice/ground. Almost 60% of the respondents are illiterate, followed by an educational attainment of 5-grade (20%) and 8-grade (5%). Only about 2% of them are college/university graduates. The respondents' most dominant occupation is crop agriculture (30.5%) followed by day labourers (18.6%) and fishing (17.2%). Other available occupations are petty trade, business, transport-work, formal jobs and various off farm and on farm economic activities. The yearly average (mean) income of the respondents' families is 141 438 BDT (2065 US$; 1 US$ = 68.5 BDT). The average farm size is 0.36 ha. Nevertheless, one fourth (24.6%) of the families do not have any farmland, meaning that farmland holdings is highly skewed. Almost 83% of the respondents do not belong to any social group. The remaining 17% are members of various social and economic groups/organizations.

21.3.2 Respondent's concern over livelihood security and forced migration

In general, respondents are not very familiar with the term 'sea level rise'. However, they are very familiar with periodic/occasional inundation of their farmlands by rushing water from the nearby sea, the Bay of Bengal. They experience such inundation due to recurrent exposure to tidal floods, storm surges, breach of embankments and high tides. Considering their low level of familiarity with SLR, practical means were used to give them an understanding of it. For example, permanent inundation of their farmland next to their homestead with knee-deep water (30–40 cm) or surge heights a few metres higher than they experienced in the past. When the essence of SLR is communicated using this kind of iconic image they were able to understand the impact of various effects of SLR on their livelihood security.

Three story lines that narrate likely scenarios of CC-SLR for the years 2020–30, 2050–75, and 2080–2100 were presented to the respondents. These scenarios were adopted from the National Adaptation Programme of Actions (NAPA) for Bangladesh (GOB, 2005). Accordingly, the respondents were asked what they or their descendents would probably do if farmlands next to their homesteads went permanently 'almost ankle-deep (10–15 cm)', 'half knee-deep (20–25 cm)' and 'knee-deep (30–40 cm)' under saline water by the years 2020–30, 2050–75 and 2080–2100 respectively? Altogether eight types of responses were identified that indicate their spatial mobility and adaptive responses to livelihood security. These are:

- building homesteads on higher stilts and continuing with the same occupation anyway;
- raising homestead and continuing agriculture with saline and flood tolerant varieties;
- raising homestead and use of saline affected land for aquaculture;
- shifting to non-farm occupations;
- evacuating and settling near safer localities;
- evacuating and taking refuge nearby urban centres and major metropolises, especially in Dhaka;
- emigrating from the homestead without predefined destination.

As we are interested in the possible occurrence of large-scale emigration from the vast coastal areas, we, therefore, separated their spatial mobility and adaptive responses into three more distinct and meaningful categories by collapsing similar categories into one. These are: *in situ* adaptation with same occupation, *in situ* adaptation with changed occupation, and (forced) emigration (Table 21.1). Our main interest concerns the forced migrants to whom we referred as climate refugees who evacuate their current locations and settle permanently elsewhere, mostly to nearby cities, metropolises and to the capital city, Dhaka. Our finding suggests that in the immediate term (2020–30), no occurrences of evacuation/forced emigration will take place. However, almost 30 to 60% of the respondents believe that they or their offspring might prefer emigration in the medium term (2050–75) and distant term (2080–2100) scenarios of SLR and its associate events if no appropriate measures are taken beforehand.

Currently, 35 million people live on the coast of Bangladesh; almost 17 million people inhabit the

Table 21.1 Likely trend of emigration from the coast for three plausible scenarios of SLR

Intention for in situ adaptation or forced emigration if farmlands next to their homesteads were permanently inundated by saline water:	Ankle height (10–15 cm) by 2020–2030	Half knee height (20–25 cm) by 2050–2075	Almost knee height (30–40 cm) by 2080–2100
i. Adaptation *in situ* with same occupation	69.5% (198)	46.3% (132)	23.2% (66)
ii. Adaptation *in situ* with possibly changed occupation	30.5% (87)	23.9% (68)	17.2% (49)
iii. Permanent evacuation (forced migration)	0% (0)[a]	29.8% (85)	59.6% (170)
Total	100% (285)	100% (285)	100% (285)

Note: [a] Figure in parenthesis indicates frequency/absolute response. Respondents did not consider a 10 cm inundation as a problem and did reject the idea of permanent evacuation.

coastal low-lying areas and by 2050, the population of Bangladesh will increase by 130 million people many of whom will live in that fragile coastal area (GOB, 2008). Assuming no population growth in the coastal areas, one could expect that by 2050 almost five million people would migrate from the low-lying parts of the coast alone. However, if we consider emigration from the whole coastal tract the number would be double, that is, about 10 million population. How this huge number of people shall be relocated if two-thirds of them, eventually (before the end of this century), evacuated the low-lying areas due to SLR is a really critical concern. Massive relocation of the coastal population could only be avoided by careful anticipatory adaptation planning for which information about the main causes of people's livelihood security concerns and about the measures that, in people's estimation, would minimize their livelihood vulnerability in the face of future SLR and its associated events is needed.

21.3.3 Respondents' concerns about the sources of livelihood insecurity

To capture the respondents' stated concerns about their possibly forced emigration they were asked to identify the key sources of livelihood vulnerability that might force them eventually to end up as climate refugees. The respondents had the option of giving multiple responses to this open-ended question. Almost half of the respondents, that is, 48% (137 out of total 285), with their current knowledge and understanding have the firm intention not to evacuate (i.e. ready to adopt anticipatory adaptation) no matter what additional measures, beyond their own, would be realized in future. The other half (148 out of total 285), however, pointed out that they or their offspring might permanently leave their current homesteads and might settle elsewhere due to one or more causes of livelihood vulnerability.

Table 21.2 reports that shrinking income, as it is the key livelihood security concern, has the potential to trigger mass emigration. 35.8% of the respondents belonging to rank one consider this as the key cause of possible emigration. The second most important stated concern is the likely permanent disruption of people's access to various services including local healthcare, marketplaces or school. About 35% of the respondents would probably emigrate permanently to avoid this kind of vulnerability. Similarly, 30.4% stated the concern of losing the free or low cost access to salt-free water for drinking and other uses, which may eventually trigger mass emigration. Likewise, food insecurity (production and availability) as a stated concern is

Table 21.2 Respondents' stated concerns about livelihood insecurity which may trigger mass emigration from the coast

Respondents' stated concerns about livelihood insecurity due to SLR and its associate events which may force them to emigrate from the coastal areas N = 148**, total responses = 208*):	Count	Percent of Responses	Percent of Cases	Rank
i. If current main sources of income are likely to encounter irrecoverable loss due to events associated with SLR	53	25.5	35.8	1
ii. If current physical accesses to services – local health care, market place, schooling are likely to be severely affected due to events associated with SLR	52	25.0	35.1	2
iii. If current free/low cost accesses to potable water are likely to be diminished due to events associated with SLR	45	21.6	30.4	3
iv. If food securities (production/availability) are likely to be severely affected due to events associated with SLR	39	18.8	26.4	4
v. If most relatives are likely to quit/evacuate due to perceived threat of events associated with SLR	19	9.1	12.8	5

Note: * Multiple responses.
** Remaining 135 (285–148) respondents do not consider emigration.

expressed by almost one fourth of the respondents. It is a bit strange however, that more people are likely to emigrate because of a perceived lack of availability of salt-free drinking water than a lack of food production and availability. Only about one in ten persons might prefer emigration due to the emigration of their close relatives. The latter reason is not really important as about 5% emigration is normal from this coastal belt.

21.3.4 Vulnerability reduction measures to avoid the occurrence of climate refugees

To avoid the occurrence of climate refugees there is a need for livelihood vulnerability reduction measures that would encourage a private anticipatory adaptive response which eventually would reduce if not completely stop emigration from the fragile coastal region. At the very basic level, three types of measures that is, exposure and sensitivity minimization, and adaptive capacity maximization are adopted to address the vulnerability that relates to livelihood security of natural resources-dependent community (Klein et al., 2001). Respondent's preference for various measures are presented in the following sections.

Measures to minimize the exposure

Exposure is the nature and degree to which a system experiences environmental or socio-political stress. The characteristics of these stresses include their magnitude, frequency, duration and areal extent of the hazard (Burton, 1997). In the context of hydro-meteorological disasters on the coast, the exposure could be the number of natural hazards per year, the extent of inundated land, the height of inundation, the duration of inundation, and so on. Due to natural settings people in some areas are more exposed to recurrent hazards than people living in some other areas. Usually people on an exposed coast are more vulnerable than people living on a coast that is protected by natural vegetation such as green belt, coastal mangrove, or manmade protection such as sea walls or embankments.

As we noted earlier the respondents were provided with a list of exposure minimization measures and were asked to identify one or more measures that would encourage them or their offspring to go for anticipatory adaptation rather than evacuation of their homestead in the case of a future CC-SLR. To mini-

mize exposure the construction of more multipurpose cyclone/flood shelters has been given their highest priority (Table 21.3). In fact, all of the respondents consider construction of new multipurpose cyclone/flood shelters as essential measures; not simply a necessity. Maintenance of existing embankments/levees seems to be another important requirement to encourage them anticipatory adaptation. Almost 75% of the responses put emphasis on the construction of effective floodgates. Similarly, about 65% think that the construction of new flood walls/embankments/levees will encourage anticipatory adaptation instead of emigration. Mangrove protection and forestation got identified as exposure minimization measures by 29% respondents (Table 21.3). It was a bit surprising for us to observe very low demand for mangrove afforestation programmes. However, close consultation with the community reveals that most respondents do not support the idea of mangrove protection and afforestion for two reasons. The first reason is attributed to their fear of losing access to Sundarbans mangrove forests and its adjoining rivers if mangrove protection programmes are consolidated. In fact a large majority of the respondents earn their livelihood directly from the Sundarbans mangrove forest and rivers located in and around the forests. Secondly, it is often the case that when a mangrove plantation programme is launched the landless poor and marginalized families who build their homestead on *khas* land (government owned waste-land) have to vacate the occupied land. Therefore, they do not prefer mangrove afforestation programmes although such programmes could provide a natural defence against exposure to cyclonic hits.

Measures to minimize the sensitivity

Sensitivity is the degree to which a system can be modified or affected by perturbations of certain types (McCarthy et al., 2001; Adger, 2006). In the context of coastal management sensitivities of both the system and the elements of the system need to be considered. For example, some families live in such remote locations that a high-tide alert gets through to them too late and, as a result, they cannot get under shelter effectively. It implies that those families are sensitive to late warnings due to their locational disadvantage. Similarly, some families can survive a fortnight's flood with stock-piled food before they can no longer cope.

It means that the family is susceptible /sensitive to the duration of a flood. Therefore, it is sometimes very hard to distinguish between sensitivity and threshold. In general, sensitivity increases after reaching a certain threshold.

The respondents were asked to identify measures for minimization of sensitivity that would encourage them or their relatives to follow anticipatory adaptation. Responses regarding the sensitivity minimization are very much centreed around measures such as enhanced/quicker and better access to cyclone/flood shelters (75.6%), followed by an establishment of emergency rescue and recovery units at community level (73.0%), and early dissemination of warning information (62.6%). Other measures include the establishment of a community food security programme (51.1%) or activities that come under integrated coastal management programmes (37.8%) (Table 21.3). The findings suggest that a significant portion of the respondents would prefer anticipatory adaptation to emigration if there were initiatives for the minimization of a current as well as a future sensitivity.

Measures to maximize the adaptive capacity

Adaptive capacity is the ability of a system to adjust to climate change, to moderate potential damages, to take advantage of opportunities, or to cope with the consequences (Fankhauser, Smith & Tol, 1999; Brooks, Adger & Kelly, 2005; Smit & Wandel, 2006; Fussel, 2007). In general, societies which have the

Table 21.3 Measures to minimize exposure and sensitivity, as well as to maximize adaptive capacity to deal with climate change induced emigration issues

The respondents were asked to identify the abovementioned three measures that would restrain people from emigration and encourage anticipatory adaptation *in situ*. Their responses are as follows.	Count	Percent of Responses	Percent of Cases
Exposure minimization measures (N=285, total responses = 1187):*			
i. Construction of new flood walls/embankments/levees	187	15.80	65.6
ii. Maintenance of existing embankments/levees	226	19.0	79.3
iii. Construction of new floodgates	214	18.0	75.1
iv. Re-excavation of illegally occupied networks of canals	192	16.2	67.4
v. Protection of mangroves and forestation	83	7.0	29.1
vii. Construction & maintenance of cyclone/flood shelters	285	24.0	100.0
Sensitivity minimization measures (N=270, total responses = 810):*			
i. Early dissemination of warning information	169	20.9	62.6
ii. Emergency rescue and recovery unit at community level	197	24.3	73.0
iii. Enhanced/quicker & better access to cyclone/flood shelter	204	25.2	75.6
iv. Community food security programme	138	17.0	51.1
v. Others – as part of integrated coastal management	102	12.6	37.8
Adaptive capacity maximization measures (N=285, total responses = 1187):*			
i. Awareness campaign on CC & SLR	285	20.2	100.0
ii. Strengthening the sense of cohesiveness	140	9.9	49.1
iii. Help in upgrading indigenous coping and adaptation measures	128	9.1	44.9
iv. Low-cost innovation for salt-free water	170	12.0	59.6
v. Cash incentive for post disaster rehabilitation	285	20.2	100.0
vi. Special social safety net for coastal communities	185	13.1	64.9
vii. Support for coastal resource-based adaptive livelihood, (e.g. brackish water shrimp and mud-crab culture, stress tolerant cultivars)	221	15.6	77.5

Note: * Multiple responses.

ability to respond to and cope with changes instantaneously are easily considered to have a high adaptive capacity. Although there is no all-sizes-fit indicator for adaptive capacity, Smit and Wandel (2006) have identified a possible set of indicators that are assumed to be useful in many contexts. These indicators are related to resources and their distribution, human capital, social capital, risk spreading systems, information management, technological options, institutional structure and others. Though most of them are macro-indicators, some could be used at the micro-level after a contextualization.

The respondents were asked to identify measures for a maximization of adaptive capacity that would encourage people or their offspring to adopt anticipatory adaptation. Responses regarding the maximization of adaptive capacity are very much centred around measures, such as awareness campaigns about the interconnected issues of climate change and sea level rise, or providing cash incentive for post disaster rehabilitation. In fact, 100% of the respondents go along with the abovementioned measures. Other measures include new initiatives for coastal resource-based adaptive livelihood, such as brackish water shrimp and mud-crab culture, stress tolerant crops and cereals production and so on, (77.5%) followed by special social safety nets for coastal communities (64.9%) and programmes for strengthening the sense of cohesiveness (49.1%) (Table 21.3).

It is surprising to note that only 44.9% identified an updating of their indigenous coping and adaptation capacity as necessary for encouraging themselves in anticipatory adaptation. The latter one gives a clear message that most of the respondents take it for granted that their traditional/indigenous coping and adaptation measures are effective for future adaptation against the impacts of SLR as well. If this attitude was not the case, in reality it would be problematic in the future to deal with the impacts of SLR and its associated events.

21.3.5 Measuring the influence of vulnerability reduction measures to discourage people from emigrating

Measures for minimization of exposure and sensitivity and maximization of adaptive capacity can have different degrees of influence on people's adaptation prospects and who might otherwise emigrate due to

the four causes of concern related to livelihood security. Therefore, four separate Binomial Logistic Regression models are built to predict the amount/extent of influence of three categories of measures on people's likely emigration behaviour. Here the three sets of 14 measures are independent variables which are binary coded; if a particular measure is considered (by the respondent) as appropriate then it is coded as 1 and otherwise coded as 0. Emigration associated with a particular cause is a dependent variable. If the respondent believes emigration could be prevented it is coded with 1 otherwise coded with 0. Therefore, the first model, for example, predicts the likely influence of vulnerability reduction measures on dissuading people from emigrating for perceived irrecoverable loss of income from current (and known) main source due to future CC-SLR. Similarly, the second model predicts the likely influence of vulnerability reduction measures to deter people likely to migrate out of the area for perceived permanent loss of physical access to community services/infrastructures such as hospital, marketplace, school and so on, due to future CC-SLR. The third and fourth models computed the likely influence of vulnerability reduction measures on halting the future trend of outmigration more likely to be caused due to loss of free/low cost access to saline-free water for drinking and other uses, and threatened food security which are all elaborated on in the following couple of sections.

Preventing emigration associated with a fear of income loss from main source

The respondents were asked to identify vulnerability reduction measures that can stop people's emigration and who are afraid of an income loss from main and known sources. Based on their responses, how much influence various vulnerability reduction measures can have on people's will to emigrate is predicted with the use of binomial logistic regression. The results show that implementation of exposure minimization measures and adaptive capacity maximization measures will have significant influence (at $p < 0.05$) to arrest the likely trend of income-loss induced future emigration. However, sensitivity minimization measures do not have a significant influence in this respect.

Among the respondents who believe implementation of appropriate measures could stop future emigration there is a very strong preference for 'construction of new flood walls/embankments/levees' followed by 'special social safety nets for coastal communities', 'support for coastal resources based adaptive livelihood and 'community food security programme'. For instance, the probability of adaptation *in situ* would increase 24 times if new flood walls/embankments/levees were constructed on the exposed coast. Similarly, the probability of adaptation *in situ* would increase 5.17, 3.34 and 1.9 times if 'special social safety nets', 'coastal resource-based adaptive livelihood opportunities' such as brackish water shrimp and mud-crab culture were intensified and a 'community food security programme' was implemented.

Overall, the findings reveal that a programme of 'construction of new flood walls/embankments/levees' for exposure minimization is the most dominant factor followed by 'special social safety nets for coastal communities' for adaptive capacity enhancement. They can significantly contribute to the arrest or slowing down of the likely future trend of emigration. However, finally, it is worth noting that despite the rate of various measures the contribution/influence of 'construction of new flood walls/embankments/levees' outweighed the total contribution of the four other measures in arresting the trend of emigration. In fact, people primarily put too much emphasis on structural measures like flood-walls/embankments/levees as most of the people are engaged in agriculture and allied occupations. For them it counts more that the agriculture and allied production systems are not exposed to the risk of inundation by rising sea levels. If that is the case they are then able to deal with other hydro-meteorological disasters of shorter duration such as temporary coastal flooding, storm surges and tidal inundation. Thus, it is suggested that the construction of 'new flood walls/embankments/levees' need to rank first.

Preventing emigration associated with permanent impairment of physical access to various facilities

The respondents have identified various vulnerability reduction measures that can stop people's emigration and who are afraid of impairment of access to various physical and social services and facilities. Based on their responses, how much influence various vulnera-

bility reduction measures can have on people's will to emigrate are predicted with the use of binomial logistic regression. The results show that although implementation of exposure and sensitivity minimization measures and adaptive capacity maximization measures have significant influence (at $p < 0.05$) on the likely trend of 'physical-mobility affects' induced future emigration, sensitivity minimization measures do have a significantly higher influence in this respect.

Among the respondents who believe implementation of appropriate measures could stop future emigration some have very strong preference for 'construction of new flood walls/embankments/levees' followed by 'integrated coastal management measures' such as climate compatible robust planning of physical infrastructures, 'timely dissemination of early warning' 'special social safety nets for coastal communities' For instance, probability of adaptation *in situ* would increase 25 times if new flood walls/embankments/levees are constructed on the exposed coast. Similarly, probability of adaptation *in situ* would increase 5.56, 5.54, 3.28 and 2.75 times if 'special social safety nets', 'integrated coastal management measures' such as climate compatible robust planning of physical infrastructures, 'timely dissemination of early warning' and 'operation of emergency rescue and recovery units at community level' are properly implemented. Finally it could be said that although 'construction of new flood walls/embankments/levees' for exposure minimization is the most dominant factor to halt the trend of emigration that might result from people's physical mobility problems, but implementation of 'integrated coastal management measures' such as climate compatible robust planning of physical infrastructures, 'timely dissemination of early warning' and 'operation of emergency rescue and recovery units at community level' are important as well.

Preventing emigration associated with the fear of loss of free/low-cost access to salt-free water

The respondents have identified the influence of various measures on the prevention of people's emigration due to the fear of loss of free/low-cost access to salt-free water for drinking and other uses. The results show that all three measures, such as exposure minimization, sensitivity minimization and adaptive capacity maximization have a statistically significant ($p < 0.05$) impact on hindering prospective emigration.

Among the exposure minimization measures at least three measures have the potential to prevent people from emigration due to a perceived loss of free/low-cost access to salt-free water. The probability of adaptation in situ would increase 5.72, 3.36 and 2.48 times if programmes such as 'construction of new flood walls/embankments/levees', 'construction of new floodgates' and 're-excavation of illegally occupied canal networks' were implemented. Similarly integrated coastal management measures such as river training and fresh-water reservoir management may have significant influence on stopping emigration. For instance implementation of these measures would increase up to 4.61 times the probability of adaptation in situ. Similarly, implementation of a new programme for 'upgrading indigenous coping and adaptation capacities of people' and 'low-cost innovation of salt-free potable water' would increase the probability of adaptation *in situ* 3.58 and 2.39 times respectively.

The overall findings reveal that programmes of 'construction of new flood walls/embankments/levees' for exposure minimization, which is the most dominant factor, followed by 'integrated coastal management' for sensitivity minimization and 'help for upgrading indigenous coping and adaptation capacities' for adaptive capacity maximization can significantly contribute to the slowing down of the likely future trend of emigration due to the fear of loss of free/low-cost access to salt-free water for drinking and other uses.

Preventing emigration associated with threatened food security

The respondents were asked to identify vulnerability reduction measures that could stop emigration due to the fear of loss of food security (production and availability). The results show that all three measures, such as exposure minimization, sensitivity minimization and adaptive capacity maximization have a statistically significant ($p < 0.05$) impact on retaining prospective emigrants. In fact, all three measures demand almost equal importance to ensure future food security.

Among the four measures of exposure minimization implementation of 'construction of new flood walls/embankments/levees', 're-excavation of illegally occupied canal networks' and 'construction of new floodgates' could increase the probability of adaptation *in situ* 19.52, 7.08 and 3.89 times by preventing people likely to emigrate because of a perceived threat in food security. That means, exposure minimization measures, more precisely, construction of new flood walls as well as new floodgates, and freeing and re-excavation of illegally occupied canals, can significantly arrest the likely trend of emigration. However, the protection of mangroves and forrestation has a negative impact on the food security issue even if only negligible. Despite its public ecological benefit people do not preserve it because their past experiences showed that only influential people profit economically from it while the native people were often evicted.

The 'Community food security programme' as a sensitivity minimization measure has an influence on dissuading people from likely emigration due to the fear of loss of food security. Implementation of this programme has 3.15 times higher probability of slowing down likely emigration. Similarly among the adaptive capacity enhancement programmes, implementation of 'special social safety nets for coastal communities', and 'upgrading indigenous coping and adaptation capacities' could increase the probability of adaptation *in situ* up to 20.67 and 3.27 times.

The overall lessons learned are that the introduction of 'special social safety nets for coastal communities' for adaptive capacity maximization is the most dominant factor followed by 'construction of new flood walls/embankments/levees' for exposure minimization and 'community food security programme' for sensitivity minimization, which can all significantly contribute to slow down the likely future trend of emigration due to the fear of loss of food security.

21.4 Discussion and concluding remarks

In the context of rural Bangladesh circular migration or seasonal displacement is not a new phenomenon for people living in the poverty stricken and flood affected areas. However, permanent emigration is the last expedient for the people who have lost their farmlands and settlements due to natural calamities, for instance massive riverbank erosion. In the coastal low-lying areas people rarely go for permanent displacement due to the fertile soil. As the study findings show, almost half of the respondents believe that many people might go for emigration which means that the perceived threats related to CC-SLR and identified in four different ways are deep-rooted in their mindset. Unless they were assured that substantial initiatives shall be directed towards a more resilient community building, it would be very hard to stop their permanent displacement. Their emigration may even create panic among others that might intensify the future trend of emigration from the low-lying areas. At worst, resettlement of the displaced population will create a financial burden for individuals, society and the nation at large. Further, socio-psychological costs of resettlement, which are often ignored, may even have a longstanding effect on the displaced people's life-course. In addition, in a land-scarce country like Bangladesh, the area to be vacated and the areas where people would be resettled are equally vulnerable to most of the impacts of CC-SLR. Given this reality experts already opined that resettlement should be the last option. Adaptation *in situ* should be the first option to be explored against the impacts of SLR.

Adaptation *in situ* requires measures to minimize exposure and sensitivity as well as measures to maximize adaptive capacity. A balanced and coordinated effort to minimize exposure and sensitivity, and maximize adaptive capacity could deterr people from emigrating. To deal with this evacuation issue this study has, as a starting point, mapped out the particular causes of concern about the vulnerability that relates to livelihood security which is, in the respondents' point of view, crucial for emigration. These true causes of emigration are identified as:

• permanent income loss from current main sources;
• permanent disturbance in / loss of physical access to various social services/facilities such as local healthcare, market, school, and so on;
• loss of free/low-cost access to salt-free water for drinking and other uses; and
• threatened food security (production and availability).

Most of the respondents tagged one or two while a few others went up to four. The responses constantly showed their belief in people's helplessness to deal with disastrous events that originate from the sea (i.e. Bay of Bengal). They considered people would feel helpless against impacts likely to be associated with future CC-SLR. Many people think like that because of their previous experiences of higher exposure and sensitivity in a low adaptive capacity context. It should be a strong signal that there is a need to devise ways in which to deal with each cause of livelihood vulnerability and to remove people's fears. In line with this the study attempted to explore the measures that the respondents think could be initiated to dissuade people from emigrating as for each of the four identified causes. The following passages will deal with the respondents' thoughts about these four causes.

Respondents think people might emigrate due to a perceived income loss from current (and known) main sources in the event of future CC-SLR, and who would not emigrate if a few actions were realized, such as the construction of flood walls/embankments/levees, or the introduction of community food security programmes, a strengthening of the community's sense of cohesiveness, an introduction of special social safety nets for coastal regions and also an implementation of coastal resources based adaptive livelihood initiatives such as brackish water shrimp, mud-crab culture, stress tolerant cultivars. As the key sources of livelihood are agricultural and allied economic activities followed by fishing, it is not unusual that the respondents put more emphasis on coastal resources-based adaptive livelihood for longer term adaptation. In the same vein they put a higher emphasis on measures of social insurances and cushion against temporary/seasonal income loss as a short-term or medium-term coping and adaptation measure. This is possibly the case for people, like day labourers, off-farm workers, or subsistence fishermen who really lack year-round permanent sources of income.

Respondents would also like to see initiatives for exposure minimization such as the construction of sea walls, floodwalls/embankments/levees, and so on. In fact, income losses of certain occupation groups are related to the extent of exposure. An agricultural farming community is at greater risk of suffering from the intrusion of saline water than people engaged in aquaculture who are more familiar with dealing with salinity intrusion. For both of these occupational groups a minimization of exposure to storm surges, salinity intrusion and coastal inundation is very crucial. Our analysis further reveals that despite the demand for exposure and sensitivity minimization, the development of adaptive capacities has the highest priority among those who think income loss and emigration are interconnected. Among such measures three are directly related to the development of adaptive capacity and one is related to each of either exposure minimization or sensitivity minimization.

Respondents who think people might emigrate due to the perceived permanent disturbance in / loss of physical access to social services/facilities, such as local healthcare, marketplaces, schooling, and so on, in the event of future CC-SLR and would stop evacuation if programmes such as the construction of new sea walls, flood walls/embankments/levees, the dissemination of warning information well before the onset of the event, the establishment of emergency rescue and recovery units at a local level, the integration of coastal management measures (such as climate compatible robust planning of physical infrastructure), the initiation of programmes for helping people to upgrade their indigenous coping and adaptation and special social safety-nets for coastal communities and the introduction of coastal resources-based adaptive livelihood were realized. In fact, this group seems very hard to satisfy as respondents of this category demanded a long list of measures to keep people in the coastal area. Of course, many of these measures are equally applicable to respondents of other categories. The universal measures must be initiated on a basis of top priority to hold prospective emigrants back and to encourage adaptation *in situ*. However, according to the analysis results it is evident that even for the rural coastal people secured access to essential service facilities is no less important than their secured access to chances of earning money.

Respondents who think people might out migrate due to perceived loss of free/low-cost access to saline-free water in the event of future CC-SLR may stop migrating out if a few programmes, such as new sea walls, that is, flood walls/embankments/levees are constructed, more new sluice gates are operated, illegally occupied canals are re-excavated, other measures as part of an integrated coastal management are taken, programmes for helping people to upgrade their indigenous coping and adaptation are initiated and a low-cost innovative source of saline-free potable water are ensured.

The coastal region as a whole and the study area in particular suffer a shortage of salt-free potable drinking water. While in most parts of the country the fresh-water aquifer is 50–100 metres below the ground level (surface), in the study area this depth ranges from 400–500 metres which entails higher installation and maintenance costs for tube-wells. The reasons for the presence of salt water throughout the subsurface water column could be attributed to lateral percolation of sea water along with infiltration of salt water from surface sources such as seasonal encroachment of high salinity waters from rivers and canals as well as from brackish aquaculture that pollutes channels and lowland with salt water. The construction of sea walls almost certainly will protect against the penetration of salt water during a high tide. Judicial regulation of sluice gates can allow penetration of fresh water during monsoon and the post monsoon period which may help the community to store fresh water in their surface water bodies. The remaining fresh water may help refill the water table and thus, reducing soil salinity. Similarly, re-excavation of illegally occupied canals can also be used to trap fresh water during monsoon and post monsoon for use in the dry season. Further, their current localized practice of rainwater harvesting for domestic use could be upgraded so that they can increase water harvesting efficiency with minimal costs. Apart from these, new initiatives for ensuring sustained free/low-cost potable water need to be initiated to see long-term prospects of anticipatory adaptation among the people likely to emigrate due to shortage of salt-free water for drinking and other uses.

Even for the current climatic conditions concerns about food security exist and may amplify with respect to the likely scenarios of CC-SLR. In the study area most farmlands are put out of operation during winter crops. First of all, that is because throughout the winter season the farming community lacks salt-free water for irrigation. It is not like there is no fresh water in the lowlands or in the canal network or even in the river system. There is plenty of water indeed. But, the influential people, the *'gheer'* (shrimp farm) owners, control the sluice gates and the canal network inside the existing floodwalls and let through mostly salt water that cannot be used for agricultural crop production as the salinity level often exceeds 10 parts per thousand (PPT) or soil salinity exceeds 10 deciSiemens per metre (dS/m). For the marginalized crop farming community it is reasonable to think about increased production of winter crops to ensure food security. They would have to assume control of the sluice gates to store fresh water in the canal network and lowlands during monsoon or post monsoon. On the downside, the influential shrimp farm owners lock the sluice gates during monsoon or post monsoon to stop penetration of fresh water from the river system to the embanked areas.

Exposure minimization measures such as the construction of new seawalls, sluice gates and the re-excavation of canals could be very instrumental to future food security. Along with these exposure minimization measures, social safety nets and community food security programmes could have a positive synergy effect on ensuring food security which may eventually encourage people to think about anticipatory adaptation rather than emigration.

The effectiveness of various vulnerability reduction measures to dissuade people from emigrating due to the identified causes is neatly presented in a matrix and will be discussed in the following paragraphs (Table 21.4). Among the exposure minimization measures, solely the 'construction of new flood walls/embankments/levees' had an effect on the four identified reasons. Likewise, measures like 'construction of new floodgates' and 're-excavation of illegally occupied canals' could hinder a significant proportion of emigration that results from a loss of free/low-cost access to salt-free potable water and the loss of food security.

Table 21.4 Matrix of intervention strategies to restrain people from emigrating and encourage anticipatory adaptation against the identified causes of vulnerability related to CC-SLR

Vulnerability reduction measures	Causes of livelihood vulnerability that are significantly influenced by measures cited in the left column			
	1	2	3	4
Measures to minimize exposure:				
Construction of new flood walls/embankments/levees	✓	✓	✓	✓
Maintenance of existing embankments/levees				
Construction of new floodgates			✓	✓
Re-excavation of illegally occupied canals			✓	✓
Protection of mangroves and forestation				✓***
Measures to minimize sensitivity:				
Early dissemination of warning information		✓		
Emergency rescue & recovery units at local level		✓		
Community food security programme	✓			✓
Others – as part of integrated coastal management		✓	✓	
Measures to maximize adaptive capacity:				
Help for upgrading indigenous coping & adaptation capacities		✓	✓	✓
Strengthening the community's sense of cohesiveness	✓	✓		✓
Low-cost innovation for salt-free potable water			✓	
Special social safety nets for coastal communities	✓	✓		✓
Coastal resources based adaptive livelihood (e.g. brackish water shrimp and mud-crab culture, stress tolerant cultivars)	✓	✓		

Note: *** Indicates negative influence.
1. People who otherwise would go for emigration due to loss of income from main sources.
2. People who otherwise would go for emigration due to impairment in access to service facilities.
3. People who otherwise would go for emigration due to loss of free/low-cost access to potable water.
4. People who otherwise would go for emigration due to threatened food security (production/availability).

Demand for the construction of new cyclone/flood shelters and maintenance of the existing ones are universal in the study area. The construction of new seawalls or embankments with sufficient height can minimize people's exposure to impacts associated with SLR. Thus, their ultimate vulnerability in terms of all four causes will be minimized if not eliminated altogether. While the benefit of exposure minimization measures such as sea walls or embankment and floodgates is apparent, however, costs might be substantial as well. An implementation of the named exposure minimization measures is beyond the scope of private adaptive means. It requires a public planning adaptation strategy to ensure a monetary distributive justice and a balanced allocation of responsibilities.

Among the sensitivity minimization measures, 'early dissemination of warning information' can motivate a significant number of people to think about anticipatory adaptation, because access to reliable and timely information about an imminent hazard can minimize the likely damage potential or even avoid any kind of damages. However, over time it will help people to familiarize with such events and to change their attitudes (i.e. from fatalistic to objective) which will eventually contribute to developing a more resilient community (Schmuck, 2000; Grothmann & Patt, 2005; Leal Filho, 2009). This finding is consistent with the existing literature which states that people will become familiar with the identification of their 'escape routes' even without the assistance of others if they get the warning information just in time before the onset of the event. The introduction of a 'community food security programme' could be instrumental in stopping those people's future emigration if they are afraid of an income loss from current main sources and the threat of food insecurity (production and availability). In fact, community food security programmes would play a central role in helping people in cases of hazardous events and they will become more important as these events might be magnified due to CC-SLR.

Likewise, an establishment of 'emergency rescue and recovery units at a local level' and some other activities related to integrated coastal management can help minimize sensitivity and, consequently, ban emigration. Nevertheless, this study could not yield any substantive ground to provide cash incentives for rehabilitation even if that was urgent for each of the respondents. Any attempt to implement such measures may even have a boomerang effect causing influx of population from poverty stricken inland areas to the coastal areas. Many of the sensitivity minimization programmes the respondents were in favour of, are beyond the scope of private adaptive means. Therefore, support from other actors, such as the government, civil societies and research organizations are indispensable.

Among the adaptive capacity enhancement measures, 'assistance for upgrading indigenous coping and adaptation capacities' has a significant impact on lowering the future trend of emigration. In particular, the people who intended to emigrate due to the loss of access to basic services, a loss of free/low-cost access to salt-free water, a loss of food security, and so on, will be more likely to adopt anticipatory adaptation if they are provided with the knowledge to upgrade their existing coping and adaptation strategies. Similarly, the introduction of 'special social safety nets for coastal communities' and programmes for 'strengthening the community's sense of cohesiveness' could halt future emigration. In fact, these two measures, along with an 'initiative for coastal resource based adaptive livelihood' such as brackish water shrimp mud-crab culture and stress tolerant cultivars might help to lay the foundation for adaptive capacity development, an essential element of private anticipatory adaptation.

From an overall analysis it is evident that coastal inhabitants in general believe the construction of new sea walls, that is, flood walls/embankments/levees and the upgrading of their indigenous coping and adaptation capacities, special social safety nets for coastal communities and a few activities such as river training/creation of a micro-reservoir for freshwater and climate compatible physical infrastructure planning that related to integrated coastal management are the key instruments that will be able to deter people from emigrating and can encourage anticipatory adaptation. Although most adaptive capacity enhancement measures require fewer technicalities, less budgetary allocation and less execution time than exposure and sensitivity minimization measures, experience, however, shows people would rarely go for adopting these reasonable measures unless cost and time intensive exposure and sensitivity minimization measures are initiated by other actors such as the public body, donors and civil society organizations. Thus, there is a need for advance planning whereby exposure and sensitivity minimization measures have to remain visible while launching adaptive capacity formation programmes to encourage people to accept anticipatory adaptation *in situ* against the impacts of future CC-SLR.

References

Adger, W.N., Huq, S., Brown, K., Conway, D. and Hulme, M., 2003. Adaptation to climate change in the developing world. *Progress in Development Studies*, 3 (3), pp.179–195.

Adger, W.N., 2006. Vulnerability. *Global Environmental Change*, 16, pp.268–281.

Agrawala, S., Ota, T., Ahmed, A.U., Smith, J. and van Aalst, M., 2003. *Development and Climate Change in Bangladesh: Focus on Coastal Flooding and Sundarbans*. Paris: OECD.

Ahmed, A.U. and Alam, M., 1998. Development of climate change scenarios with general circulation models. In: Huq, S., Karim, Z., Asaduzzaman, M. and Mahtab, F., eds. 1998. *Vulnerability and Adaptation to Climate Change for Bangladesh*. Dordrecht: Kluwer Academic Publishers, pp.13–20.

Ali, A., 1999. Climate change impacts and adaptation assessment in Bangladesh. *Climate Research*, 12 (2/3), pp.109–116.

Ali, A. and Chowdhury, J.U., 1997. Tropical cyclone risk assessment with special reference to Bangladesh. *MAUSAM*, 48, pp.305–322.

Ali Khan, T.M., Singh, O.P. and Rahman, M.S., 2000. Recent sea level and sea surface temperature trends along the Bangladesh coast in relation to the frequency of intense cyclones. *Marine Geodesy*, 23 (2), pp.1–14.

Assan, J.K., Caminade, C. and Obeng, F., 2009. Environmental variability and vulnerable livelihoods: Minimising risks and optimising opportunities for poverty alleviation. *Journal of International Development*, 21 (3), pp.403–418.

Azad, A.K., Jensen, K.R. and Lin, C.K., 2009. Coastal aquaculture development in Bangladesh: Unsustainable and sustainable experiences. *Environmental Management*, 44 (4), pp.800–809.

Brooks, N., Adger, W.N. and Kelly, P.M., 2005. The determinants of vulnerability and adaptive capacity at the national level and the implications for adaptation. *Global Environmental Change*, 15, pp.151–163.

Bryman, A. and Cramer, D., 2001. *Quantitative Data Analysis with SPSS Release 10 for Windows: A guide for social scientists*. East Sussex: Routledge.

Burbridge, P. and Humphrey, S., 2003. Introduction to special issue on the European demonstration programme on integrated coastal zone management. *Coastal Management*, 31 (2), pp.121–126.

Burton, I., 1997. Vulnerability and adaptive response in the context of climate and climate change. *Climatic Change*, 36, pp.185–196.

Cannon, T., 2002. Gender and climate hazards in Bangladesh. *Gender and Development*, 10 (2), pp.45–50.

CARE, 2003. *Report of a Community Level Vulnerability Assessment Conducted in Southwest Bangladesh. Reducing Vulnerability to Climate Change (RVCC) Project*. Dhaka: CARE Bangladesh.

Castro-Ortiz, C.A., 1994. Sea level rise and its impact on Bangladesh. *Ocean and Coastal Management*, 23 (3), pp.249–270.

Choudhury, A.M., Neelormi, S., Quadir, D.A., Mallick, S. and Ahmed, A.U., 2005. Socio-economic and physical perspectives of water related vulnerability to climate change: results of field study in Bangladesh. *Science and Culture*, 71 (7–8), pp.225–238.

Cutter, S.L., Boruff, B.J. and Shirley, W.L., 2003. Social vulnerability to environmental hazards. *Social Science Quarterly*, 84, pp.242–261.

Fankhauser, S., Smith, J.B. and Tol, R.S.J., 1999. Weathering climate change: some simple rules to guide adaptation decisions. *Ecological Economics*, 30, pp.67–78.

Faruque, H.S.M. and Ali, M.L., 2005. Climate change and water resources management in Bangladesh. In: Mirza, M.M.Q. and Ahmad, K.Q., eds. 2005. *Climate Change and Water Resources in South Asia*. Leiden: Balkema Press, pp.231–254.

Fussel, H.-M., 2007. Vulnerability: A generally applicable conceptual framework for climate change research. *Global Environmental Change*, l (17), pp.155–167.

GOB (Government of Bangladesh), 2005. *National Adaptation Programme of Action (NAPA), Final report: November*. Dhaka: Ministry of Environment and Forests, Government of Bangladesh.

GOB, 2006. *Bangladesh Climate Change Impacts and Vulnerability: A Synthesis, Ministry of Environment and Forests*. Dhaka: Climate Change Cell, Department of Environment.

GOB, 2008. *Bangladesh Climate Change Strategy and Action Plan 2008*. Dhaka: Ministry of Environment and Forests.

Grothmann, T. and Patt, A., 2005. Adaptive capacity and human cognition: the process of individual adaptation to climate change. *Global Environmental Change*, 15, pp.199–213.

Haddad, B.M., 2005. Ranking the adaptive capacity of nations to climate change when socio-political goals are explicit. *Global Environmental Change*, 15, pp.165–176.

Huq, S., Karim, Z., Asaduzzaman, M. and Mahtab, F. eds., 1998. *Vulnerability and Adaptation to Climate Change for Bangladesh*. Dordrecht: Kluwer Academic Publishers.

Ingham, A., Ma, J. and Ulph, A., 2005. Climate change, migration and adaptation with uncertainty and learning. *Energy Policy*, 35 (11), pp.5354–5369.

IPCC (Intergovernmental Panel on Climate Change), 1996. *Climate Change 1995: The Science of Climate Change. Contribution of Working Group I to the Second Assessment Report of the Intergovernmental Panel on Climate Change*. Cambridge: Cambridge University Press.

IPCC WG I, 2001. *Climate Change 2001: The Scientific Basis. Contribution of Working Group I to the Third Assessment Report of the Intergovernmental Panel on Climate Change*. Cambridge: Cambridge University Press.

IPCC WG I, 2007. *IPCC Fourth Assessment Report: Climate Change 2007. Contribution of Working Group I to the Fourth Assessment Report of the Intergovernmental Panel on Climate Change*. Cambridge: Cambridge University Press.

IPCC WG II, 2007. *Climate Change 2007: Impacts, Adaptation and Vulnerability. Contribution of Working Group II to the Fourth Assessment Report of the Intergovernmental Panel on Climate Change*. Cambridge: Cambridge University Press.

Islam, S.M.R., Huq, S. and Ali, A., 1998. Beach erosion in the eastern coastline of Bangladesh. In: Huq, S., Karim, Z., Asaduzzaman, M. and Mahtab, F., eds. 1998. *Vulnerability and Adaptation to Climate Change for Bangladesh*. Dordrecht: Kluwer Academic Publishers, pp.72–93.

Islam, S.M.R., Ahmad, M., Huq, H. and Osman, M.S., 2006. *State of the Coast 2006*. Dhaka: Program Development Office for Integrated Coastal Zone Management Plan Project, Water Resources Planning Organization, Government of Bangladesh.

Kelly, P.M. and Adger, W.N., 2000. Theory and practice in assessing vulnerability to climate change and facilitating adaptation. *Climatic Change*, 47 (4), pp.325–352.

Khan, M.S.A., 2008. Disaster preparedness for sustainable development in Bangladesh. *Disaster Prevention and Management*, 17 (5), pp.662–671.

Khondker, H.H., 1996. Women and floods in Bangladesh. *International Journal of Mass Emergencies and Disasters*, 14 (3), pp.281–292.

Klein, R.J.T., Nicholls, R.J., Ragoonaden, S., Capobianco, M., Aston, J. and Buckley, E.N., 2001. Technological options for adaptation to climate change in coastal zones. *Journal of Coastal Research*, 17 (3), pp.531–543.

Leaf, M., 1997. Local control versus technocracy: the Bangladesh Flood Response Study. *Journal of International Affairs*, 51 (1), pp.179–200.

Leal Filho, W., 2009. Communicating climate change: challenges ahead and action needed. *International Journal of Climate Change Strategies and Management*, 1 (1), pp.6–18.

Leggett, J., Pepper, W.J. and Swart, R.J., 1992. Emissions scenarios for the IPCC: an update. In: Houghton, J.T., Callander, B.A. and Varney, S.K., eds. 1992. *Climate Change 1992: the Supplementary Report to the IPCC Scientific Assessment*. Cambridge: Cambridge University Press, pp.69–95.

Matin, N. and Taher, M., 2001. The changing emphasis of disasters in Bangladesh NGOs. *Disasters*, 25 (3), pp.227–239.

McCarthy, J.J., Canziani, O.F., Leary, N.A., Dokken, D.J. and White, K.S. eds., 2001. *Climate Change 2001: Impacts, Adaptation and Vulnerability, Intergovernmental Panel on Climate Change (IPCC), Work Group II Input to the Third Assessment Report*. Cambridge: Cambridge University Press.

Middleton, N., 1999. *The Global Casino: An Introduction to Environmental Issues*. 2nd ed. London: Arnold.

Mimura, N., 1999. Vulnerability of island countries in the South Pacific to sea level rise and climate change. *Climate Research*, 12, pp.137–143.

Mirza, M.M.Q., 2002. Global warming and changes in the probability of occurrence of floods in Bangladesh and implications. *Global Environmental Change*, 12 (2), pp.127–138.

Nicholls, R.J., 1995. *Synthesis of vulnerability analysis studies. Proceedings of WORLD COAST '93*. Rijkswaterstaat: Coastal Zone Management Centre.

Nicholls, R.J., Leatherman, S.P., Dennis, K.C. and Volonte, C.R., 1995. Impacts and responses to sea-level rise: qualitative and quantitative assessments. *Journal of Coastal Research*, 14, pp.26–43.

Nicholls, R.J. and Lowe, J.A., 2004. Benefits of mitigation of climate change for coastal areas. *Global Environmental Change*, 14, pp.229–244.

Paul, B.K., 2005. Evidence against disaster-induced migration: the 2004 tornado in north central Bangladesh. *Disasters*, 9 (4), pp.370–385.

Paul, B.K., 2010. Climate refugees: the Bangladesh case. *The Daily Star*, 6 March.

Patt, A.G., Daze, A. and Suarez, P., 2009. Gender and climate change vulnerability: what's the problem, what is the solution? In: Ruth, M. and Ibarraran, M.E., eds. 2009. *Distributional Impacts of Climate Change and Disasters: Concept and Cases*. Massachusetts, USA: Edward Elgar Publishing Limited, pp.82–102.

Rashid, S.F. and Michaud, S., 2000. Female adolescents and their sexuality: notions of honour, shame, purity and pollution during the floods. *Disasters*, 241, pp.54–70.

Saroar, M.M. and Routray, J.K., 2010a. Adaptation *in situ* or retreat? A multivariate approach to explore the factors that guide the peoples' preference against the impacts of sea level rise in Bangladesh. *Local Environment*, 15 (7), pp.663–686.

Saroar, M.M. and Routray, J.K., 2010b. *In situ* adaptation against sea level rise (SLR) in Bangladesh: does awareness matter? *International Journal of Climate Change Strategies and Management*, 2 (3), pp.321–345.

Saroar, M.M. and Routray, J.K., 2012. Impacts of climatic disasters in coastal Bangladesh: why does private adaptive capacity differ? *Regional Environmental Change*, 12 (1), pp.169–190.

Schmuck, H., 2000. An act of Allah: religious explanations for floods in Bangladesh as survival strategy. *International Journal of Mass Emergencies and Disasters*, 18 (1), pp.85–95.

Singh, O.P., Ali Khan, T.M., Murty, T.S. and Rahman, M.S., 2001. Sea level changes along the Bangladesh coast in relation to southern oscillation phenomenon. *Marine Geodesy*, 24, pp.65–72.

Singh, O.P., 2001. Cause-effect relationships between sea surface temperature, precipitation and sea level along the Bangladesh coast. *Theoretical and Applied Climatology*, 68, pp.233–243.

Smit, B., Burton, I., Klein, R. and Wandel, J., 2000. An anatomy of adaptation to climate change and variability. *Climatic Change*, 45, pp.223–251.

Smit, B. and Skinner, M., 2002. Adaptation options in agriculture to climate change: a typology. *Mitigation and Adaptation Strategies for Global Change*, 7 (1), pp.85–114.

Smit, B. and Wandel, J., 2006. Adaptation, adaptive capacity, and vulnerability. *Global Environmental Change*, 16 (3), pp.282–292.

Smith, J.B., 1997. Setting priorities for adapting to climate change. *Global Environmental Change*, 7 (3), pp.251–264.

Smithers, J. and Smit, B., 1997. Human adaptation to climatic variability and change. *Global Environmental Change*, 7 (2), pp.129–146.

Sterr, H., 2008. Assessment of vulnerability and adaptation to sea-level rise for the coastal zone of Germany. *Journal of Coastal Research*, 24 (2), pp.380–393.

Stern, N., 2006. *The Stern Review on the Economic Effects of Climate Change (Report to the British Government)*. Cambridge: Cambridge University Press.

Tol, R.S.J., Klein, R.J.T. and Nicholls, R.J., 2008. Toward successful adaptation to sea-level rise along Europe's coast. *Journal of Coastal Research*, 24 (2), pp.432–442.

UNDP (United Nations Development Programme), 2004. *Reducing Disasters Risk: A Challenge for Development. Bureau for Crisis Prevention and Recovery*. New York: UNDP.

UNISDR (United Nations International Strategy for Disaster Reduction Secretariat), 2009. *Global Assessment Report on Disaster Risk Reduction*. Geneva, Switzerland: United Nations.

World Bank, 2000. *Bangladesh: Climate Change and Sustainable Development*. Dhaka: Rural Development Unit, South Asia Region.

22 Promoting Risk Insurance in the Asia-Pacific Region: Lessons from the Ground for the Future Climate Regime under UNFCCC

S.V.R.K. Prabhakar[1], Gattineni Srinivasa Rao[2], Koji Fukuda[1] & Shinano Hayashi[1]

[1]Institute for Global Environmental Strategies, Japan
[2]eeMausam, Weather Risk Management Services, India

22.1 Introduction

An increase in the number of catastrophic disasters and related insured and uninsured losses has been reported (Munich Re, 2010) undermining the developmental gains across the world. The Asia-Pacific Region is one of the regions most vulnerable to a range of primary hydro-meteorological and geological natural hazards such as earthquakes, storms, floods, tsunamis, landslides and droughts. The Emergency Events Database (EM-DAT) of the Center for Research on the Epidemiology of Disasters (CRED) suggests that specifically the number of hydro-meteorological disasters over the 2000–09 period was 10 times more than the number of disasters reported during 1947–56 (CRED, 2010). In the Asia-Pacific Region, the hydro-meteorological disasters have claimed the lives of 0.22 million people with estimated total economic damage costs of 285 million US\$ during 2001–12 (CRED, 2012).

The region's high vulnerability to natural hazards compared to other regions in the world, is primarily due to range of geophysical, socioeconomic and developmental conditions. These include a long coastline of 187 193 km, a highly variable monsoon system, high volcanic and tectonic activity, high poverty both within and outside of urban areas, high population densities associated with massive immigration to cities, poorly planned urban development, absence of proper disaster risk mitigation mechanisms and institutional/regulatory frameworks including prevalence and enforcement of structural standards such as building- and land-use planning regulations, as well as the poor development of risk spreading instruments such as risk insurance systems. The data available since 1900 show a steady increase in economic losses and a plateauing trend in loss of lives from disasters caused by natural hazards in the Asia-Pacific Region (CRED, 2012). During 1960 and 2010, the average per capita deaths and average per capita economic losses were significantly higher in developing countries than in developed countries in the Asia-Pacific Region (CRED, 2012). A disaster of the same intensity can lead to a greater number of deaths and higher economic damage in developing countries (e.g. Bangladesh and

Climate Change Adaptation in Practice: From Strategy Development to Implementation, First Edition.
Edited by Philipp Schmidt-Thomé and Johannes Klein.
© 2013 John Wiley & Sons, Ltd. Published 2013 by John Wiley & Sons, Ltd.

Philippines) than in developed countries (e.g. Japan) (Mechler, 2004). This clearly indicates differences in exposure and vulnerability between developed and developing countries of the Asia-Pacific Region. What is noteworthy as well is that the loss of assets and related livelihoods significantly limit returning the affected population to their normal life irrespective of the developmental state of a country and hence, the protection of assets deserves greater attention (Vatsa, 2004).

Climate change has brought an additional dimension to disaster risks in the Asia-Pacific Region as it is projected to exacerbate the intensity and magnitude of various natural hazards such as storms, high-intensity rainfall events, heat waves, floods and droughts (IPCC, 2007; Kunreuther and Michel-Kerjan, 2007). Especially, the projections suggest high probability for an increasing trend in the high-intensity and low probability events. These increased catastrophic risks will further undermine the developmental gains already made in the Asia-Pacific Region.

Take, for example, the case of the agricultural sector which is one of the sectors in the region that is highly vulnerable to climate change. Farming communities in particular are at greater risk to weather-related crop failures. Often, farmers borrow loans from local banks prior to the cropping season. However, farmers, banks and governments are exposed to higher financial risks due to an increasing frequency of crop failures, and in many cases the governments are forced to waive the loans. In the case of India, estimates suggest that the government waived crop loans totaling 14.4 billion US$ in 2008 (Kanz & Robert, 2011). Similar incidences are observed across other countries in the Asia-Pacific Region (e.g. bailout of Thai farmers in 2010) (Sompo Japan Insurance Inc., 2010; Kanz & Robert, 2011).

Hence, in order to address additional risks brought by the impact of climate change, there is a need to reassess and reframe the current risk reduction strategies especially in terms of development and utilization of risk-spreading instruments within the Asia-Pacific Region. Keeping this in mind, this chapter reviews the current status of risk insurance and identifies emerging issues and experiences. Those are compared with various risk insurance proposals made by the COP to the UNFCCC for assessing the extent to which they can promote the risk insurance.

22.2 Risk Insurance and Climate Change Adaptation

The concept of risk transfer or risk spreading entails that the individual (the insured) risks are reduced by spreading or transferring the risks from the insured to the insurance provider (the insurer) since the insurer is in a stronger financial position than the insured (Njegomir & Maksimovic, 2009). The insurance provider is able to insure the risks of the insured to a great extent due to the fact that the insurer obtains premiums from a large number of insured who are at different levels of risks and by making sure that the total amount of premiums collected exceeds the underwriting of risks (termed as the law of large numbers). Insurance agencies in turn underwrite some of these risks with reinsurance firms that provide the needed buffer against losses related to catastrophic events. In sum, the risk insurance scheme functions as part of the social safety net through risk transfer mechanisms and thereby contribute to an enhancement of the resilience of societies.

Risk transfer has been widely advocated as one of the best means of risk mitigation across the world (Siamwalla & Valdes, 1986; Arnold, 2008; Swiss Re, 2010a) and as a result of several advantages it:

- Promotes emphasis on risk mitigation compared to the current response-driven mechanisms.
- Provides a cost-effective way of coping financial impacts of climate- and weather-induced hazards.
- Supports the climate change adaptation by covering the residual risks uncovered by other risk reduction mechanisms such as building regulations, land-use planning and disaster risk management plans.
- Stabilizes rural incomes and hence reduces the adverse effects on income fluctuation and socio-economic development.
- Provides opportunities for public-private partnerships.
- Reduces burden on government resources for post-disaster relief and reconstruction.
- Helps communities and individuals to quickly renew and restore the livelihood activity.
- Addresses a wide variety of risks emanating from climatic and non-climatic origin, depending on the way the insurance products are designed.

22.3 Current state of risk insurance in the Asia-Pacific Region

The prevailing insurance widely observed in the Asia-Pacific Region could be broadly classified into health- and non-health-based insurances which are offered both by the governmental insurance programmes and by private insurers. The most popular form of insurance is the life insurance where the insurance companies pay for the insured party's death or other risks such as critical terminal illness. Other forms of insurances cover for health, vehicles, properties, liability, credit, housing and crop among others. Though both life and non-life insurances have a stake in disaster risk reduction, promoting the non-life insurances is of paramount importance in the region due to its poor spread compared to the life insurance.

Among the world regions compiled by Swiss Re, the non-life insurance penetration indicated by premium volumes is highest in North America followed by Western Europe and South and East Asia (Figure 22.1; Swiss Re, 2010b). Within Asia, the non-life insurance penetration is highest in Japan followed by China, South Korea, Taiwan and India. In general, the spread of health insurances in the region is much higher than that of the non-health insurances overlooking the premiums, though the magnitude varies between developed and emerging economies. Car insurances and insurances for industrial and commercial establishments are among the dominant forms of non-life insurances in the region.

It should be noted that most insurance mechanisms have been conceptualized and developed largely in the developed country markets and are being adapted in the developing countries. While most high-income households in the developing countries pay their own insurance premiums, most of the premiums of the low- and middle-income families are often enrolled by their employers (O'Donnell et al., 2008).

The poor spread of the insurances remains a concern for the Asia-Pacific Region, especially in the non-health disaster risk insurance sector, and it is attributed to the following factors:

1. **Affordability:** The issue of affordability could be put at the top of all the bottlenecks limiting the spread of risk insurance in the developing Asia-Pacific. Though insurance premiums in the majority of those countries are lower than in the developed countries, the annual insurance premium costs are still not affordable for most of the income groups. Part of the high insurance premium costs emerge from the high residual risks and the low number of insured persons (i.e. poor development of the insurance portfolio).

2. **Residual risks:** High residual risks are one of the major causes for the poor risk insurance coverage in the region. The high residual risks exist due to poor disaster risk mitigation mechanisms as well as the poor enforcement and inadequacies of laws, respectively, such as building regulations, structural codes and laws pertaining to land-use planning.

3. **Presence of insurers and reinsurers:** One of the reasons for the poor penetration of insurances as well as insurance prices above affordability is the limited presence of private insurers and reinsurers. Reinsurers play an important role in providing shock-absorbing capacity to the insurers. To date, very few national (e.g. General Insurance Corporation in India,

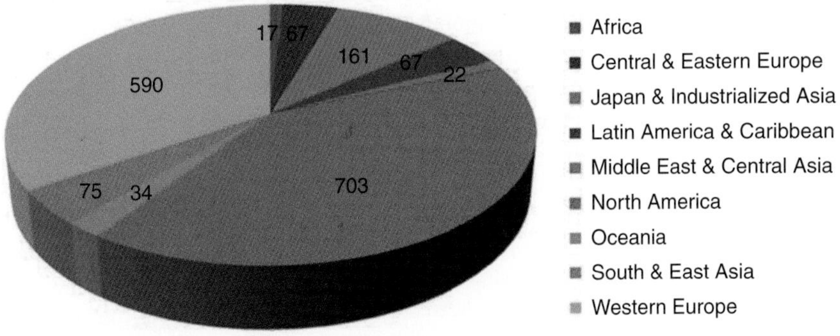

Figure 22.1 Penetration of non-life insurances indicated by premium volumes in different world regions (in billion US$).
Source: Swiss Re, (2010b)

- Africa
- Central & Eastern Europe
- Japan & Industrialized Asia
- Latin America & Caribbean
- Middle East & Central Asia
- North America
- Oceania
- South & East Asia
- Western Europe

China Reinsurance Company in China, Zenkyoren or *Zenkoku Kyousai Seikatsukyoudoukumiai Rengou Kai* in Japan) and international reinsurers (e.g. Munich Re, Swiss Re, Toa Re, Axis Re) operate in the region. Hence, there is a high potential for the expansion of the reinsurance sector. Insurers and reinsurers cannot afford to operate in the region unless there is a sufficient enabling environment including efforts to reduce the residual risks.

4. High premium costs: The high residual risks, lack of optimum number of insurers, low competition, and low number of insured parties all lead to the premium costs being higher than what they could be in the Asia-Pacific Region.

5. Policy environment: Though risk insurance is a 'market instrument', its dynamics are determined or governed by the principles of an open market, government policies and regulatory guidelines act as precursors for a flourishing of the sector and ensure the effectiveness of the instrument. Hence, the role of the government in promoting the culture of risk mitigation through awareness-raising activities, as well as designing and implementing structural and non-structural disaster risk mitigation codes/laws, which include institutional mechanisms and conducive regulations, is paramount.

Though there has already been a significant improvement of policy support for the insurance sector, as apparent from the high growth rates of the insurance sector in the region, the support is still not comprehensive enough. For example, currently, most developing countries in the Asia-Pacific Region are at the nascent stages of formulating national disaster risk mitigation plans and policies (GFDRR, 2009) and they have not fully utilized the potential of risk insurance in promoting risk reduction. Traditionally, most governments propagate disaster response rather than mitigation to hinder the public participation in risk insurance schemes (Yucemen, 2008). Limited financing is the major reason behind the poor emphasis of disaster risk mitigation in the region.

6. Cultural and perceptional issues: A general lack of awareness and misplaced perceptions about dealing with risk in general and risk insurance in particular among the common people and the business sector is also an obstacle (Yazici, 2007; Yucemen, 2008). Sociological research has indicated the existence of behavioural patterns that can be characterized

as 'lethal attitude', that is, things will happen whatever is done and things are beyond ones' control. As a consequence, the individual willingness to mitigate risks is limited.

7. Lack of data: Infrastructure for collecting and managing systematic and comparable data on past risks, vulnerabilities, disasters and the nature of disaster losses provides important information on designing risk insurance schemes. In fact, this infrastructure is neither fully developed nor readily available and accessible to the risk insurance industry as well as to the general public in most of the developing countries in the Asia-Pacific Region.

Another important challenge, which has not gained much attention in the region yet and which could undermine the implementation of an effective insurance facility, is that of liability. Insurers will have to deal with it when failing to report their climate-related risks to their shareholders (O'Connor, 2005; Kunreuther & Michel-Kerjan, 2007). Besides, the probability of high insurance payouts increases due to the greater uncertainty and higher frequency of occurrence of extreme weather events in a changing climate that could lead, for example, to crop failures/harvest losses at increasing intervals (Iizumi, Yokozawa, Hayashi & Kimura, 2008). As a result of these limitations, most of the initiatives could not be effectuated and there are still large, sometimes even important regions as well as socio-economic groups that could not benefit from insurance-related instruments.

Thus it appears that most of the above factors are interlinked and that the situation is akin to the 'chicken and egg' dilemma. In order to promote the risk insurance in the Asia-Pacific Region, there is a need to overcome these limitations. In this regard, drawing lessons from some of the existing examples of implementing risk insurance in the Asia-Pacific Region and elsewhere can provide insights into how to overcome these limitations.

22.4 Case study of current experiences

At present, several pilot projects exist within and outside the Asia-Pacific Region that provide practical knowledge of promoting risk insurance (Table 22.1). One of the features of existing examples is that most

Table 22.1 Selected cases of risk insurance mechanisms from the Asia-Pacific Region and elsewhere

S No	Case	Geographical coverage	Hazards covered	Direct benefactor	Payment trigger
1	Caribbean Catastrophe Risk Insurance Facility	Caribbean (Regional)	Hurricane and earthquakes	National governments	Parametric
2	Mexico cat bonds	Mexico	Earthquakes and hurricanes	Government	Parametric
3	Turkish catastrophic insurance pool	Turkey	Multi-peril (Currently earthquake only)	Building owners	Indemnity
4	BASIX-ICICI Lambard microinsurance	Andhra Pradesh, India	Monsoon failures	Farmers	Index
5	Indian National Agricultural Insurance Scheme	All over India	Crop failure due to a range of conditions	Farmers	Indemnity
6	Agricultural weather index insurance	Thailand	Crop failure (Maize and rice)	Farmers	Index
7	Crop insurance in Japan	Japan	Crop failure (Rice)	Farmers	Indemnity
8	Microinsurance for cooperatives	Philippines	Protect loan portfolio from typhoons	Cooperatives and farmers	Parametric

Sources: Abousleiman, Zelenko and Mahul, 2011; Ghesquiere, Mahul, Forni and Gartley, 2006; Manuamorn, 2007; Munich Re, 2011; Sompo Japan Insurance Inc., 2010; Yazici, 2007.

of these experiences emanate from efforts to promote disaster risk reduction funded by the multi- and bilateral assistance organizations implemented at the local, regional and national level.

The Caribbean Catastrophic Risk Insurance Facility (CCRIF) is probably the epitome. It is the only insurance facility implemented and with premiums pooled on the regional level in which national governments pay the premiums for the insurable risks assessed at the national level. There are a number of examples for national level insurance facilities (e.g. Mexico cat bonds, Turkish catastrophic insurance pool, and Indian national agricultural insurance scheme, Japanese rice insurance) and numerous examples at the local level mostly implemented by non-governmental organizations (e.g. BASIX-ICICI Lambard microinsurance in India). Among the local level experiences, India and Mexico are reported to have well developed weather-based insurance programmes (Barnett & Mahul, 2007).

22.4.1 Weather index insurance is the way: Experiences from India

National policy environment

Around 70% of Indian agriculture is susceptible to the vagaries of the monsoon and other factors beyond the control of farmers. As a result, Indian agriculture has always been affected by nature's caprices. Each agro-climatic region requires different cropping plans as well as a distinct policy regime. With this in mind, the Government of India has initiated several policy initiatives to address various risks faced by farmers in the country:

(i) Programme based on 'individual' approach (1972–78): The first-ever crop insurance programme was introduced in 1972 to cover H-4 cotton in Gujarat and was later extended to a few other crops and states.

(ii) Pilot Crop Insurance Scheme – PCIS (1979–1984): PCIS was introduced on the basis of a report by Prof. V. M. Dandekar (Dandekar, 1976) presenting the

'Homogeneous Area' approach. The scheme covered food crops (cereals, millets and pulses), oilseeds, cotton and potato and was confined to borrowing farmers on a voluntary basis.

(iii) Comprehensive Crop Insurance Scheme – CCIS (1985–99): The scheme was an expansion of PCIS and has made insurance compulsory for borrowing farmers.

(iv) National Agriculture Insurance Scheme – NAIS (1999): NAIS (area yield index based crop insurance programme) replaced CCIS in the year 2000. Despite it being ideally suited for Indian conditions, the scheme has some shortcomings. The most important one is 'basis risk' as the area (insurance unit) is rarely homogenous. As the index is based on yield, the insurance covers, primarily, the processes between sowing and harvesting, but pre-sowing and post-harvest losses are not reflected in the yield index. Another challenge is the infrastructure and manpower required to conduct millions of crop cutting experiments (CCEs) across the country to estimate the yield of crops. The process also contributes to a delay in the settlement of indemnities as the CCEs can take several months. Moreover, yield index based insurances can be designed only for those crops for which historical yield data for at least 10 years (at insurance unit level) exist. Despite these shortcomings, the area yield index crop insurance operational in India is still regarded as one of the most illustrious crop insurance programmes in the world.

(v) Modified National Agricultural Insurance Scheme (MNAIS): The government announced a pilot project on an experimental basis in selected states and districts which is an improved version of NAIS titled 'Modified NAIS' (MNAIS). The new version bridges to a large extent the gaps of the existing NAIS.

The following are a few salient features of MNAIS:

a) Insurance unit for major crops is the village Panchayat or other equivalent units.

b) In case of prevented / failed sowing, claims up to 25% of the sum insured are payable immediately, while insurance cover for subsequent periods gets terminated.

c) Post-harvest losses caused by cyclonic rains are assessed at farm level for the crop harvested and left in 'cut and spread' condition up to a period of two weeks.

d) Individual farm level assessment of losses in the case of localized calamities like hailstorm and landslide.

e) Payment on account of up to 25% of likely claim in advance, for providing immediate relief to farmers in the case of severe calamities.

f) Threshold yield based on average yield of past seven years excluding up to two years of declared natural calamities.

One of the major issues in implementing MNAIS is that the insurance unit for major crops has been lowered to village Panchayat which is good for the farmers but increases exponentially the workload of CCEs. Many states are moving away from the pilot because of the enormity of the workload. Some states are requesting that the federal government shares parts of the costs of CCEs. From the insurer's point of view, accurate and timely data are needed to price the product accurately and to make timely payouts.

Risk insurance experiences in India

Significant experimentation and pilot projects have been taken up since 2003 in various states by all the major insurance service providers (Table 22.2). To provide risk cover to farmers, weather index insurances are better placed. Advocates of index-based insurances argue that it is transparent, inexpensive to administer, enables quick payouts and it minimizes moral hazard and adverse selection problems associated with other risk-coping mechanisms and insurance programmes (Giné, Townsend & Vickery, 2007; Hellmuth et al., 2009). Most importantly there are many low-income countries for which no historical data are available, except for weather data, affording an opportunity to try out index insurances of some kind. As a result, weather index-based insurances caught the imagination of policy makers at the beginning of the 21st century. Development institutions like the World Bank initiated pilot projects of this form of crop insurance in low-income countries where traditional crop insurances could not take off for various reasons that include unavailability of historical yield and/or loss data. The underlying principle for weather index insurances is the quantitative relationship between weather parameters and crop yields. There are various crop modeling and statistical techniques to estimate the impact of deviations in weather parameters on the crop yields (Rao, 2011).

Table 22.2 Comparison of various local initiatives of risk insurance in India

S No	Case	Geographical coverage	Hazards covered	Direct benefactor	Payment trigger	Benefits accrued
1	Weather Insurance by ICICI Lombard General Insurance Company (BASIX as local partner)	Mahabubnagar District, Andhra Pradesh during kharif 2003	Rainfall	Farmers	Excess and deficit rainfall	Claim amount to be adjusted against the crop loan
2	Mausam Bima Yogana by IFFCO Tokio General Insurance Company	Coimbatore District, Tamil Nadu during rabi 2008	Rainfall	Farmers and State Cooperatives	Excess rainfall	
3	Varsha Bima by Agriculture Insurance Company of India Limited (AIC)	Selected districts in 15 states in the Country	Rainfall	Farmers	Sowing failure and deficit rainfall during various phenophases	Partial payments as per the case during the crop cycle
4	Weather Based Crop Insurance Scheme (WBCIS) by AIC	Whole Rajasthan along with various other states	Rainfall, temperature, frost, heat wave and relative humidity	Farmers	Excess and deficit rainfall, deviation from the normal temperature and relative humidity	i. Trigger events can be verified independently; ii. Quick settlements of indemnities; iii. Covered all ranges of farmers
5	Weather Insurance by ICICI Lombard General Insurance Company	Nagapattinam, Tamil Nadu	Rainfall	Gujarat Heavy Water Chemicals Limited for Salt Industry	Rainfall disruption for salt preparation	Hedging against rainfall
6	Weather Based Crop Insurance Scheme (WBCIS) by AIC	Nashik, Maharashtra	Rainfall, temperature and relative humidity	Vine Making Industry	Disease and pest incidences considering weather as proxy	Hedging against weather for selected period of risk
7	Bajaj Allianz microinsurance	All over India	Life, asset damage, accidents etc	Rural communities	On maturity or damage	Accidental death benefit & accidental permanent total / partial disability benefit

The first pilot project of weather index insurance in India was carried out in 2003 by ICICI Lombard General Insurance Company Limited which was followed by projects of Agriculture Insurance Company of India Limited and IFFCO-Tokio General Insurance Company Limited, both during 2004. An impressive repository of historical weather data, high dependence on rainfed agriculture and a huge pool of scientific resources place India at the forefront in piloting different models of weather index insurance. The government's realizing of the need for encouraging pilot projects of this risk management tool has supported weather index insurance programmes from 2007 onwards by providing financial support in the form of premium subsidy paid up-front. The weather parameters that have been incorporated so far in weather index insurance include rainfall (deficit, excess, number of rainy days, consecutive dry days and wet days), temperature (minimum for frost, hourly chilling units, maximum for heat wave, mean etc.), humidity and wind speed.

Rao (2011) reported that, from 2007 onwards, both borrowed and non-borrowed farmers are covered under this scheme. Between 2010 and 2011 as many as 15 states have implemented the project Weather Based Crop Insurance Scheme (WBCIS) in over 100 districts covering more than 800 blocks/tehsils (administrative subdivision of a district). As per estimates, it insured nearly eight million farmers in an area of more than 12 million hectares for a sum insured of approximately Rs. 96 350 million at a premium of Rs. 8830 million. The cumulative number of Indian farmers covered under WBCIS during 2010 and 2011 is estimated to have reached 9.27 million, in 13.23 million hectares and risk exposure of Rs. 143 000 million at a premium of Rs. 12 900 million.

Lessons from the ground

Despite several risk insurance experiences in India, the farmers' loyalty has not been won completely (Singh & Jogi, 2011). There are various reasons for it and the most important are basis risk, either no or delay in payment of claims, lack of knowledge and awareness of various contracts, lack of historical weather data and hence high premiums.

The two major challenges of the present weather risk index-based insurance products are (i) designing a proxy weather risk index with predictive capability to measure crop losses realistically and (ii) basis risk. Basis risk results if the actual experience of weather risk (rainfall) in the neighbourhood significantly differs from the data recorded at the weather station. The two aspects lead to compounding of the problem for all parametric triggered insurance products: both may not trigger a payout despite the occurrence of damages at an individual farm, or these may trigger a payout when loss did not occur.

The State of Knowledge Report by the Global AgRisk (2010) has brought out a few important observations on using a weather index for small, moderate and large losses. When rainfall is around the optimal level for a crop, many other important factors affect crop yields (e.g., soil quality, fertilizer use, pesticide use, crop husbandry practices, etc.). Around this level, the correlation between rainfall and crop yields is likely to be not very strong. When rainfall is extremely low, however, the relationship between rainfall and yields is expressed more strongly. Other variables such as use of fertilizers and pesticides have very little effect on yields at low levels of rainfall.

Due to high transaction costs, insurance is perceived to be rather an expensive financial instrument and is mostly designed to protect against low probability and extreme loss events. However with increasing awareness, penetration and efficiency, the unit cost is going down rapidly. The schemes like WBCIS are in fact more desirable as they have the ability to mitigate even small to moderate losses and also provide extended coverage for pre-sowing periods and quality of output which are difficult to cover under other schemes.

On the other hand, catastrophic events affect not just yields but assets and long-term income. A 'generic' insurance product (in place of a sophisticated product), therefore, can do well for mitigating such losses (Rao, 2011). Compared with a weather index insurance a catastrophic coverage could be an alternative as it is designed to cover the insured party in case of a fire, flood, earthquake, tornado, or other major accidents. The data requirements for designing catastrophic coverage insurance products are relatively low and hence the basis risk is lower. The cost of administration is also lower for catastrophic coverage. Premiums for covering catastrophic risks through catastrophic coverage are generally affordable which leads to availing insurance for almost all important assets, which in turn can lead to increased demand for insurance and ultimately a high level of insurance penetration.

Insurers have to find a way to offer a technically sound product that is simple and easily accessible to farmers. Farmers must be able to understand the products sufficiently in order to calculate claims and expect realistic payouts. The lack of benchmarking for weather index insurance products erodes the value of financial support provided by the state governments under WBCIS.

By their very nature, weather index insurance products are difficult to comprehend, especially by a typical Indian farmer who has limited capacities and experience. The multitude of weather index insurance products offered by various insurance providers necessitates the need for benchmarking the various products to enable the farmer to make an informed choice. Through benchmarking it may be ascertained whether the products offered by the different insurance companies carry at least comparable benefits (Protection vis-a-vis Premium). The complex weather index insurance products may be disintegrated into the constituent covers for different perils.

The opinion of WBCIS beneficiaries on 16 different aspects of weather index insurance was assessed by the Government of India (2010). Eighty per cent of the respondents highlighted high basis risk (location of weather station), 57% were not satisfied with the grievance redress mechanism, an equal number have reported inconvenience in enrolment, 17% were not satisfied with the transparency, 19% were not satisfied with the reliability of weather data and 25% were not satisfied with a weather index as a substitute for yield index insurance. From this exercise, it is clear that weather risk index-based insurance is rated well on data accuracy, transparency and quick claims settlement, which are very attractive to both farmers and the reinsurance market.

Though the WBCIS programme is perceived by states as a good alternative to NAIS, there are some key challenges to be overcome for the scaling up of the scheme (Rao, 2011).

1. Scope of WBCIS is limited to parametric weather exigencies like rainfall, temperature, humidity and so on. In addition to weather-related impacts, often crops suffer due to hailstorms, floods, pests and diseases, which to a large extent are difficult to cover under the scheme. However, over the years with increasing understanding between weather parameters and effects on crops, indices have been designed which provide cover against pests and diseases by considering weather as a proxy.

2. Product design under WBCIS is challenging as crop yield and weather relationship is not only complex, but also influenced by various factors such as cultural practices, date of sowing, soil type and crop variety. It requires focused research by agricultural scientists to fine-tune the weather-yield relationship.

3. The growth of WBCIS demands that every village has a weather station so that basis risk in weather index insurance is minimized. Nevertheless, with con-

sistent increase in coverage under WBCIS the penetration of weather stations is also increasing. The weather stations are now available at about a radius of 15 km for locations where they were available at more than 30 km earlier. The acceptable radius for insuring rainfall is about 5 km and for other parameters is 10 km. For achieving these levels, nearly 50 000 weather stations are required as against about 5000 stations which are presently available including both public and private stations in the country (Milesi et al., 2011).

4. Calibration of sensors and data at weather stations is another challenge as presently weather data providers are using stations of different make and quality. This would require third party accreditation and calibration services to vouch for reliability and accuracy of the data.

5. There is a need to develop location specific and crop specific insurance contracts by making use of local historical weather data.

Recognizing the problems being faced in creating and delivering weather index based insurance products around the world, systems such as Terrestrial Observation and Prediction System (TOPS) have been developed to organize disparate streams of information into a cohesive framework to serve a variety of societal needs (Nemani et al., 2009). The need for data synthesis for producing actionable information is greatest in rural India where nearly 70% of the population lives and works.

TOPS is a modeling software system designed to produce ecological forecasts. TOPS brings together advances in information technology, weather/climate forecasting, ecosystem modeling and satellite remote sensing to enhance management decisions related to floods, droughts, crop condition, human health, forest fires, forest production and so on. TOPS provides a suite of ecosystem 'nowcasts' (measures of current conditions) and forecasts. These data products include measures of vegetation condition and productivity, snow dynamics, soil moisture and meteorological conditions and forecasts (Milesi et al., 2011). Another key feature of TOPS is an automated system for ingesting climate observations from local, regional and global networks of meteorological stations in real-time to produce spatially continuous gridded meteorological fields. This capability allows TOPS to provide continuous estimates of ecosystem conditions for any location in the country, including remote and sparsely instrumented regions.

Information from TOPS can benefit the risk insurance society in a number of ways. By blending data from a few weather stations with satellite data available for over 30 years, TOPS creates high-quality information at village level. Similarly the integrated information from TOPS allows one to verify fraudulent claims, in a way acting as third party verification. Using TOPS capabilities for long-term simulations of vulnerabilities, key insights about the potential consequences of climate change in a variety of sectors can be generated and disseminated. The weather stations are not totally tamper-proof and their maintenance costs are high. The concept of a 'virtual weather station network' based on TOPS platform can produce the daily weather data at a scale of one kilometre grid for the past 10 years which is valuable for agriculture meteorological risk reduction.

22.4.2 Financial markets play a vital role: Experiences from Japan

Although Japan has made large investments in infrastructure, agriculture, fisheries and allied sectors, the great Tohoku earthquake in 2011 has revealed that these sectors are still vulnerable to natural hazards. The damage caused by the Tohoku earthquake reaffirms that Japan is potentially one of the countries in the world that holds the highest vulnerability to natural hazards. According to UNU-EHS's and Munich Re's (2007) Natural Hazard Risk Index (Table 22.3) potential disaster losses in Japan's megacities are significant.

Based on statistics issued by the Government of Japan, both direct and indirect social and economic damages caused by Hanshin-Awaji earthquake (also known as the Great Hanshin or the Kobe Earthquake) was approximately 2.6% of national GDP in 1995 (Kato, 2009). This quake also had surprisingly low insured losses compared to the very high economic losses (see also Figure 22.1). The recent Tohoku earthquake has affected wider areas leading to significant economic and financial damage to the Japanese economy. It was estimated that the damage could be to the tune of 16.9 trillion Japanese Yen, 1.7 times of Hanshin-Awaji earthquake in 1995 (Cabinet Bureau of Japan, 2011). The earthquake has worsened the nation's fiscal balance.

Having passed a depression era in the 1990s, the so-called 'lost decade', Japan issued deficit-covering government bonds in large proportions (OECD, 2011). Due to Japan's economic situation, which includes fiscal imbalance, it is not easy for the national and prefectural governments to issue bonds for disaster reconstruction, since this could affect fiscal consolidation and fiscal deficit. Hence, the Tohoku disaster damage cannot be covered by the reconstruction bonds issued by the national and prefectural governments and it is essential to consider other financial instruments (OECD, 2011).

Though Japan's individual financial asset is the biggest in the world (1 450 trillion Yen) (Bank of Japan, 2012) consisting of bonds and bank deposits (nearly 80%), only a small contribution comes from risk assets such as equities and mutual funds. This shows a huge potential for the growth of market oriented risk management schemes such as equities and mutual funds, replacing public financing assumes importance in Japan.

Table 22.3 Natural Hazard Risk Index for Megacities

City	Natural Hazard Risk Index	City	Natural Hazard Risk Index
Tokyo/Yokohama	710.0	Hong Kong	41.0
San Francisco	167.0	London	30.0
Los Angeles	100.0	Beijing	15.0
Osaka/Kobe/Kyoto	92.0	Dhaka	7.3
New York	42.0	Mumbai	5.1

Source: UNU-EHS and Munich Re, 2007.

National policy environment

In Japan, the risk insurance is mainly represented in the form of earthquake insurance. Earthquake insurance issued by private insurance companies is designated to financially support lives of earthquake victims and hence is limited to cover residential houses and personal properties. Until Niigata Earthquake in 1964, damages on assets caused by earthquakes in Japan were not covered by fire insurance since the fire insurance schemes in place have regarded the earthquake damages as legally immune to compensatation even though fire could have led to secondary disaster in the wake of an earthquake. The Government of Japan has decided to support the earthquake insurance as an exclusive reinsurer since private insurance companies could not fully afford to compensate without government support. As a result, the Government of Japan has established Japan Earthquake Reinsurance Co. Ltd. as a Special Purpose Vehicle through 'The Law Related to Earthquake Insurance, 1966' to give impetus to the earthquake insurance in Japan (Ministry of Internal Affairs and Communications of Japan, 1966). Despite this, the earthquake insurance could not become popular and the major spread of insurance has limited to fire insurance with coverage of 30–50% in the present value of the asset (with an upper limit of 50 million Yen). Due to the high premium costs and low compensation levels, the earthquake insurance has not been popular until Japan experienced Hanshin Earthquake in 1995. After 1995, the number of earthquake insured has steadily increased up to 23.7% of entire households in 2010. Such an inadequate spread of earthquake insurance could be a financial burden in the wake of a major earthquake (General Insurance Association of Japan, 2011).

Due to limited demand for risk insurance products and limited volume of Japan's reinsurance market, only very few other forms of risk insurance, other than the government supported earthquake insurance, could be observed in Japan (Financial Centre Futures, 2011). During recent years, big corporations have started using alternative risk transfer products such as captive, finite risk insurance and cat bonds. For these firms, risk hedging using the financial schemes helps to get higher evaluation by credit rating agencies. However, these practices are hardly contributing to risk insurance for vulnerable citizens.

In early 2008, the Government of Japan has introduced Japanese Sarbanes-Oxley Act (J-SOX) to strengthen corporate internal control (Kato, 2009). The act stresses firms' risk management and ensures risk mitigation against physical and financial damages. Since then, Japanese businesses, from small to large scale, have found encouragement to minimize the disaster risks and realized that the government reinsured earthquake insurance is inadequate to address the range of risks these firms face.

The Government of Japan has begun amending Private Finance Initiative Act (PFI Act) to diverge risks on public infrastructure between public and private sectors (Kato, 2011). J-SOX require companies to disclose various risks that companies face in their financial reporting. With this amendment, private corporations can also get involved in large scale public infrastructure projects such as agricultural community sewerage projects with government subsidies. This requires corporations to invest in weather derivatives to deal with unexpected hydro-meteorological extreme events.

Although the Government of Japan has recognized the importance of aggressively promoting the risk financing, the related processes are still in their infancy. The current rigid financial law does not allow captive insurance firms in Japan to be established and this needs to be addressed at the policy level. The direction of Government of Japan's policy is to emphasize various risk mitigation provisions for both public and private sectors. With this, the application of risk insurance and alternative risk transfer is being increasingly recognized in Japan.

Risk insurance experiences in Japan

Currently, a major part of alternative risk transfers in Japanese market is through cat bonds,[1] captive,[2] finite risk insurance,[3] and weather derivatives[4] (METI, 2006). These form the greater part of risk transfer

---

[1] A debt instrument that is usually insurance linked and planned to raise money in case of a catastrophe such as a typhoon or earthquake.
[2] A subsidiary that is designed to provide financing to customers by purchasing the parent company's product.
[3] An insurance contract that shifts the risk of loss from an insured to an insurer during a given years.
[4] A financial commodity used by companies is designed to hedge against the risk of weather-related losses.

strategy for big companies and large associations such as Zenkyoren, the National Mutual Insurance Federation of Agricultural Cooperatives. Zenkyoren in turn will compensate the financial losses faced by its members. After the Hanshin-Awaji earthquake in 1995, Oriental Land, which runs Tokyo Disney Land, issued two different cat bonds to cover reconstruction costs.

Big corporate bodies can create Special Purpose Vehicle (SPV) for captive insurance and they can afford to accumulate sufficient funds for finite risk insurance. Small businesses and households can purchase risk insurance issued by associations like Zenkyoren. However, these forms of insurance are inadequately covered by reinsurance and hence pose financial threat to the insurer in the wake of a large disaster (Froot, 1999). In addition, it is extremely unlikely for small businesses to utilize these financial risk transfer mechanisms due to the need for a large portfolio for these instruments to work effectively.

Munich Re issued a cat bond called 'Muteki', which covers Zenkyoren's earthquake risk and transfer to the capital market (Munich Re Group Risk Trading Unit, 2011) (Table 22.4). The bond is well recognized in the market since it is independent of the stockmarket and practical to put in theportfolio. According to Steve Evans (2011a), Muteki was triggered during the Tohoku earthquake and made a loss to the tune of US$300 million. However, despite these losses,

the investors may not have lost their confidence in Japanese cat bonds (Evans, 2011b). The earthquake also took a toll on the Zenkyoren which faced a reported loss of US$ 11.2 billion (Evans, 2012a). If the insurer (Zenkyoren) cannot bear the financial damage, small businesses and households that are insured cannot expect to receive full compensation.

The Midori bonds are designated for the JR East, the largest railway company in Japan since 2007. This five-year bond is expected to cover loss of public transportation services and the infrastructure when a significant earthquake hits within a 70 km radius of Tokyo.

Despite rising reinsurance costs, Kibou, another cat bond, has been issued on behalf of Zenkyoren (Evans, 2012b). According to Swiss Re Global Cat Bond Performance Index, Index Exposure by Peril (shown in Evans, 2011b), the global cat bond market is still dominated by the US hurricane risk bonds and the share of typhoon/earthquake risk related bonds in Japan are on the decline. Since capital markets welcome diversification of the investable, there is enough room for cat bonds in Japan.

Initiatives 3 to 5 in Table 22.4 are examples of weather derivatives in Japan which are emerging as a significant financial tool for farmers and small businesses (Yokouchi, 2007). The weather derivative sales have increased significantly recently (Yamada, 2010;

Table 22.4 Comparison of various local initiatives in risk insurance in Japan

No	Case	Geographical coverage	Hazards covered	Direct benefactor	Payment trigger	Benefits accrued
1	Muteki Ltd (cat bond)	Japan	Earthquake	Zenkyoren	Richter scale	Investors
2	Midori Ltd (cat bond)	Japan	Earthquake	JR East	Richter scale	Investors
3	Typhoon Derivative (Tokio Marine Co.)	Japan	Typhoon	Farmers Union, Hotels, Leisure industry,	Number of typhoon passed	None
4	Warm Winter Derivative (Sompo Japan)	Japan	Climate Change	Farmers Union, Energy retailers, Fashion Industry	Temperature	None
5	Winter Preparation aka Fuyu no Sonae (Aioi Insurance)	Japan	Climate Change	Farmers Union, Energy retailers, Fashion Industry, Hotels	Temperature, Rainfall, Snowfall	None

Sources: Aioi Nissey Dowa Insurance Co. Ltd., 2011; Sompo Japan Insurance Inc., 2010b; Munich Re Group – Risk Trading Unit, 2011; Tokio Marine and Nichido Fire Insurance Co. Ltd., 2011.

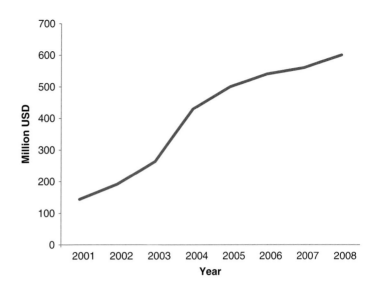

Figure 22.2 Growing market volume of weather derivatives (million USD) in Japan. *Source*: Compiled by the author, based on Yamada, 2010.

Figure 22.2) as local small banks and credit unions played an intermediate role in selling weather derivatives for agriculture and allied sectors (Figure 22.3). Weather derivatives in Japan are designated for specific clients since the derivatives are not publicly offered. Therefore, the premium is usually over one million Yen (approximately US$ 1250 at an exchange rate of 80 JPY per US$) and this is too expensive for individuals and small businesses. However, the local financial institutions (FIs) such as local banks and credit unions with a strong local network played a significant role in selling weather derivatives benefiting commission from insurance companies. FIs know

the need for weather derivatives and how to access the potential market. These FIs can even accommodate farmers and business owners with a loan for the derivative since FIs have a strong local network through which to sell financial commodities unlike insurance companies.

The strong ties of farmers and business owners with FIs have helped them to get familiar with the sophisticated financial products such as weather derivatives. As a result, this has emerged as a key model in disseminating alternative risk transfer. The mediation by the local FIs has stimulated their sales and has drastically reduced the cost of designing the financial instrument. The growing number of weather derivative sales helped to reduce the premium costs (0.3 million Yen, approximately US$ 3750) and helped in their spread. No significant impact of the financial crisis could be seen on the derivatives market in Japan since people are sensitive enough and aware of weather risks.

Although the weather derivatives market is growing, their volume in Japan is much smaller than in the US (Bank for International Settlements, 2010). To avoid holding domestic risk insurance within the country, it is necessary to transfer the risk abroad by increasing trade volume and risk transfer will be accelerated if foreign investors hold more Japan issued derivatives. Weather derivatives in Japan are only negotiated over-the-counter and are not traded on exchanges. This is due to the Commodity Exchange Act (METI, 1950) that allows only trading of 'tangible

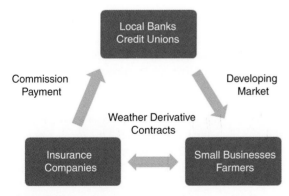

Figure 22.3 Weather derivative sales scheme using local financial institutions. *Source*: Based on Ono, 2004.

objects'. Without a provision in the law, the weather derivatives in Japan cannot be traded in domestic and international markets. However, exchanges such as Chicago Mercantile Exchange, London International Financial Futures and Options Exchange (LIFFE), Intercontinental Exchange (ICE) and the Catastrophe Risk Exchange Inc. (CATEX) are listing standardized weather derivatives (Amazaki et al., 2003). Such trading of weather derivatives will have a positive impact on price discovery, volume and market liquidity.

The weather derivatives sold by Japanese insurance companies are only intended for corporate bodies and unions, since selling the financial products to an individual can be a violation of the Consumer Contract Act in Japan (Itabashi, Iwazawa & Watanabe, 2007). Individuals, mostly farmers, purchase the derivatives through local agricultural associations. Each derivative has a different trigger for the payment and hence it is not rare to purchase multiple financial products. The strong demand from businesses and industries affected by weather change has stimulated the insurance companies to develop numbers of weather derivatives, and expansion of the market has reduced the premium prices. With this trend, the insurance companies are now able to provide various weather derivatives on a small to big scale.

Lessons for scaling up

It is clear that Japan needs multiple sources to finance disaster risk reduction in the future. As described in the previous section, financial risk transfer in Japan is mostly used by small businesses and big corporations leaving individuals out of the risk insurance market. From the analysis, it can be said that the risk finance market is segmented and the market should be consolidated to adjust needs from various sectors such as households, public and industry, if risk transfer is to be strengthened. The government supported reinsurance is a strong backbone for the earthquake insurance system, but alternative risk transfers also need to be encouraged.

One of the reasons why the market cannot respond to the growing need for risk financing is the immaturity of Japan's capital market infrastructure in terms of capability of domestic FIs. It is indicated by the existence of a few Japanese reinsurance companies with small gross billings compared to top reinsurers elsewhere. Consequently, most of the funds flow

to Europe, the United States and other regions such as Bermuda as the Japanese market cannot provide much needed financial services meeting domestic investors' demand. If Japan and countries in Asia need to utilize abundant funds in Asia for the region's risk reduction, it is necessary to establish sound capital market and reinsurance market supported by domestic agents. By doing so, disaster risk will be shared not only in Japan and the region, but also in the world through a flow of funds.

The following points emerge clearly, considering the current fiscal condition and discussion above: a) existing risk mitigation mechanisms such as investments in infrastructure are not sufficient to mitigate disaster risks; b) there is a huge potential for increasing the risk insurance in Japan in terms of insurance from the individual subscribers as against institutional and corporate insurance and in terms of insurance for specific natural hazards such as earthquakes, tsunami and floods; c) the role of financial markets can be further strengthened by linking the domestic risk insurance market with that of the regional and international financial markets; d) there is a diminishing role of governments (national or prefectural) in promoting risk insurance in the country, but it is expected to support establishing generic risk finance market in the future; e) consolidating risk insurance and alternative financial risk transfer markets is not only complementary of Japan's current risk mitigation but also provides an opportunity to disperse various risks to natural hazards in the region; and f) it is highly desired to establish transparent and openly accessible risk finance market in Japan for farmers, small business owners and investors.

22.5 Proposals to the UNFCCC for the Future Climate Regime

The future climate regime can facilitate promoting the climate risk insurance in the Asia-Pacific Region through providing the additional finances required which is one of the major limitations in promoting disaster risk mitigation (GFDRR, 2009). Mention of the risk insurance can be found in the negotiated text of the UNFCCC and the Conference of Parties. The Article 4, paragraph 8 of the UNFCCC text refers to the risk insurance as a funding mechanism to meet the needs of the developing countries arising from the adverse effects of climate change (UNFCCC, 1992) 'including actions related to funding, insurance and the transfer of technology, to meet the specific needs and concerns of developing country Parties arising from the adverse effects of climate change and/or the impact' (p.8). The UNFCCC text also characterizes countries eligible for financing and insurance mechanisms. The Bali Action Plan goes further and explicitly states that the risk insurance mechanisms should be used in promoting adaptation (UNFCCC, 2007).

Various proposals have been submitted by the Parties to the Convention as well as by those outside the Convention for promoting the risk insurance under the Convention (Table 22.5). The Alliance of Small Island States (AOSIS), the most rigorous promoter of such a risk insurance scheme, has proposed that an International Insurance mechanism and Solidarity Funds addresses catastrophic risk and collective loss sharing. Cook Islands proposed the International Insurance Scheme where it emphasized the collective burden sharing, subsidy elements to maintain the fund as a compensation for unavoidable impacts, and funding risk reduction initiatives (Harmeling, 2008). A Swiss proposal to the UNFCCC on promoting risk insurance includes prevention and insurance pillars with funds coming from global CO_2 levy with greater benefit to low income countries (Government of Switzerland, 2008).

Munich Climate Change Initiative (MCII) made a proposal consisting of two tracks or pillars, one for supporting risk reduction through mitigation activities and the other supporting the insurance (Bals, Burton & Butzengeiger, 2008). The insurance component was divided into two tiers with tier I consisting of climate insurance pool to cover the high level risks in non-Annex I countries and the tier II consisting of public safety nets and insurance systems through public-private partnerships covering medium level risks.

Recent negotiations have anchored the risk insurance subject in the Paragraph 14 of Decision 1/CP.16 of the Cancun Adaptation Framework (CAF) in Cancun Agreements. Under the Cancun Agreements, a work programme to consider approaches to address the loss and damage associated with climate change impacts in vulnerable developing countries was established by the COP. These discussions have stressed the need for a climate risk insurance facility. The

related submissions by Parties display the general convergence among Parties and relevant organizations for risk insurance mechanisms to be included under the work programme (UNFCCC, 2011). However, divergence of views can also be observed in terms of its form: for instance, AOSIS envisages establishment of an international risk insurance mechanism under the UNFCCC framework whereas countries such as the United States and EU prefer a climate risk insurance facility at national and regional level, taking into account country differences and respecting country-driven approaches. The related discussions in March 2012 in Tokyo have emphasized the need for improved knowledge sharing among various UNFCCC processes; the need to standardize damage assessment and reporting, and identify entry points and facilitate the engagement of the finance sector in disaster risk reduction (UNFCCC, 2012).

22.6 Messages for the future climate regime

Several lessons and best practices emerge in terms of what should be the essential design elements for promoting risk insurance under the future climate regime.

1. Keep the price of the insurance premium affordable: The price of the insurance premium is one of the major determinants for enrolling a maximum number of insured and hence keeping its price affordable is an important aspect of the overall design of the insurance system. In the case of Japan, the premiums were heavily subsidized (over 50%) to make the premiums affordable (Tsuji, 1986). Since the amount of residual risks and premium prices are directly correlated, other insurance programmes such as the Turkey catastrophe insurance pool have combined to promote the risk mitigation measures such as enforcing seismic resistance codes along with the insurance programme. In the Philippines, the prices of premiums were able to be kept at an affordable level by linking microinsurance with the cooperatives (Munich Re, 2011). However, there is a limit to which the insurance agencies can reduce the insurance premium prices since the premium prices would have to cover capital costs, reinsurance costs and admin costs and profit margins. Any substantial reduction in

Table 22.5 Summary of Selected Country/Consortium Proposals on Disaster Risk Insurance Mechanisms at UNFCCC Negotiations

Characteristics	Proposals			
	AOSIS	**MCII**	**Cook Islands**	**Switzerland**
Target group (governments/ individuals)	National Governments of SIDS, LDCs and other developing countries	Governments and individuals	National governments of SIDS	Regional authorities, governments, and individuals
Geographical coverage (national/local/ regional)	Regional/National	National and regional	National	• Regional and sub-regional (insurance pillar); • National (prevention pillar)
Source of funding	• Convention Adaptation Fund • Kyoto Protocol Adaptation Fund (existing) • Other bilateral and multilateral sources	Financial mechanism of the Convention channeled through CIP, CIAF, and CRMF	Internationally-sourced pool of funds (subsidy in establishing establish-ing/maintaining fund)	• Global Carbon Tax • Insurance pillar funded through MAF
Promotion of re-insurance	Yes, through conventional risk sharing and transfer instruments	Yes, through CIP	No reference to re-insurance	Yes, through public-private partnership
Targets premium prices	No indication for premium prices	No indication for premium prices	No indication for premium prices	Provides funding for premiums
Inclusion of risk mitigation component	Yes, through technical and financial support for risk reduction efforts	Yes, through the prevention pillar	Yes, mechanism funds risk reduction initiatives	Yes, through the prevention pillar
Reference to guidelines for implementation	No reference to guideline	Yes, under the authority and guidance of COP	No reference to guideline	Yes, defines eligible extreme events and insured damage
Reference to awareness	No reference to awareness	No reference to awareness	No reference to awareness	Yes, awareness generation is financed by NCCF
Addressing the risk data gaps	Yes, though improved risk management tools, collection and analysis of data	No reference to addressing data gaps	No reference to addressing data gaps	Yes, through small budget under the insurance pillar
Sustainability issues if any	No reference to sustainability	No reference to sustainability	No reference to sustainability	No reference to sustainability

Notes: **AOSIS**: Alliance of Small Island States; **MCII**: Munich Climate Insurance Initiative; **SIDS**: Small island developing states; **LDC**: Least developed countries; **CIP**: Climate Insurance Pool; **CIAF**: Climate Insurance Assistance Facility; **CRMF**: Chronic Risk Management Facility; **MAF**: Multilateral Adaptation Fund; **NCCF**: National Climate Change Fund.

Sources: AOSIS, 2008; Cook Islands on behalf of AOSIS, 2008; The Munich Climate Insurance Initiative, 2009; Government of Switzerland, 2008.

insurance costs can only be possible by a combination of approaches such as efficient management by the insurance firms, reducing basis risks through risk mitigation measures such as enforcing structural standards and land-use planning regulations, and subsidies from the national governments.

2. Generate public awareness: Apart from the issue of the price of the premium, the lack of awareness among various stakeholders is a major hurdle in spreading the risk insurance. This hurdle was mostly overcome by incorporating the grassroots level awareness generation activities. For instance, such an effort can be seen in agricultural weather index insurance, im Thailand; and in various locally implemented insurance programmes (e.g. BASIX-ICICI Lambard microinsurance; Turkey catastrophe risk insurance pool). Through insurance agencies working closely with farmer associations, the Japanese example provides a good case for increasing public awareness and overcoming other attitudinal barriers.

3. Avoid the moral hazard: One of the major problems with the traditional insurance programmes including the crop insurance programmes has been the moral hazard, i.e. unfair insurance claims leading to a higher risk for the insuring agencies (Giné, 2009). This limitation has largely been overcome by the advent of index based insurance systems where payment is triggered by factors that are extraneous to human control, i.e. the actual incidence of the particular intensity level of the hazard (e.g. 60% reduction in rainfall). One factor that needs to be taken into consideration, however, is the weather data required for developing such indexes. The India case provides a good example of overcoming this barrier.

4. Link with reinsurers and investment in financial markets: Support by reinsurers is one of the important considerations for putting in place robust risk insurance systems as reinsurers provide needed financial backup to the insurers. In addition, insurance facilities created may also consider investing, in part or total, in international financial markets using the support of the international reinsurance facilities. Such an example is epitomized by the current agricultural weather index programme in Thailand (Sompo Japan Insurance Inc., 2010a) and the Caribbean catastrophe risk insurance facility (Ghesquiere, Mahul, Marc & Ross, 2007). The structure of current financial markets is only favourable for large corporations and busi-

nesses and does not seem to benefit direct risk reduction for the individuals. Efforts should be made so as to ensure that the financial markets provide greater risk reduction benefits to individuals by giving right price signals encouraging greater participation in risk insurance.

5. Enhance availability of risk information: Availability of reliable rainfall data and associated crop losses is a prerequisite for designing a robust index-based insurance facility. Similarly, comprehensive information on the physical characteristics of the infrastructure such as buildings, warehouses and so on to be insured is needed for estimating the risk from hazards such as floods, droughts and earthquakes. Such a robust information infrastructure is still not readily available on a large-scale in most of the countries, including the Asia-Pacific Region, hindering expansion of the risk insurance facilities.

For example, the lack of widespread historical data to assess the relationship between weather parameters and crop losses has limited the implementation of risk insurance facility to the area where historical weather information is available in Thailand (Sompo Japan Insurance Inc., 2010a). Risk insurance facilities have overcome this limitation by investing the resources to collect and analyse the available information, employing simulation modelling, interpolation and extrapolation techniques as well as by increasing the risk margin while calculating the price of the premium (United Nations, 2007; O'Connor, 2005). Nevertheless, in all the cases, the availability of risk information determined the feasibility and success of an insurance facility.

Comparing these experiences with the issues identified at the beginning of this section, the insurance initiatives did not translate in terms of scaling up and sustainability of these initiatives which are areas where the future climate regime could play an important role.

22.7 Conclusions and way forward

This chapter has identified existing limitations in promoting risk insurance by drawing lessons from within and outside of the Asia-Pacific Region and looking into how the future climate regime could help overcome these limitations.

Numerous risk insurance experiences show that risk spreading is a way forward for dealing with a variety of climate and non-climate related risks. However, feasibility and sustainability of implementing an insurance facility at the global, regional, national and local level could face several barriers, as identified in this chapter, which include limited knowledge among stakeholders about the benefits of risk insurance systems, limited expertise to design and implement insurance products, challenges in keeping the premium prices sustainable, lack of good quality data on risks and historical losses and limited presence of reinsurers. Addressing these limitations is essential in enhancing readiness to accept insurance as a risk reduction tool.

While divergent positions are observed between Annex I and non-Annex I parties on the fundamental need to support an insurance mechanism, it is crucial for parties to consider and assess the opportunities that insurance mechanisms provide in reducing risks at different levels in line with the role of the UNFCCC as a catalyst to promote collective actions. It is important for the Annex I parties to recognize the fact that any risk reduction promoted in Non-Annex I countries would benefit the Annex I countries as well due to the role these countries are playing in terms of production of goods and services.

To adapt to the future climate, it may be necessary to consider adopting a convergence approach consisting of the lessons drawn from regional models such as CCRIF as well as from local models, for example, numerous microinsurance schemes which have proven to be suitable in particular for developing countries. In this regard, further assessment is needed to identify the best mix or combination of such tools for each region concerned, including Asia-Pacific. The proposals to the Convention should aim at promoting public awareness on risk insurance, putting in place robust and transparent systems to collect, analyse and disclose risk information, providing for continuous evaluation of the performance of the risk insurance systems, encouraging greater private sector participation, and most importantly, helping to keep the premium prices at affordable levels. The latter objective could be achieved by a combination of approaches, such as targeted subsidies or enforcing structural and land-use planning regulations. In addition, the proposals should make clear how the regional and local insurance mechanisms are to be governed and sustained while improving the existing risk governance systems at the national level. The ultimate metric for the real impact of these proposals should be in terms of the scaling up of insurance leading to substantial risk reduction on the ground.

Acknowledgements

We thankfully acknowledge the funding support from the Ministry of Environment Strategic Environment Research Project (S8) led by Professor Mimura, Ibaraki University and the Asia Pacific Adaptation Network, Bangkok.

References

Abousleiman, I., Zelenko, I. and Mahul. O., 2011. *Mexico MultiCat Bond – Transferring Catastrophe Risk to the Capital Markets*. Washington DC: Global Facility for Disaster Reduction and Recovery, The World Bank.

Aioi Nissey Dowa Insurance Co. Ltd., 2011. *Aioi Nissey Dowa Insurance Co. Ltd.* [online] Available at: <http://www.aioinissaydowa.co.jp/> [Accessed 21 November 2011].

Amazaki, Y., Okamoto, H., Shiihara, K., Niimura, N. and Hirose, N., 2003. *Weather Derivatives*. Tokyo, Japan: Tokyo Denki University Press.

AOSIS, 2008. *Multi-Window Mechanism to Address Loss and Damage from Climate Change Impacts: Submission to the AWG-LCA*. [pdf] Alliance of Small Island States, UN. Available at: <http://unfccc.int/resource/docs/2008/awglca4/eng/misc05a02p01.pdf> [Accessed 12 November 2010].

Arnold, M., 2008. *The Role of Risk Transfer and Insurance in Disaster Risk Reduction and Climate Change Adaptation*. Stockholm: Commission on Climate Change and Development.

Bals, C., Burton, I., Butzengeiger, S. et al., 2008. *Insurance-related options for adaptation to climate change*. [pdf] Bonn: The Munich Climate Insurance Initiative. Available at: <http://germanwatch.org/rio/c11insur.pdf> [Accessed 12 November 2010].

Bank for International Settlements, 2010. *Triennial Central Bank Survey – Report on global foreign exchange market activity in 2010*. [pdf] Available at: <http://www.bis.org/publ/rpfxf10t.htm> [Accessed 12 December 2011].

Bank of Japan, 2012. *Flow of Funds – Overview of Japan, US, and the Euro area*. [pdf] Available at: <http://www.boj. or.jp/en/statistics/sj/sjhiq.pdf> [Accessed 15 April 2012].

Barnett, B.J. and Mahul, O., 2007. Weather index insurance for agriculture and rural areas in lower income countries. *American Journal of Agricultural Economics*, 89 (5), pp.1241–1247.

Cabinet Bureau of Japan, 2011. *Estimate of damage by Northeast Japan Earthquake*. [pdf] Available at: <http://www.bousai.go.jp/oshirase/h23/110624-1kisya.pdf> [Accessed 28 November 2011].

Commodity Exchange Act 1950. (Act No. 239 of August 5, 1950), Tokyo: METI. Available at: <http://www.meti.go. jp/policy/commerce/b00/pdf/b0000008.pdf> [Accessed 22 May 2012].

Cook Islands on behalf of AOSIS, 2008. *Advancing adaptation through finance and technology, including National Adaptation Programmes of Action – Views of AOSIS*. Presentation material at the Workshop on advancing adaptation through finance and technology, including National Adaptation Programmes of Action. [pdf] Available at: <http://unfccc.int/files/adaptation/application/pdf/cookislands_awgcla2_adaptation_workshop.pdf> [Accessed 12 November 2010].

CRED (Centre for Research on the Epidemiology of Disasters), 2010. *Emergency Disasters Data Base, 2010*. [online] OFDA/CRED International Disaster Database, Data Version 06.06. Brussels: Université Catholique de Louvain. Available at: <www.em-dat.net> [Accessed 15 December 2010].

CRED, 2012. *EM-DAT Database, 2012*. [online] OFDA/CRED International Disaster Database, Data Version 12.07. Brussels: Université Catholique de Louvain. Available at: <www.emdat.be> [Accessed 14 December 2011].

Dandekar, V.M., 1976. Crop insurance in India. *Economic and Political Weekly*, 11 (26), pp.A61–A80.

Evans, Steve, 2011a. Muteki Ltd. catastrophe bond triggered by Japan earthquake confirmed as total loss. *Artemis.bm – The Alternative Risk Transfer Portal*, [blog] 7 May. Available at: <http://www.artemis.bm/blog/2011/05/07/muteki-ltd-catastrophe-bond-triggered-by-japan-earthquake-confirmed-as-total-loss/> [Accessed 22 May 2012].

Evans, Steve, 2011b. Catastrophe bond market still largely dominated by U.S. hurricane risk. *Artemis.bm – The Alternative Risk Transfer Portal*, [blog] 29 December. Available at: <http://www.artemis.bm/blog/2011/12/29/catastrophe-bond-market-still-largely-dominated-by-u-s-hurricane-risk/> [Accessed 18 May 2012].

Evans, Steve, 2012a. Zenkyoren loss creep makes Muteki loss seem small by comparison. *Artemis.bm – The Alternative Risk Transfer Portal*, [blog] 6 January. Available at: <http://www.artemis.bm/blog/2012/01/06/zenkyoren-loss-creep-makes-muteki-loss-seem-small-by-comparison/> [Accessed 18 May 2012].

Evans, Steve, 2012b. Kibou Ltd. catastrophe bond from Zenkyoren doubles in size. *Artemis.bm – The Alternative Risk Transfer Portal*, [blog] 7 February. Available at: <http://www.artemis.bm/blog/2012/02/07/kibou-ltd-catastrophe-bond-from-zenkyoren-doubles-in-size/> [Accessed 22 May 2012].

Froot, K.A., 2001. The market for catastrophe risk: A clinical examination. *Journal of Financial Economics*, 60, pp.529–571.

General Insurance Association of Japan, 2011. *Numbers of earthquake insurance consumers*. [pdf] Available at: <http://www.sonpo.or.jp/useful/insurance/jishin/pdf/reference/jishin_suii.pdf> [Accessed 12 December 2011].

GFDRR, 2009. *Disaster Risk Management Programs for Priority Countries*. Geneva, Washington DC: United Nations International Strategy for Disaster Reduction (ISDR), The World Bank.

Ghesquiere, F., Mahul, O., Forni, M. and Gartley, R., 2006. *Caribbean Catastrophe Risk Insurance Facility: A solution to the short-term liquidity needs of small island states in the aftermath of natural disasters*. IAT03-13/3, Washington, D.C.: The World Bank.

Giné, X., Townsend, R. and Vickery, J., 2007. Statistical analysis of rainfall insurance payouts in Southern India. *American Journal of Agricultural Economics*, 89 (5), pp.1248–1254.

Giné, X., 2009. Innovations in insuring the poor: Experience with weather index-based insurance in India and Malawi. In: R. Vargas Hill and M. Torero, 2009. *Innovations in Insuring the Poor*. Washington DC: International Food Policy Research Institute, pp.17–18.

Global AgRisk, 2010. *State of Knowledge Report – Data Requirements for the Design of Weather Index Insurance*. [pdf] Lexington, US: Global AgRisk, Inc. Available at: <http://www.globalagrisk.com/Pubs/2010_GlobalAgRisk_State_of_Knowledge_Data_sept.pdf> [Accessed 15 November 2011].

Government of India, 2010. *Report on Impact Evaluation of Pilot Weather Based Crop Insurance Scheme (WBCIS)*. New Delhi, India: Ministry of Agriculture, Government of India.

Government of Switzerland, 2008. *Funding Scheme for Bali Action Plan: A Swiss Proposal for global solidarity in financing adaptation*. [pdf] Available at: <http://unfccc.int/files/kyoto_protocol/application/pdf/switzerlandfinancebap091008.pdf> [Accessed 12 November 2010].

Harmeling, S., 2008. *Adaptation under the UNFCCC – the road from Bonn to Poznan 2008.* [pdf] Briefing paper, Bonn: Germanwatch. Available at: <http://www.germanwatch.org/klima/bonnadapt08e.pdf> [Accessed 12 November 2010].

Hellmuth, M.E., Osgood, D.E., Hess, U., Moorhead, A. and Bhojwani, H. eds., 2009. *Index insurance and climate risk: Prospects for development and disaster management.* Climate and Society No. 2. Columbia University, New York, USA: International Research Institute for Climate and Society (IRI).

Iizumi, T., Yokozawa, M., Hayashi, Y. and Kimura, F., 2008. Climate change impact on rice insurance payouts in Japan. *Journal of Applied Meteorology and Climatology*, 47 (9), pp.2265–2278.

IPCC, 2007. Summary for policymakers. In: Solomon, S., Qin, D., Manning, M., Chen, Z., Marquis, M., Averyt, K.B., Tignor, M. and Miller, H.L., eds. 2007. *Climate Change 2007: The Physical Science Basis.* Contribution of Working Group I to the Fourth Assessment Report of the Intergovernmental Panel on Climate Change. Cambridge, New York: Cambridge University Press.

Itabashi, C., Iwazawa, Y. and Watanabe, Y., 2007. *Recommendation of Weather Derivatives.* Working Paper 12. Aizu, Japan: University of Aizu.

Kanz, M. and Robert, C., 2011. What does debt relief do for development? Evidence from a large scale policy experiment. In: Indira Gandhi Institute of Development Research, *The Emerging Markets Finance Conference.* Bombay, India 20–21 December 2011, Bombay, India: Indira Gandhi Institute of Development Research.

Kato, N., 2009. Reports of practicians: Post-disaster recovery and risk finance. *Studies in Disaster Recovery and Revitalization*, 1 (1), pp.143–166.

Kato, N., 2011. *Utilization of Private Fund for Disaster Finance.* [pdf] Nishinomiya, Japan: Kwansei Gakuin University. Available at: <http://www.fukkou.net/e-japan/contribution/files/contribution_03.pdf> [Accessed 21 Novemebr 2011].

Kunreuther, H. and Michel-Kerjan, E., 2007. *Climate Change, Insurability of Large-Scale Disasters and the Emerging Liability Challenge.* Cambridge, USA: National Bureau of Economic Research.

Manuamorn, O.P., 2007. *Scaling up Microinsurance: The Case of Weather Insurance for Smallholders in India.* Agriculture and Rural Development Discussion Paper 36, Washington D.C.: The World Bank.

MCII (Munich Climate Insurance Initiative), 2009. *Climate Risk Management Mechanism including Insurance, in the context of Adaptation to Climate Change.* [pdf] Available at: <http://unfccc.int/resource/docs/2009/smsn/ngo/132.pdf> [Accessed 12 November 2010].

Mechler, R., 2004. *Natural Disaster Risk Management and Financing Disaster Losses in Developing Countries.* Karlsruhe, Germany: Verlag Versicherungswirtschaft.

METI, 2006. *Report of Risk Finance Study Group: Toward the Prevalence of Risk Finance.* [pdf] Tokyo: Ministry of Economy, Trade, and Industry of Japan, Government of Japan. Available at: <http://www.meti.go.jp/english/report/downloadfiles/0607riskfinancereport.pdf> [Accessed 12 December 2011].

Milesi, C., Geethalakshmi, V., Rao, G.S. et al., 2011. Producing valid meteorological data using Terrestrial Observation and Prediction System (TOPS) and ecological forecasting. In: ASCI (Administrative Staff College of India), *Weather Insurance: Addressing the Risk Mitigation Needs of Weather Sensitive Industries in India.* Hyderabad, India 28 – 29 January 2011. *Journal of Management*, 41 (1), pp.102–107.

Ministry of Internal Affairs and Communications of Japan, 1966. Law related to Earthquake insurance. [online] Available at: <http://law.e-gov.go.jp/htmldata/S41/S41HO073.html> [Accessed 12 December 2011].

Munich Re, 2010. *Topics Geo – Natural Catastrophes 2009: Analyses, Assessments and Positions.* Munich, Germany: Munich Re.

Munich Re, 2011. Protecting cooperatives and their low-income members in the Philippines against extreme weather events through microinsurance. [press release] 11 October 2011, Available at: <http://www.munichre.com/en/media_relations/press_releases/2010/2010_10_11_press_release.aspx> [Accessed 13 June 2012].

Munich Re Group – Risk Trading Unit, 2011. *Muteki: Japanese Earthquake Protection for Zenkyoren.* [pdf] Available at: <http://www.munichre.co.jp/public/PDF/Topics_MutekiLtdAsia1.pdf> [Accessed 12 December 2011].

Nemani, R., Hashimoto, H., Votava, P. et al., 2009. Monitoring and forecasting ecosystem dynamics using the Terrestrial Observation and Prediction System (TOPS). *Remote Sensing of Environment*, 113 (7), pp.1497–1509.

Njegomir, V. and Maksimovic, R., 2009. Risk transfer solutions for the insurance industry. *Economic Annals*, 54 (180), pp.57–90.

O'Connor, P.M., 2005. Recent trends in the catastrophic risk insurance / Reinsurance Market. OECD Publishing, doi: 10.1787/9789264009950-20-en.

O'Donnell, O., van Doorslaer, E., Rannan-Eliya, R.P. et al., 2008. Who pays for health care in Asia? *Journal of Health Economics*, 27 (2), pp.460–475.

OECD, 2011. *Economic Outlook Statistic Database.* [online] Available at: <http://www.oecd.org/document/0,3746,

en_2649_201185_46462759_1_1_1_1,00.html> [Accessed 12 December 2011].

Ono, M., 2004. *Weather Derivatives*. Japan: Sigmabase Capital Publishing.

Rao, K.N., 2011. Crop insurance and mitigation tool. In: Institute of Development Studies, *National Seminar on Agriculture at Crossroads: Issues and Challenges*. Jaipur, India 28 – 29 September 2011. Jaipur, India: Institute for Development Studies.

Siamwalla, A. and Valdes, A., 1986. Should crop insurance be subsidized? In: P. Hazell, C. Pomareda and A. Valdez, eds. 1986. *Crop Insurance for Agricultural Development: Issues and Experience*. Baltimore: John Hopkins University Press.

Singh, S. and Jogi, R.L., 2011. *Managing Risk for Indian Farmers: Is Weather Insurance Workable?* IDSJ Working Paper 157, Jaipur, India: Institute of Development Studies.

Sompo Japan Insurance Inc., 2010a. *Weather Index Insurance Launched for Drought Risk in Northeast Thailand: Provision of adaptation measure for climate change utilizing insurance*. Tokyo, Japan: Sompo Japan Insurance Inc.

Sompo Japan Insurance Inc., 2010b. *Weather Derivative*. [online] Available at: <http://www.sompo-japan.co.jp/hinsurance/art/weather_derivative/> [Accessed 21 November 2011].

Swiss Re, 2010a. *Weathering Climate Change: Insurance Solutions for More Resilient Communities*. Zurich, Switzerland: Swiss Reinsurance Company Ltd.

Swiss Re, 2010b. *World Insurance in 2009: Premiums Dipped, but Industry Capital Improved*. Zurich, Switzerland: Swiss Reinsurance Company Ltd.

Tokio Marine and Nichido Fire Insurance Co., Ltd., 2011. *Weather Derivatives*. [online] Available at: <http://www.tokiomarine-nichido.co.jp/hojin/risk/weather/index.html> [Accessed 21 November 2011].

Tsuji, H., 1986. An economic analysis of rice insurance in Japan. In: P. Hazell, C. Pomareda and A. Valdez, eds. 1986. *Crop Insurance for Agricultural Development: Issues and Experience*. Baltimore: John Hopkins University Press.

UNFCCC, 2007. *Bali Action Plan*. [pdf] Bali, Indonesia: United Nations Framework Convention on Climate Change. Available at: <http://unfccc.int/files/meetings/cop_13/application/pdf/cp_bali_action.pdf> [Accessed 10 October 2010].

UNFCCC, 2011. Views and information on elements to be included in the work programme on loss and damage. [pdf] In: UNFCCC, *Subsidiary Body for Implementation: Thirty-fourth session*. Bonn, Germany 6 – 16 June 2011. Bonn: UNFCCC. Available at: <http://unfccc.int/resource/docs/2011/sbi/eng/misc01.pdf> [Accessed 14 June 2012].

UNFCCC, 2012. Report on the expert meeting on assessing the risk of loss and damage associated with the adverse effects of climate change. [pdf] In: UNFCCC, *Subsidiary Body for Implementation: Thirty-sixth session*. Bonn, Germany 14–25 May 2012. Bonn: UNFCCC. Available at: <http://unfccc.int/resource/docs/2012/sbi/eng/inf03.pdf> [Accessed 14 June 1012].

United Nations, 1992. *United Nations Framework Convention on Climate Change*. [pdf] Available at: <http://unfccc.int/resource/docs/convkp/conveng.pdf> [Accessed 28 October 2010]

United Nations, 2007. Developing index-based insurance for agriculture in developing countries. *Sustainable Development Innovation Briefs*, 2. New York: United Nations.

UNU-EHS/Munich Re, 2007. *Social Vulnerability: Summer Academy 2007 – Megacities as Hotspots of Risk*. [pdf] Available at: <http://www.munichre-foundation.org/NR/rdonlyres/5FC116A6-B33F-4370-A047-71AE170339B3/0/PosterLoewNaturalHazardRiskIndex.pdf> [Accessed 12 December 2011].

Vatsa, S.K., 2004. Risk, vulnerability, and asset-based approach to disaster risk management. *International Journal of Sociology and Social Policy*, 24 (10/11), pp.1–48.

Yamada, R., 2010. *Risk Finance 2: Role of Reinsurance Market, 'Risuku fainansu 2 [Saihoken Shijo no Yakuwari]'*. [Online via internal VLE], Keio University Faculty of Commerce. Available at: <http://www.fbc.keio.ac.jp/~tyabu/risk4.pdf> [Accessed 12 December 2011].

Yazici, S., 2005. The Turkish Catastrophe Insurance Pool (TCIP) and Compulsory Earthquake Insurance Scheme. In: OECD, 2005. *Catastrophic Risks and Insurance*. Paris: OECD Publishing, doi: 10.1787/9789264009950-20-en.

Yeandle, M., 2011. *The Global Financial Centres Index 10*. [pdf] Available at: <http://www.zyen.com/PDF/GFCI%2010.pdf> [Accessed 12 December 2011].

Yokouchi, A., 2007. *Introduction of Weather Risk Management Technology in Farm Management*. Tokyo, Japan: Agricultural Information Research, Japan Weather Association.

Yucemen, M.S., 2008. Turkish catastrophe risk insurance pool. [pdf] In: Organization for Economic Co-operation and Development, *International Conference on Financial Education*, Washington D.C., USA 7 – 8 May 2008. Washington D.C., USA: Organization for Economic Co-operation and Development. Available at: <www.oecd.org/dataoecd/16/40/40607615.pdf> [Accessed 21 November 2011].

Index

Climate Change Adaptation in Practice: From Strategy Development to Implementation, First Edition.
Edited by Philipp Schmidt-Thomé and Johannes Klein.
© 2013 John Wiley & Sons, Ltd. Published 2013 by John Wiley & Sons, Ltd.

Index compiled by Terry Halliday